The Country Life Gu

Shells of the World

A.P.H. Oliver

Illustrated by
James Nicholls

Country Life Books

Preface

The information in this book is gleaned from many sources apart from personal observation and the author would like to acknowledge his indebtedness to many conchologists past and present, and in particular the authors of the books listed on page 8. While most of the shells illustrated are from his own collection, some have been kindly lent by the National Museum of Wales, Mr and Mrs T. Pain, and Messrs W. Karo, F. Mayer and K. R. Wye.

Thanks are also due to Dr A. J. Rundle for his assistance with the bivalves, and to Miss G. M. Warland without whose help and patience in turning barely legible manuscript into typed copy, the project would never have been completed. Readers will not need to be told of the author's very sincere appreciation of the work done by the artist, Mr J. Nicholls. The encouragement and help from Mr S. P. Dance, Mr R. P. Scase and other members of the Conchological Society of Great Britain, and the patient and helpful attitude of the staff of the Hamlyn Group who have handled this book are also gratefully acknowledged.

A. P. H. Oliver

Published by Country Life Books,
an imprint of Newnes Books,
a Division of The Hamlyn Publishing Group Limited,
84-88 The Centre, Feltham, Middlesex, England.

Fifth impression 1984
First published in hardback 1980
ISBN 0 600 34397 9 (paperback)
ISBN 0 600 35354 0 (cased)

Printed in Spain by Printer Industria Gráfica S.A.
Barcelona D.L.B. 34598-1983

Contents

Introduction

A Guide to Shells—common shells, popular shells, spectacular shells? To choose any group would be difficult and inevitably subjective, so I have given up the struggle and selected those which I should most like to see illustrated. I believe my tastes are fairly typical of the average collector and hope that I shall, therefore, please most readers while accepting that all of them will say 'Why has this shell been included?' and 'Why has that been left out?'.

I have devoted the majority of this book to Gastropoda as I believe most collectors find this the most fascinating class. I have included some of the more interesting Bivalvia, and have given a short reference to each of the other classes of molluscs, Cephalopoda, Amphineura, Scaphopoda and Monoplacophora. I have devoted more space to the most popular families, such as the cowries, cones, volutes and strombs. As to species, I have tried to concentrate on the more common shells but some rare, or at best very uncommon, ones are included because they are known and coveted by so many collectors, because of their beauty, or even because they are particular favourites of mine. A typical example of a species has normally been illustrated, but occasionally a more unusual variety has been included for interest.

Wherever possible the illustrations are life size, but in many cases it has been necessary to alter the scale. However, shells on any one page are in proportion to each other. In addition, the average large size of each species is given in the text. Distribution areas can only be approximate as every month there is news of one or more species being found in a new location.

Classification

Classification of living things is essential to any study of animals and plants to enable their relationships to be determined. It is usually based, according to Linnaeus, on anatomical similarities and differences. The first subdivision is into the animal kingdom and plant kingdom. The animal kingdom is comprised of phyla. The phylum Mollusca, with which this book is concerned, is second only to Arthropoda in number of species. Within each phylum, members are grouped into classes according to gross anatomical differences. Mollusca comprises six classes—Gastropoda, Bivalvia, Scaphopoda, Amphineura, Cephalopoda and Monoplacophora—the main differences between them being in the structure of the shell and foot. Each class is then divided into orders, again according to anatomical differences, though less obvious ones. Members of an order are always grouped into families and then into genera. Intermediate grades such as subclass, suborder, superfamily, subfamily and subgenera are sometimes used. A genus is divided into species, members of a particular species being similar enough to allow interbreeding to result in fertile offspring.

The scientific name of each species consists of its generic and specific names. This is usually followed by that of the person who first described and named it (the author) and the date, so that the relevant work can easily be referred to. It has not been possible to specify the dates of the authors' descriptions of some of the species mentioned in this book and a space has been left for readers to fill these in when they can. Authors' names have not been abbreviated except for Linné or Linnaeus who has been denoted by L.

The classification of such a large phylum as Mollusca is inevitably complex, and despite the many books which are available on the subject today, the naming of shells can at times be very difficult even if one is fortunate enough to be able to refer to a well-curated collection at a museum. Taxonomists are constantly finding that species were described under an earlier name than that in current use, and the earliest name given is considered to be the valid one. Also, further study of shells and their animals may show that what have been thought to be two separate species are only varieties of the same species and vice versa.

A shell has been included on page 175 as an example of the difficulties which may be encountered. I acquired it as *Murex radiatus* from west Central America. There does not appear to be a shell so named. It is not *Murexiella radicata* Hinds as that has five varices and mine only three and looks like a species of the genus *Chicoreus*. Dr Myra Keen, an expert on the west Central American area, does not list any *Chicoreus* species from there, but in *An Illustrated Catalog of the Recent Species of the Rock Shells* by Maxwell Smith there is an illustration of a shell not unlike mine, which he calls *Murex (Chicoreus) palmarosae mexicanus* Stearns 1897, giving as the locality the Gulf of California. Dr Myra Keen in *Marine Shells of Tropical West America* 1st edition puts this under 'Muricidae of doubtful status' and says 'This is to be rejected for two reasons: first the type specimen seems to be an Indo-Pacific form erroneously recorded as from the American coast; second, the name *Murex mexicanus* Petit 1852, was used for a Caribbean form.'!

Dr E. H. Vokes, an expert on Muricidae, does not list the name in her *Catalogue of the Genus* Murex *Linne; Muricidae and Ocenebridae 1971*. Next, checking with the British Museum collection the specimens most like mine are labelled *M. (Chicoreus) corrugatus* Sowerby 1841, which is found in the Indo-Pacific. To add to the confusion I have another shell which is very similar to that labelled *M. radiatus*, which I collected in Malaya and have been unable to identify.

If the problem shell is that illustrated by Maxwell Smith, then either he would appear to be correct in attributing it to that region and mine could be another record, or if Dr Myra Keen is correct—and she is a zoologist of world renown with a good knowledge of the Muricidae—then both Maxwell Smith's specimen and mine are from the Indo-Pacific, perhaps a chance introduction to America or possibly the wrong data was sent with both shells. All of which illustrates the problems facing the amateur collector; but though problems there may be, these should not be deterrents.

About Molluscs

Approximately half of the vast phylum of Mollusca, which contains about 100000 species, lives in salt water; the other half in fresh-water or on land. This book only deals with the former. These live in all seas, polar and tropical, shallow and deep water. Generally, those living in warm water are more colourful than those living in cold—cold water regions including the deep waters of temperate and tropical seas as well as the arctic and antarctic. The majority of shells covered in this book are found in warm water and any reference to, for example, the Indo-Pacific excludes the extreme south of the Indian and Pacific Oceans and the north of the latter.

All molluscs have a brain, digestive system and reproductive system. Nearly all lay eggs and some even look after their eggs after laying. The female cowrie is often found sitting on top of her pile of eggs—and should be left there! After hatching, many marine species have a free-swimming or veliger stage before changing into miniatures of their final form.

Excepting the bivalves, they have a ribbon-like set of teeth, the radula, which they use like a file to eat their food. Some of the animals are vegetarians, some are carnivorous and some are scavengers. The cones, augers and turrids kill their prey by injecting poison into it with a harpoon-shaped barb or tooth. While the augers and turrids are not harmful to man, some of the cones can cause considerable pain and a few can be lethal, so care must be taken when collecting such shells alive. The barb can pierce more than the skin so the shell should not be carried about alive in a pocket or bag next to the skin.

Molluscs are soft-bodied animals which in the vast majority of cases protect themselves by building a hard covering or shell. This may be in one piece as in the gastropods, scaphopods or tusk shells, and one genus of the cephalopods, *Nautilus*. The bivalves build a shell in two parts, as the oysters, cockles and mussels. Watering-pot shells start life as a normal bivalve and then construct a long, shelly tube, many times bigger than their original shell, which they incorporate into their new structure. The chiton's shell consists usually of eight plates which overlap and form a cover so that the animal resembles a common woodlouse, though it has no legs. Some gastropods, the nudibranchs in the sea and the slugs on land, and many of the cephalopods, including the octopus and the squid, have no external shell. Some of these have an internal shell, an example of which is the cuttlefish 'bone', often found washed up on the beach and fed to cage birds.

Shelled molluscs have a lobe or a pair of lobes, the mantle, from the edge of which is secreted the material from which the shell is built up. Bivalves grow by adding to the edges of their valves away from the beak—the place where the valves are joined together by a rubbery like ligament. Gastropods generally grow by adding to the edge of a tube which is coiled, usually dextrally. The limpets and abalones do not, however, produce tubes.

In some families, the Muricidae for example, the animal has a periodic rest in its building. When it has completed part of a whorl it constructs a strong, often heavily reinforced lip before continuing to build. The early lips show as varices on the shells. Most of the Muricidae have three varices per whorl, some up to eight, while the Bursidae often have two. Some shells, such as the members of the genus *Lambis*, do not build a strong, reinforced lip until they are fully adult, when the lip becomes extended with long, strong spines. The cowries also wait until adult before the lip turns in, thickens and develops 'teeth'. Others, such as the cones, do not modify their lip and therefore give no indication of having reached their maximum size.

Many molluscs have a skin-like covering over their shells, the periostracum, often rather furry. This will usually crack and flake off when dry, but can easily be removed by soaking the shell in ordinary household bleach for a few hours and then washing thoroughly with water. This will not damage the shell, and until the periostracum is removed the beauty of the colours will not show. A living animal under water with its periostracum intact looks very different from a specimen with its periostracum either dried out or removed. The cowrie and olive shells have no periostracum. They are as bright and beautiful alive as in a collection—usually more so as they tend to fade, some more than others, when the animal is dead.

Many gastropods have a 'trap-door' or operculum attached to the animal with which they can close the aperture of their shells to protect themselves. This may be solid and heavy as the cat's-eye of the Turbinidae and the Neritidae, or pliant and horny as in the Trochidae and Muricidae. Sometimes, as in some of the cones, it is very small and obsolete and certainly not big enough to seal the aperture. Most collectors like to keep the operculum with their specimens, either tucked inside the shell or stuck on to a piece of cotton-wool so that it can be held in its proper position. A horny operculum must not be put into bleach when cleaning off the periostracum as it will dissolve.

Hints for collectors

It should not be forgotten, especially in the tropics, that when pottering about on the reef or floating along the surface of the sea with a snorkel, one can easily get very badly sunburnt. It is wise to wear an old shirt and thin trousers to give protection from the sun and from being grazed by rock or coral and also from the stings of some of the corals and creatures such as jellyfish which can be very painful and sometimes dangerous. For the same reason never walk barefoot on a reef and always wear gloves. There are creatures apart from poisonous cones which can be harmful—the sluggish and well-camouflaged stone fish, moray eels, some of the rays, sea-snakes and sea-wasps, quite apart from sharks! A great deal has been written about sharks, and if there is one thing to be learned from this, it is that they are unpredictable. If you meet one under water, get away from it as quickly as possible, without making a lot of disturbance. It is believed that sharks are attracted by light colours and white skin, so it is wise to wear dark-coloured clothing.

Remember that the number of shells of any species is not limitless. Do not take more specimens than are needed. After turning over rocks or coral slabs to search beneath them, they should be turned back again. Many creatures lay their eggs on the underside of rock and coral, apart from living there themselves, and if left uncovered these are likely to be eaten or die of exposure. Collecting shells has become such a popular hobby in recent years that many accessible reefs have been denuded of most of their molluscan life due to irresponsible shelling.

Keep a data slip or some form of record with each specimen giving as much information as possible—the name, when found, where, depth, type of habitat and so on.

Cleaning shells can be a major operation. To extract the animal many species can be brought to the boil in water and then allowed to cool slowly; cowries and olives should not be boiled as their glaze is likely to crack. A piece of bent wire is often the final answer, and I have found that an old hypodermic syringe is helpful in producing a strong jet of water inside the shell to flush out organic remains. Burying shells in clean dry sand (but not one above the other) is effective in the tropics where ants can be helpful for a change. A poor sense of smell is also an asset!

Acid should never be used to clean off the calcium and coral deposits, as this will clean off some of the shell too, especially the delicate fronds and sculpturing on the Muricidae which seem particularly prone to such deposits. Shells should not be varnished or polished. If they become dull, a wipe over with a touch of fine oil will replace the natural oil which has dried out, and do much to restore the original colours and sheen.

Collections should not be stored where they are exposed to the light. Strong sunlight, especially, will quickly cause the shells to fade. That is why in a museum the shells on display are usually dull and faded. The main collections are always stored away in light-proof cabinets.

Many collectors who cannot visit the locality of a particular shell or cannot find the required species, obtain the shell by exchange or purchase. Addresses of suppliers can be obtained from the journals on page 9 or from a conchological society.

Finally, I recommend to a beginner that he should decide early on where to specialize. The phylum Mollusca is so vast that one is likely to get more satisfaction from limiting a collection to a few families or to the shells from a particular country or area. Apart from achieving a more 'complete' collection, there is a better chance of becoming knowledgeable in one part of a very wide field.

Bibliography

General

Dance, S. P., *Shell Collecting—An Illustrated History*, Faber & Faber Ltd, London, 1966.

Dance, S. P., *Rare Shells*, Faber & Faber Ltd, London, 1969.

Melvin, A. G., *Sea Shells of the World*, Charles E. Tuttle Co., Tokyo, 1966.

Saul, M., *Shells*, Country Life, London, 1974.

Wagner, R. J. L., & Abbott, R. T., *Van Nostrand's Standard Catalog of Shells*, Van Nostrand Reinhold Co. Ltd, New York, 1967.

Families

Burgess, C. M., *The Living Cowries*, Thomas Yoseloff Ltd, London, 1970.

Clover, P. W., *A Catalog of Popular* Marginella *Species*, P. W. Clover, New York, 1968.

Marsh, J. A., & Rippingale, O. H., *Cone Shells of the World*, The Jacaranda Press, Australia, 1968.

Weaver, C. S., & du Pont, J. E., *Living Volutes*, Delaware Museum of Natural History, 1970.

Zeigler, R. F., & Porreca, H. C., *Olive Shells of the World*, privately published, 1969.

Areas

Abbott, R. T., *American Seashells*, Van Nostrand Reinhold Co. Ltd, New York, 1955.

Abbott, R. T. (ed.), *Indo-Pacific Mollusca*, Delaware Museum of Natural History, 1959—current.

Abbott, R. T., *Seashells of the World*, Golden Press, New York, 1962.

Boss, K. J. (ed.), *Johnsonia* Monographs of the marine Mollusca of the Western Atlantic, Museum of Comparative Zoology, Harvard University, Massachusetts, 1941—current.

Habe, T., *Shells of the Western Pacific in Color* Vol II, Hoikusha, Osaka, 1964.

Hinton, A. G., *Shells of New Guinea and the Central Indo-Pacific*, Robert Brown & Associates Pty Ltd, Port Moresby, New Guinea and The Jacaranda Press, Australia, 1972.

Humfrey, M., *Sea Shells of the West Indies*, Collins, London, 1975.

Keen, A. M. & McLean, J. H., *Marine Shells of Tropical West America*, Stanford University Press, California, 1971.

Kennelly, D. H., *Marine Shells of Southern Africa*, Thomas Nelson & Son (Africa) Pty Ltd, Johannesburg, 1964.

Kensley, B., *Seashells of Southern Africa—Gastropods*, South African Museum Publication, Maskew Miller Ltd, Cape Town, 1973.

Kira, T., *Shells of the Western Pacific in Color* Vol. I, Hoikusha, Osaka, 1962.

McMillan, N. F., *British Shells*, Frederick Warne & Co. Ltd, London, 1968.

Nicklès, M., *Mollusques Testacés Marins de la Côte Occidentale d'Afrique*, Paul Lechevalier, Paris, 1950.

Nordsiek, Dr F., *Die Europäischen Meeres-Gehäuseschnecken vom Eismeer bis Kapverden und Mittlemeer*, (Prosobranchia), Gustav Fischer Verlag, Stuttgart, 1968.

Nordsiek, Dr F., *Die Europäischen Meeremuscheln vom Eismeer bis Kapverden, Mittlemeer und Schwarzes Meer*, (Bivalvia), Gustav Fischer Verlag, Stuttgart, 1969.

Powell, A. W. B., *Shells of New Zealand*, Whitcombe & Tombs Ltd, Wellington, 1957, 1961.

Rios, E. C., *Coastal Brazilian Seashells*, Fundacão Cidade do Rio Grande,

Museu Oceanográfico de Rio Grande, Brazil, 1970.
Tebble, N., *British Bivalve Seashells*, Trustees of the British Museum (Natural History), London, 1966.
Warmke, G. L., & Abbott, R. T., *Caribbean Seashells*, Livingston Publishing Co., Pennsylvania, 1961.
Wilson, B. R., & Gillett, K., *Australian Shells*, A. H. & A. W. Reed, Sydney, 1971.

Journals

Hawaiian Shell News Hawaiian Malacological Society, P.O. Box 10391, Honolulu, Hawaii.
Journal of Conchology Conchological Society of Great Britain and Ireland, London, England.
The Veliger North Californian Malacozoological Club, Berkeley, California.

Glossary

adpressed overlapping so as to converge gradually
anal canal channel or opening through which the animal passes waste matter, usually at the posterior end of the lip
anterior end pointing in the direction in which the animal moves; furthest away from the apex
aperture opening in a gastropod shell through which the animal emerges
apex first-formed part of the shell; tip of the spire, usually pointed
axis imaginary line from the apex to the base about which the whorls are formed; *adj.* axial
base anterior end of a shell, especially the flattened end of a *Trochus* type shell; also used, though strictly incorrectly, to describe the flat aperture side of cowries
beak see umbo
bifid split into two
body whorl last complete whorl formed 360° from the lip
callus thickening layer of shelly material; *adj.* callous
cancellate with sculptural cords intersecting at right angles; reticulated or latticed
carina prominent ridge or keel; *adj.* carinate
columella central pillar of a coiled gastropod shell formed by the inner or axial sides of the whorls; *adj.* columellar
cord small, round-topped ridge
coronate(d) having nodes or nodules on the shoulder
crenulate having a regularly notched edge
depressed low relative to the diameter
dorsum back of the shell, opposite the aperture
fasciole spiral band built up from the notches in the canal of successive growth stages of a gastropod
fossula shallow, linear depression on the inner lip of some cowries
fusiform shaped like a spindle—i.e., more or less equally pointed at both ends
globose more or less spherical
head scar see muscle scar
height distance from the apex to the base or anterior end of the shell
keel prominent ridge or carina
labial area surface around the inner lip or labium from the bottom of the columella to the suture
lamella thin plate or scale; *adj.* lamellate

latticed see cancellate
length in species of the Patellidae, Acmaeidae or Fissurellidae, the distance from the anterior edge to the posterior edge of the shell; in a species of the Haliotidae, the longest distance between any two points on the circumference of the shell
lip edge of the aperture. The inner lip or labium (*see* labial area). The outer lip or labrum—the edge of the aperture furthest from the columella. In the following pages the term 'lip' has been used to denote only the outer lip.
lira fine line or groove; *pl.* lirae, *adj.* lirate
malleate as if beaten with a hammer
muscle scar or **head scar** mark on the inside of a shell—e.g., bivalve or *Haliotis*—where the muscle was attached

Terminology used to describe the gastropod shell.

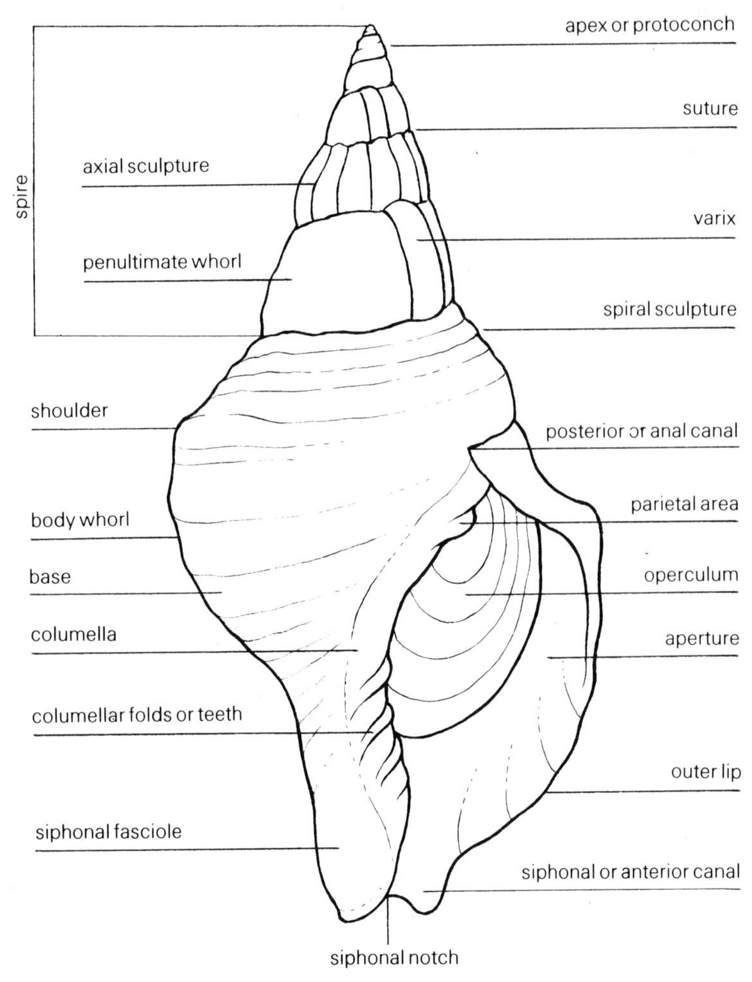

node small knob; *adj.* nodose
nodule small node; *adj.* nodulose
nucleus earliest whorls at the apex of a shell, which were formed during the egg stage of the animal; earliest formed part of the operculum
operculum shelly or horny attachment to the animal with which it can more or less close the aperture of its shell; *adj.* operculate
patulous expanded or spread
parietal area basal area of a coiled shell on the columellar side of the aperture
periostracum skin of hairy or smooth horny material (conchiolin) covering the outer surface of many shells
periphery widest part of a shell—i.e., furthest from the axis
plica plait, fold or ridge on the columella of some shells; *pl.* plicae, *adj.* plicate(d)
posterior opposite to anterior; rear or apical end of a gastropod; siphonal end of a bivalve
protoconch whorls at the apex of a shell, especially where clearly discernible from later ones.
punctate as if covered with pin-pricks
pustule small rounded elevation, smaller than a tubercle; *adj.* pustulate
pyriform pear-shaped
recurved bent backward; used especially of the anterior or siphonal canal, away from the aperture
reticulate see cancellate
rostrum beak-like extension in—e.g., *Tibia* and some *Cypraea*; *adj.* rostrate
sculpture relief pattern on the surface of a shell
shoulder angulation of a whorl forming the outer edge of the sutural ramp or shelf
siphonal canal channel at the anterior end of a shell through which the animal can extend its siphon, a tube-like part of the mantle for the passage of water
siphonal notch depression or notch in the lip through which the siphon can be extended
spatulate having broad, blunt, finger-like projections
spiral extending round the whorls more or less parallel with the suture
spire visible part of all the whorls except the last or body whorl
stria fine incised groove; *pl.* striae, *adj.* striate
sulcus furrow or slit
suture lines formed by the joining of the whorls
tubercule small rounded elevation; larger than a pustule, smaller than a nodule; *adj.* tuberculate
umbilicus axially aligned opening within the whorls of a loosely coiled shell where no columella has been formed; *adj.* umbilicate
umbo or **beak** earliest formed part of the shell of a bivalve
varix thickened ridge formed at the outer lip during a rest period in the growth of a shell; *pl.* varices, *adj.* varicose
whorl complete 360° coil of a shell
width widest part of a shell at right angles to its axis

The following maps show the positions of most of the shell locations mentioned in this book.

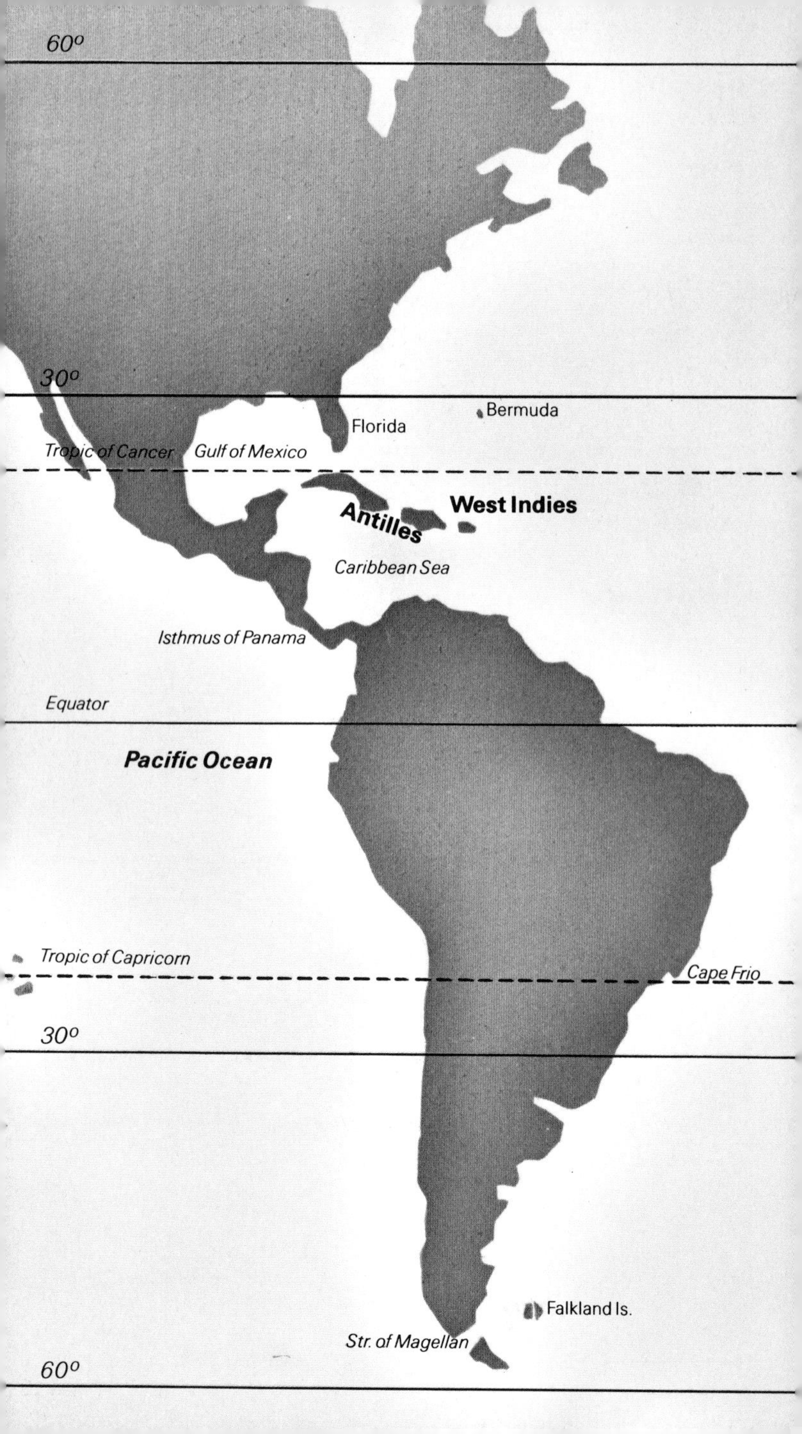

60°

30°

Bermuda

Florida

Tropic of Cancer

Gulf of Mexico

West Indies

Antilles

Caribbean Sea

Isthmus of Panama

Equator

Pacific Ocean

Tropic of Capricorn

Cape Frio

30°

Falkland Is.

Str. of Magellan

60°

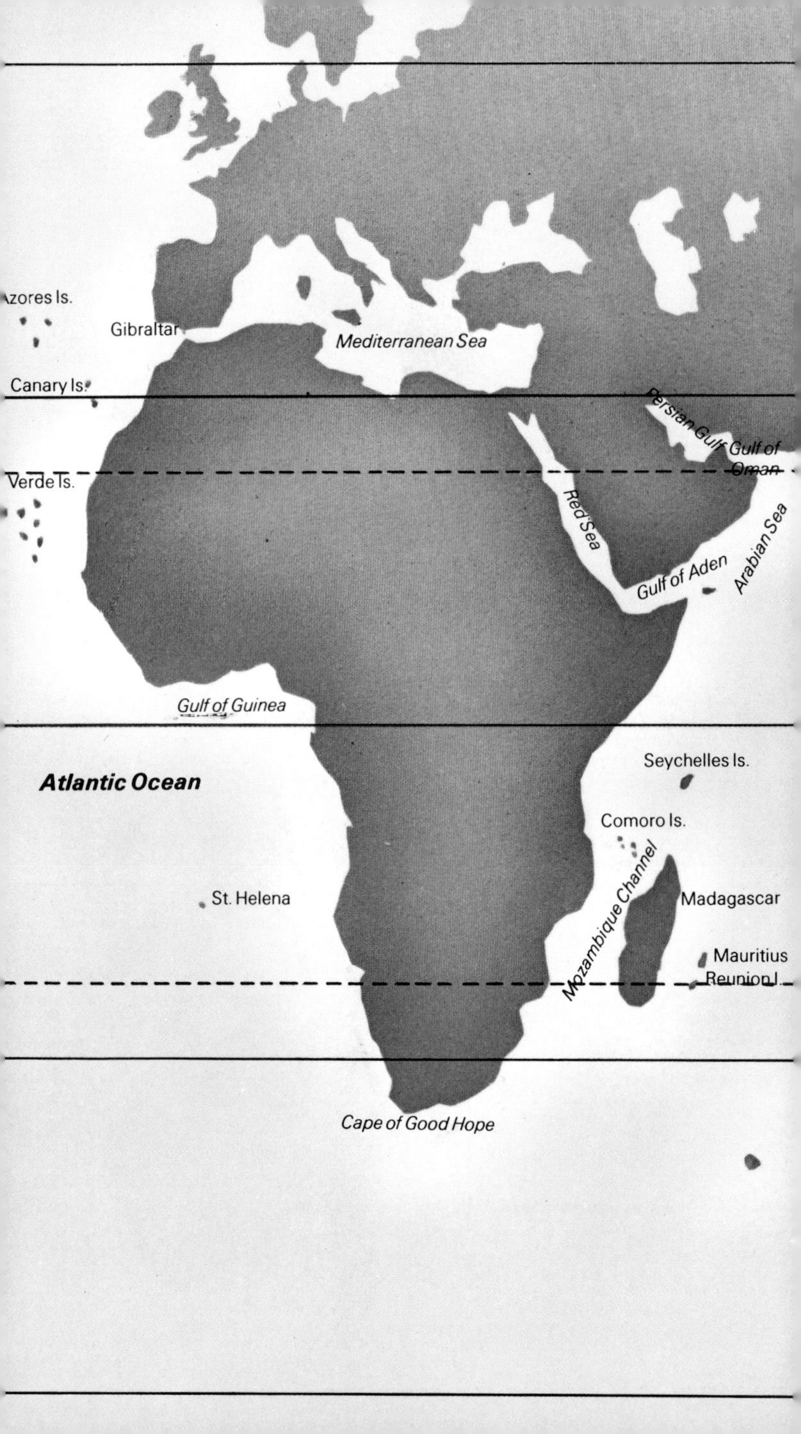

zores Is.
Gibraltar
Mediterranean Sea
Canary Is.
Persian Gulf
Gulf of Oman
Verde Is.
Red Sea
Arabian Sea
Gulf of Aden
Gulf of Guinea
Seychelles Is.
Atlantic Ocean
Comoro Is.
Mozambique Channel
St. Helena
Madagascar
Mauritius
Reunion I.
Cape of Good Hope

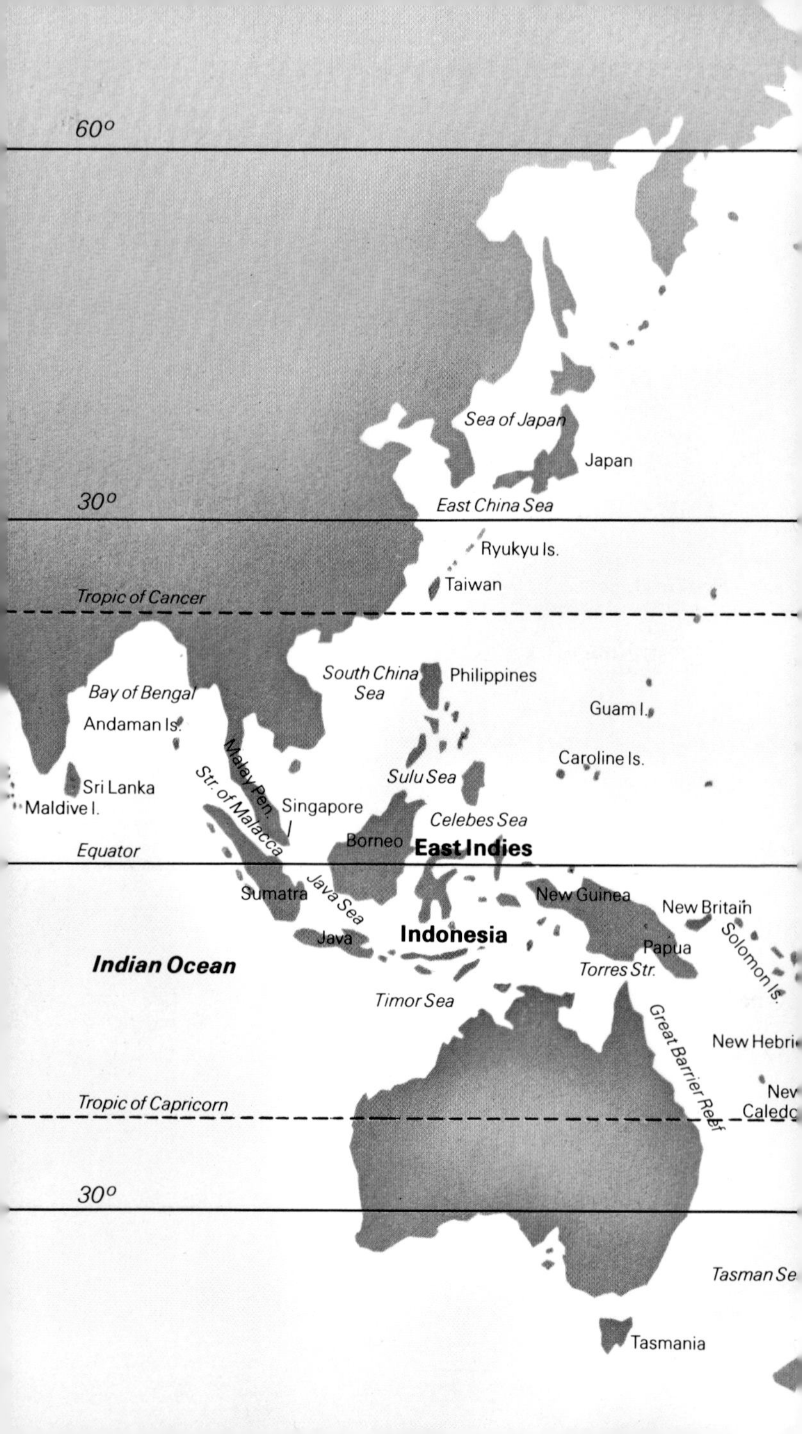

60°
Sea of Japan
Japan
30°
East China Sea
Ryukyu Is.
Taiwan
Tropic of Cancer
South China Sea
Philippines
Bay of Bengal
Guam I.
Andaman Is.
Caroline Is.
Malay Pen.
Str. of Malacca
Sulu Sea
Sri Lanka
Maldive I.
Singapore
Celebes Sea
Borneo
East Indies
Equator
Sumatra
Java Sea
New Guinea
New Britain
Solomon Is.
Java
Indonesia
Papua
Indian Ocean
Torres Str.
Timor Sea
Great Barrier Reef
New Hebri
Nev
Caledc
Tropic of Capricorn
30°
Tasman Se
Tasmania

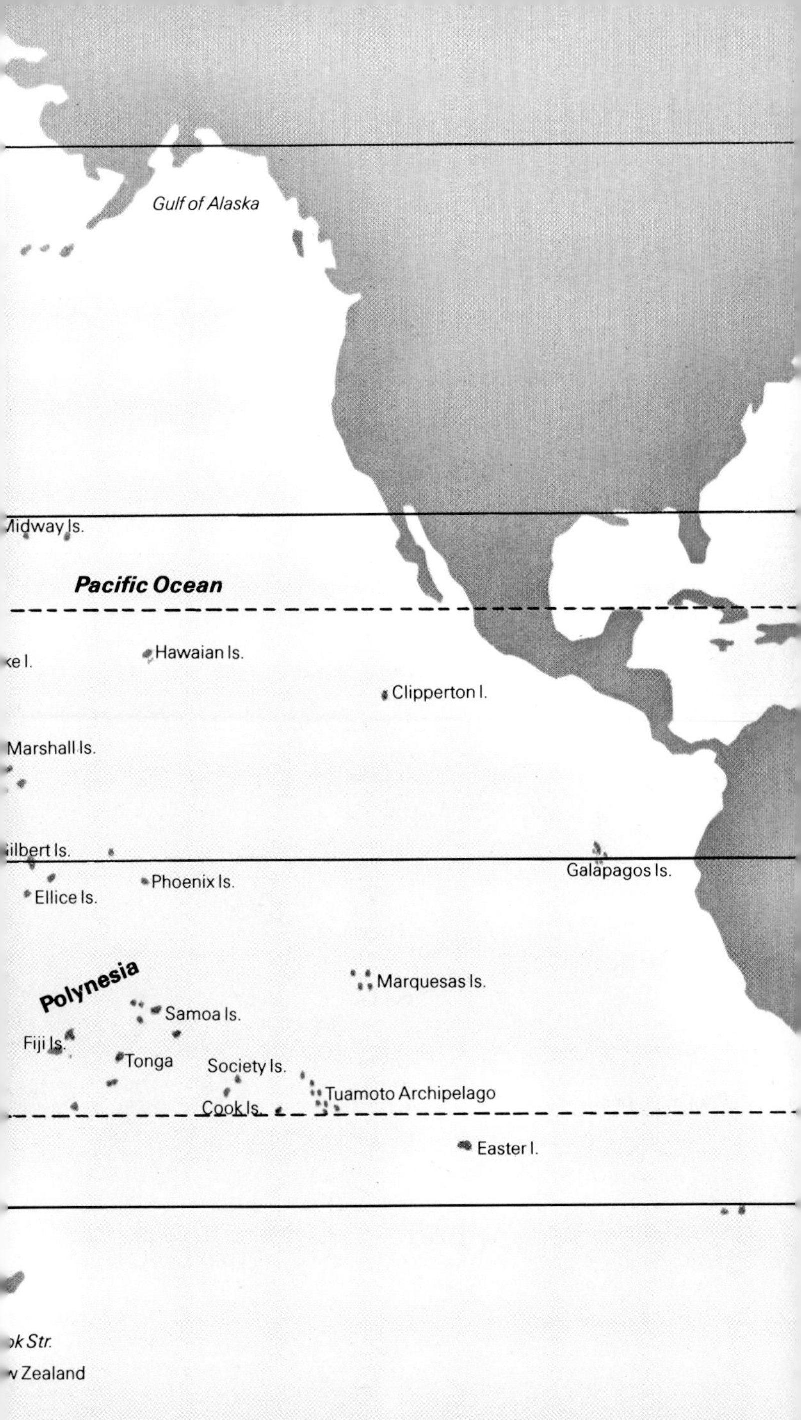
Gulf of Alaska
Aidway Is.
Pacific Ocean
Hawaian Is.
ke I.
Clipperton I.
Marshall Is.
Gilbert Is.
Galapagos Is.
Phoenix Is.
Ellice Is.
Polynesia
Marquesas Is.
Samoa Is.
Fiji Is.
Tonga
Society Is.
Tuamoto Archipelago
Cook Is.
Easter I.
ok Str.
w Zealand

Class: Gastropoda

Order: Archaeogastropoda

Superfamily: Pleurotomariacea

Family: Pleurotomariidae

Slit shells get their common name from the slit through which they pass water and waste matter. They were known from fossils and were thought to be extinct until the first recent specimen was found in the Caribbean in 1856. They are of special interest because they are probably the most primitive of all gastropods living today. Since all are deep-water shells they are rare or very uncommon. There are some seventeen known species divided among three genera, *Entemnotrochus, Perotrochus* and *Mikadotrochus*. Seven species inhabit the Caribbean area, two off east South America, seven in Japan and Taiwan waters and one off east South Africa. Most are 50mm to 125mm wide, but *Entemnotrochus rumphii* grows to over 250mm. Some are umbilicate and all have a horny operculum.

Perotrochus africanus **Tomlin 1948** the South African Slit Shell was first found off the Natal coast in 1931 at about 200 fathoms; about 125mm wide and 100mm high. A light, thin shell but like all members of the family conical with the characteristic slit. It is orange-yellow streaked with white. Only about a dozen specimens have been found.

P. hirasei **Pilsbury 1903** the Emperor's Slit Shell from depths of 50 fathoms or more off south-west Japan, 100mm. Relatively heavier than the rest of the genus, it is taller and more coarsely sculptured than *P. africanus* above. It is creamy white, heavily overlaid with diagonal streaks varying from pale salmon to a rich orange-red. Where the slit has been filled in as the shell grew bigger is a stripe which gives the effect of the diagonal lines having been pulled backward at that point. A similar effect can be seen in *P. africanus. P. hirasei* is the least rare of the family.

P. quoyanus **Fischer and Bernardi 1856** (not illustrated) Caribbean, 45mm. It was the first of the family to be found and is pale brown with yellowish streaks.

Other members of the genus *Perotrochus* are *P. teramachii* Kuroda 1955 which is very variable or is the name given to two separate species, and is from Japan and Taiwan; *P. amabilis* F. M. Bayer 1963, *P. gemma* F. M. Bayer 1965, *P. lucaya* F. M. Bayer 1967, *P. midas* F. M. Bayer 1965, *P. pyramus* F. M. Bayer 1967 all from the Caribbean area; and *P. atlanticus* Rios and Matthews 1968 from Brazil.

Entemnotrochus rumphii **Schepman 1879** (not illustrated) Japan and Taiwan, the largest of the family, up to 250mm. The only other member of this genus is *E. adansonianus* Crosse and Fischer 1861 from the West Indies and Caribbean area. It is the second largest of the family. Both are heavy and solid and have longer slits than members of the other genera. They also have a wide, open umbilicus.

Mikadotrochus beyerichi **Hilgendorf 1877** (not illustrated) Japan. Members of this genus are moderately large and heavy, and have a shorter slit than those of *Entemnotrochus,* but longer than those of *Perotrochus.* Other members of the genus are *M. schmalzi* Shikama 1961 and *M. salmianus* Rolle 1899, both from Japan, and *M. notialis* Leme and Penna 1969 from off south Brazil.

Perotrochus africanus

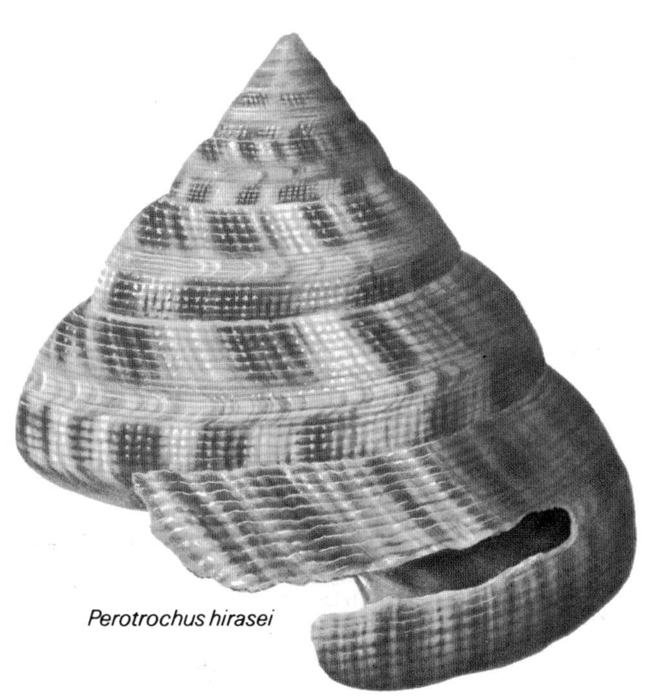

Perotrochus hirasei

Family: Haliotidae

The family contains only one genus. The shells somewhat resemble a single valve of a bivalve. Rather flat, they cling to rocks with a muscular foot like a limpet. They have a small spiral whorl at the low apex. Body whorl has a series of holes, from four to ten depending on the species, through which water and waste matter is passed, as through the slit of *Pleurotomaria*. Inside of shell is nacreous, iridescent, and used for costume jewellery. The foot is regarded as a delicacy in China, Japan, America, Channel Islands and New Zealand. Known as abalones in America, ormers in Channel Islands, paua in New Zealand, and awabi in Japan. There are some 100 species in the genus.

Haliotis asinina L. 1758 west Pacific, 120mm long and half as wide. Shape gives it the name Ass's Ear. Seven holes open. One of the few glossy members of the genus. Cream heavily overlaid with green-brown triangles and irregular blotches; interior iridescent silver.

H. iris Gmelin 1791 New Zealand, 150mm long, 115mm wide. Outer surface frequently covered with heavy calcareous deposit; interior richly iridescent silver, blue and green, dashes of gold. Muscle scar centrally.

H. discus Reeve 1846 Japan and east Asia, 100mm long, 65mm wide. Rough, rather uneven surface with bumps and hollows. Surround to holes is elevated; three to five holes open. Green-brown; interior silver. Prefers shallow water.

H. ovina Gmelin 1791 west Pacific, 60mm long, 45mm wide. More round than *H. discus*. Ribs run from apex to lip; five to six holes open. Dark green; six, radiating, cream streaks from apex to holes; twenty, narrower streaks from holes to left-hand edge; interior silver.

H. corrugata Wood 1828 California and Lower California, 180mm long. Higher than most of the genus. Nearly round with highly corrugated surface and scalloped edge; three to four holes open. Green or red-brown, often heavily covered with limey deposit, as illustrated; interior silver, gold, pink, green and blue. Muscle scar. Commonly used for food.

H. sieboldi Reeve 1846 (not illustrated) Japan, 150mm long. Oval in outline with apex close to margin. Strong radial ridges of varying widths. Holes large, somewhat tuberculate, four or five open. Brown or red-brown; interior silver.

H. ruber Leach 1814 (not illustrated) South and south-east Australia and Tasmania, 160mm long, 125mm wide. Fine, beaded spiral cords and uneven radial folds. Six or seven holes open, holes somewhat tuberculate; area between holes and margin concave. Dark red-brown; narrow green rays.

H. jacnensis Reeve 1846 (not illustrated) Philippines, south and south-west Pacific, 12mm. Very rough, irregular spiral ridges. Rather scaley. Holes, a little tuberculated and rather far apart, two or three open. Orange-red; interior silver.

Haliotis asinina

Haliotis iris

Haliotis ovina

Haliotis discus

Haliotis corrugata

H. rufescens **Swainson 1822** the Red Abalone, north and Lower California, up to 300mm. It is a large shell, oval and relatively flatter than *H. corrugata*. Its rough exterior has uneven spiral ridges and coarse, radial growth lines. It has three or four holes open. Brick red; the interior is silver and golden brown with a central muscle scar. It is popular as food, although only those over 175mm may be taken.

H. varia **L. 1758** west Pacific, up to 60mm. A common shell. The uneven spiral ridges are coarse and nodulose. It has strong growth lines and some radial folds. The holes, five or six of which are usually open, are raised. It may be an olive green colour or red-brown or mottled with white and/or black. As its name implies it is variable. The interior is silver with iridescent blues and greens.

H. lamellosa **Lamarck 1822** is recorded as coming from Gibraltar, 50mm. The illustrated specimen, however, is from the Greek Islands. It has a rather straight lip and uneven, radial corrugations looking like creases or wrinkles, and fine, spiral striae. It has four holes open. It is a blue-green, almost turquoise colour and the interior is silver, blue, green and highly iridescent. It may be a form of *H. tuberculata* L. 1758 (see next page).

H. diversicolor **Reeve 1846** western Pacific, up to 75mm. It is relatively narrower and flatter than the majority of the genus and has rather evenly marked ridges running from the spire to the lip. About nine holes are open. Dark olive green splashed with dark red, brown and lighter green.

H. fulgens **Philippi 1845** the Green Abalone, California and Lower California, up to about 200mm. It is oval with coarse, flattened radial ridges and has five or six open holes and a central muscle scar. The outside is a fairly even dull brown and the interior beautifully iridescent with blue-green, gold and dark brown on silver. *H. fulgens* is fished commercially above a size of 160mm.

H. spadicea **Donovan 1808** (not illustrated) east coast of South Africa, 75mm. Perhaps better known under its synonym of *H. sanguinea* Hanley 1841. Rather elongate shell, the exterior comparatively smooth for the genus but with uneven, radiating ribs. Rough spiral ridges, fine growth lines and about nine holes open. Red-brown; interior silver with brown stains near the apex.

H. emmae **Reeve 1846** (not illustrated) south Australia and Tasmania, 100mm. Coarse spiral riblets, uneven radial folds and six or seven holes open and raised, with a narrow channel running beneath the holes. Orange-brown with about six, uneven, radial, cream rays; interior iridescent.

Haliotis rufescens

Haliotis lamellosa

Haliotis varia

Haliotis fulgens

Haliotis diversicolor

H. gigantea **Gmelin 1791** Japan, over 200mm. Although the largest Japanese *Haliotis*, it is not as large as some others in the genus. Its holes are more elevated than any other, being tubercular in shape, and some four are open. The spiral corrugations are uneven and they cross lumpy ridges running at 90° to them. It is a dull brown or red-brown and is prized as a food. The two illustrations show the variation in colour and also how the shell becomes rounder in outline as it grows larger.

H. tuberculata **L. 1758** Mediterranean and north-east Atlantic, about 90mm long and 60mm wide. It is prized as food in the Channel Islands, where it is known as the ormer, but is very seldom found north of the English Channel. The spire is raised and growth lines clearly show. It is deeply striated spirally. There are about nine holes open. Variable in colour from dull brown, red-brown to mixtures of browns, reds and greens with pale green, zigzag waves. The inside is silver with faint red tinges.

H. midae **L. 1758** South Africa, up to 140mm long and 115mm broad. It is the largest of the South African species and is covered with deep irregular wrinkles radiating from the area of the apex. It falls very sharply from the line of holes to the left margin. The holes are numerous, small and about nine are open. It is a dull white with a red tinge; inside it has a central muscle scar and is silver with very pale greens, blues and pinks.

H. australis **Gmelin 1791** New Zealand, where it is known as the Silver Paua, up to 100mm. It has fairly even radial folds across low spiral ridges and growth lines. There are seven holes open. Pale olive green and may have faint red rays; interior silver. It is used for food.

H. cracherodi **Leach 1817** the Black Abalone, west USA and Lower California, up to 150mm. It has a smoother surface than most though the growth lines are clear. About six holes are open. Outside it is a very dark red-brown, almost black, and inside silver with a golden sheen and a central muscle scar.

H. kamtschatkana **Jonas 1845** (not illustrated) north Pacific from Japan, south Alaska to California, 150mm. Elongate, roughly corrugated and may have weak spiral cords; about four holes open. Grey brown.

H. elegans **Philippi 1899** (not illustrated) Western Australia, 100mm. It is elongate with a small spire close to the posterior edge of the shell, somewhat inflated and has strong spiral ribs. Number of open holes is variable—about eight in young shells. Orange-brown with cream rays; interior silver.

H. coccinea **Reeve 1846** (not illustrated) Cape Verde and Canary Islands, 50mm. It has close-set spiral ridges—sometimes unequal—with fine striae. Holes are close together and five or six are open. Scarlet with cream blotches or rays; interior silver.

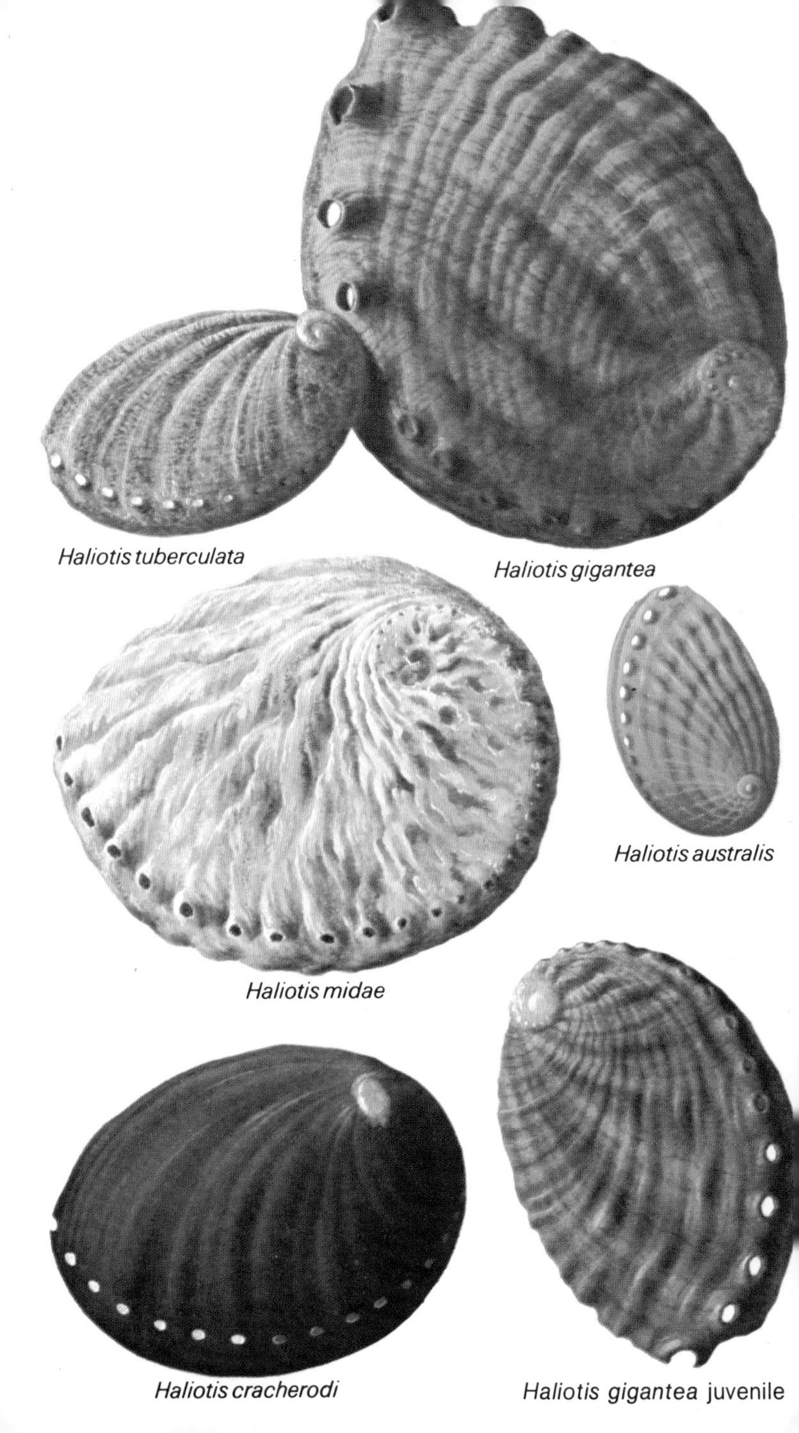

Haliotis tuberculata

Haliotis gigantea

Haliotis australis

Haliotis midae

Haliotis cracherodi

Haliotis gigantea juvenile

Superfamily: Fissurellacea

Family: Fissurellidae

The keyhole limpets are flat and round, oval or shield-shaped like true limpets, but generally have a hole in the dorsum—sometimes keyhole-shaped—and often a slit in the margin. There are many small members of the family which includes several hundred species. They cling to rock and coral, mostly in warm seas and are vegetarians.

Scutus antipodes Montfort 1810 east Australia from Queensland to Tasmania, 100mm long and about half as wide. As all the genus, it is shaped like a Roman shield from which it takes its name. It is depressed and has fine, concentric growth lines with the apex well off-centre. It has no hole and the apex points forward rather than upward. Creamy white outside and pure white within.

Glyphis elizabethae E. A. Smith 1901 east South Africa, up to 45mm long and 30mm wide. The hole is about one-third of the way from the end and there are eight prominent ridges and numerous small ones of varying sizes between. Grey; interior white.

Fissurella radiata Lamarck 1801 Caribbean, 40mm long. It has eleven irregular, radiating ribs with finer ribs between and an almost circular hole. Grey; radial ribs lighter; interior white with faint yellow tinge.

F. picta Gmelin 1790 South America including the Strait of Magellan, 50mm long and 30mm wide. It has an oval opening forward of the centre, and its growth lines give it a rough feel. Twelve dark brown rays alternate with light tan rays of about the same size; interior pure white with the ends of the rays showing round the margin.

F. grandis Sowerby 1834 west South America, 60mm long and 40mm wide. Much like *Megathura crenulata* in outline though narrowing more towards the anterior and much smoother. Very dark brown, almost black, this colour forming a narrow band round the inside edge; rest of the interior is pure white as is the immediate surround to the hole.

F. nodosa Born 1778 Florida Keys and the West Indies, 40mm long, 30mm wide and high for the genus, 20mm. It has twenty-two ribs which are nodulose, becoming more so towards the margin. The hole is in the shape of a figure eight. White inside and out and usually carries a heavy calcareous deposit.

Macrochisma producta A. Adams 1850 the Antilles, 40mm long, 20mm wide. It has much the same outline as *Scutus antipodes* but is relatively higher with a long, narrow, triangular opening running from the apex to within a few millimetres of the end. White within and without.

Megathura crenulata Sowerby 1825 California and the Pacific coast of Mexico, 125mm long and 80mm wide. A large member of the genus, it has a large oval hole a little off-centre with a white edge from which numerous fine ridges run to the crenulated margin across concentric growth lines. Grey-brown; lighter towards the margin; interior pure white.

Diodora graeca L. 1758 Mediterranean to south England, 25mm long and 18mm wide. Synonym *D. apertura* Montagu 1803. It has radial ridges, alternate ridges being larger, and very fine concentric cords, giving a slightly reticulated effect. Pink-brown; cords white.

Glyphis elizabethae

Fissurella radiata

Diodora graeca

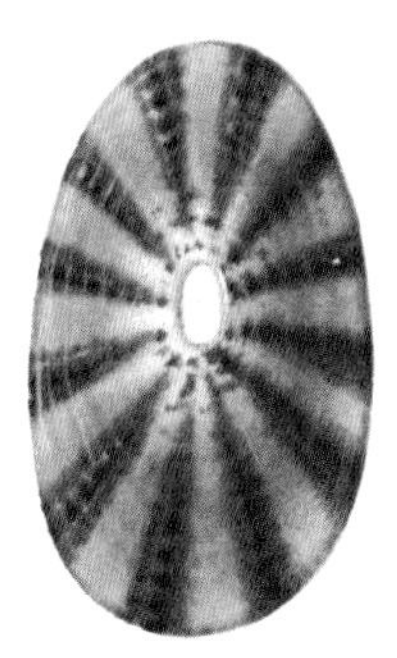

Fissurella picta

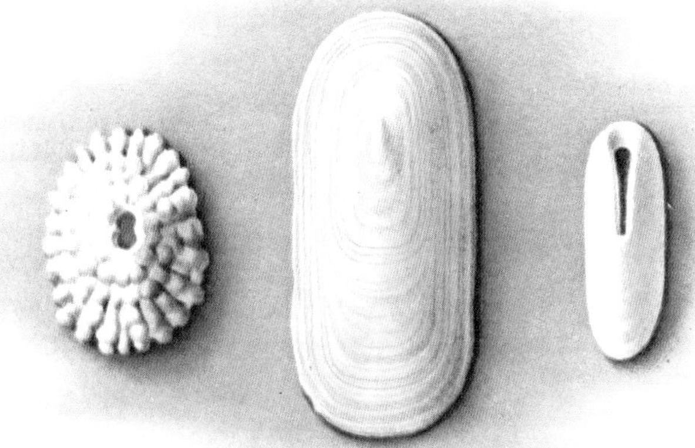

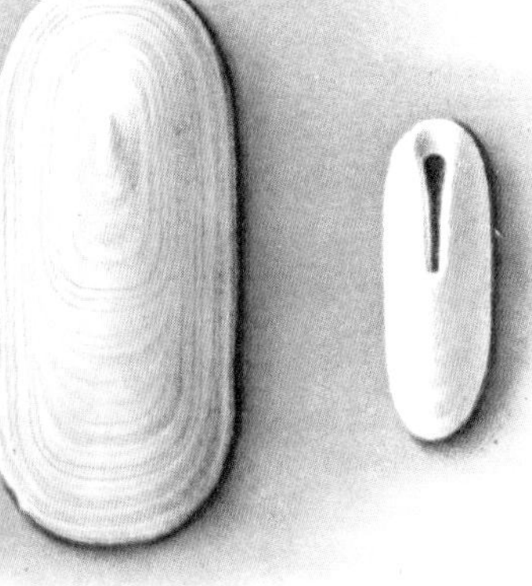

Fissurella nodosa

Scutus antipodes

Macrochisma producta

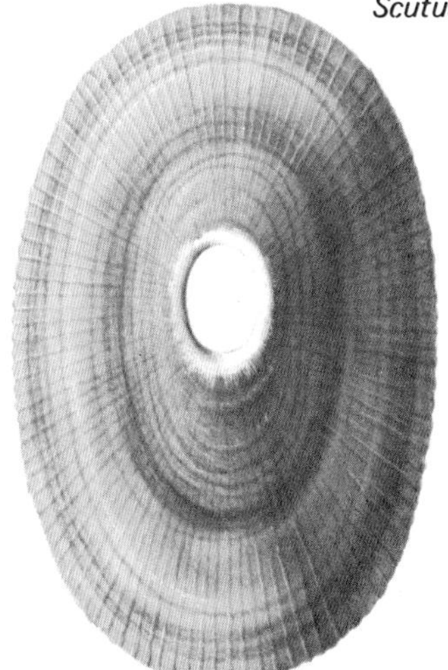

Megathura crenulata

Fissurella grandis

Superfamily: Patellacea

Some 400 species. The limpets include two main families, Acmaeidae and Patellidae and a third with few members, Lepetidae. The difference between the former two families is mainly in the animals – for example, the Acmaeidae have gills and the Patellidae have folds on the mantle edge which function as secondary gills. The interior of the Acmaeidae shells are porcellaneous whereas those of the Patellidae are iridescent. Herbivores.

Family: Acmaeidae

Acmaea patina Eschscholtz 1847 west North America, more common in the north than the south, 50mm long and slightly less wide. Smooth apart from fine growth lines. Grey-green with a blue-grey patch at apex, slightly off-centre; then an area where blue-grey colour forms a network over base colour; on outer third of shell, blue-grey forms reticulations. Interior blue-white with dark brown patch at apex and purple-brown marks around border, corresponding with ends of blue-grey rays on outer side. *A. testudinalis* Müller 1776 in the east is the same species but is smaller, more oval and has a darker brown mark inside.

A. borneensis Reeve 1855 north coast of Borneo, 40mm long and slightly less wide. Rather depressed. Finely striated radially. The apex, a little off-centre, is chestnut-coloured. The rest of the exterior is black with irregular rays of interrupted, short, white streaks. The head scar is cloudy blue, brown and white; its surround white with a blue tinge and a dark blue-black border. Until recently believed to be a *Patella*.

A. pileopsis Quoy and Gaimard 1834 north New Zealand, Cook Strait, 25mm. Numerous small ribs; apex well forward. Red-brown speckled with white; interior with dark brown head scar surrounded by blue-white, shading into dark border streaked with red.

Family: Patellidae

Patella longicosta Lamarck 1819 east South Africa, 70mm. South Africa has a particularly fine variety of large and colourful Patellidae. Rough surface. Ten strong ribs end in points like a star, projecting most at the end farthest from the apex, and smaller intermediate ribs. Dark or light brown; interior cloudy brown and white at the centre, with a sharply defined border of orange, bordered by purple then a pale orange tan, and then white. Interior margin has a narrow band of dark blue. It may also be off-white with a darker head scar.

P. saccharina L. 1758 Japan, 40mm. Like a small edition of the above. Seven major ribs and small intermediate ones. Dark brown externally; white inside with dark brown margin and dark centre.

P. oculus Born 1778 also east South Africa, 80mm long, 70mm wide. Rather thin and flat. Some eighteen ribs, not nearly as pronounced as in *P. longicosta*. Dark brown and green circular stripes. Head scar is pale pink-brown bordered with broad white band; outer third dark brown with a little pale clouding.

P. caerulea L. 1758 Mediterranean, north-west Africa, 50mm. Rather flat and thin. Irregularly sized, rather flat ribs. Green-grey; white head scar surrounded by a translucent blue-grey through white. Ribbing shows as dark blue rays.

P. testudinaria L. 1758 Philippines, 90mm. Oval, rather smooth, solid.

Patella saccharina

Patella longicosta

Patella caerulea

Acmaea pileopsis

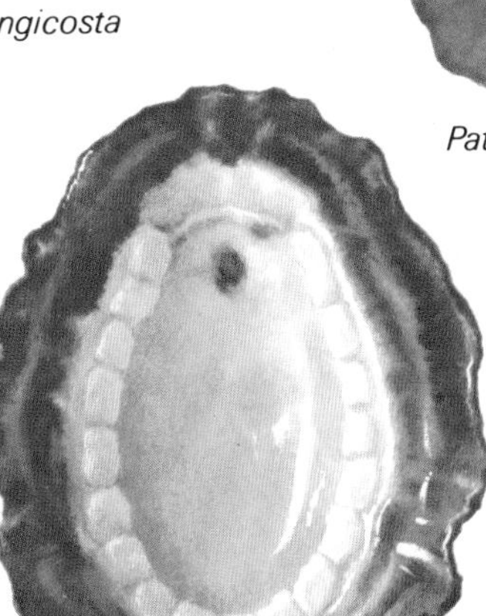

Patella oculus

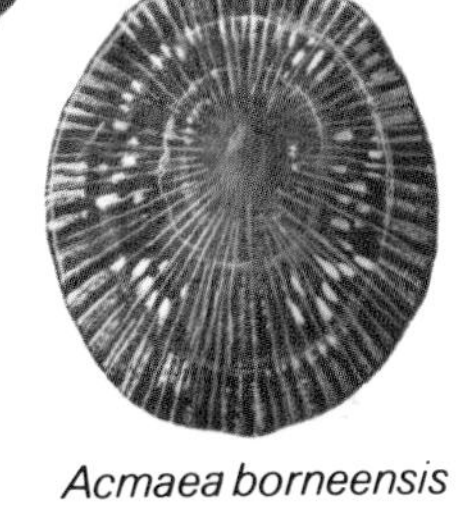

Acmaea borneensis

Patella testudinaria

Acmaea patina

Very dark, almost black, with rays of red-brown spots; white head scar. Translucent; inside the external colours show through a silvery nacreous layer.

P. granatina **L. 1758** South Africa's western seaboard, 85mm long and rather high, 30mm. It has about thirty, not very developed ribs and is grey with dark brown mottling or zigzags. The interior has a very rich, deep brown head scar surrounded by a flesh-coloured ring and this by an orange-brown or blue-white area edged with a pale purple-brown flecked band.

P. compressa **L. 1758** also from the Atlantic coast of South Africa is 110mm long, 50mm wide and 50mm high. It is compressed at the sides and the high apex is slightly forward of centre and comes to a point facing forward. When resting on a flat surface, the end margins are both clear of the ground. It is finely ribbed and smooth at the apex. Pale brown-grey or grey-pink, darker towards the margin; interior white or pearly grey edged with a broad pink band.

P. vulgata **L. 1758** the Common Limpet, west Europe, up to 70mm. Its coarsely ribbed exterior is variable in greys and light browns. Inside, the white or blue head scar is surrounded by dark blue and white rays overlaid with a translucent orange nacre, which may completely obscure the underlying colours.

P. miniata **Born 1778** South Africa, 60mm. It has many ribs of varying size, generally large and small alternately. It is grey-white with radial blue-grey streaks. These colours show through the opaque nacre lining the interior except for the white head scar.

P. barbara **L. 1758** east and west coasts of South Africa, 95mm. Variable; some flat and others high; some with strong ribs; some with a few strong but numerous small ribs. White or off-white externally; brown or red head scar.

P. cochlear **Born 1778** east and west coasts of South Africa, 65mm. Pear-shaped and the exterior is finely striated radially. White to pale tan; interior has a light chestnut head scar with a dark blue surround, a pale blue outside this, and an intermediate blue outside this again; white margin.

P. nigrolineata **Reeve 1854** Japan, 60mm long and 50mm wide. May be flat or high with the apex usually well forward of centre. Numerous, slightly prominent, narrow, radial ribs and weak, concentric growth lines. Blue-green; ribs and growth lines dark red-brown. Interior silver-blue with the ribs showing through. Head scar white, but may be stained with red or dark brown.

Patella granatina

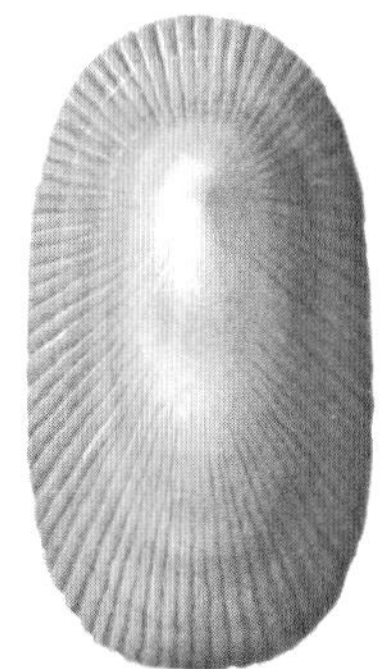
Patella compressa

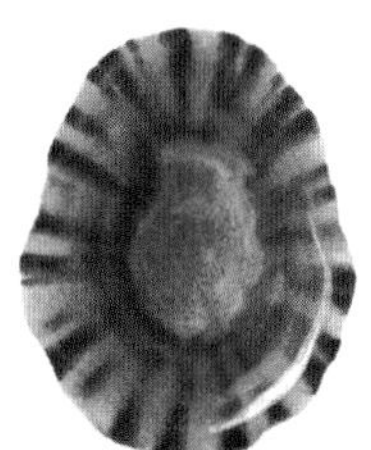
Patella vulgata

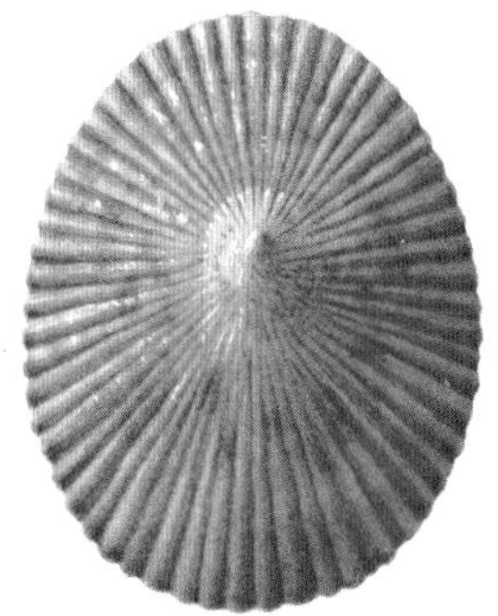
Patella miniata

Patella barbara

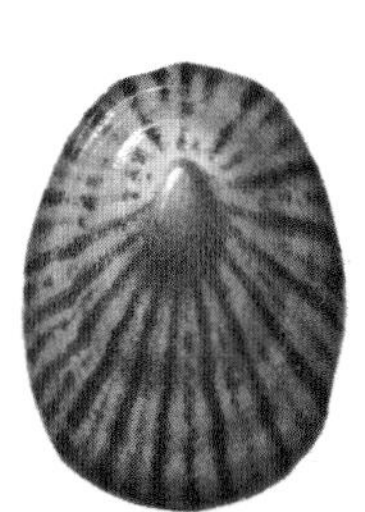
Patella nigrolineata

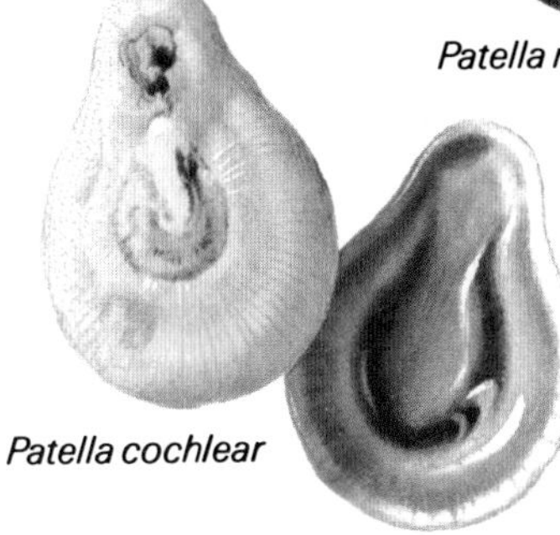
Patella cochlear

Patella vulgata

Superfamily: Trochacea

Family: Trochidae

The Trochidae or top shells, are flat-based conical shells found in tropical and temperate waters. They have a horny operculum and a nacreous interior. They inhabit the intertidal zone and shallow water and are vegetarians. Some are eaten and some used commercially.

Trochus maculatus L. 1758 the Indo-Pacific, up to 65mm high and as tall or taller than it is wide. Top-shaped, it is sculptured with rows of fine granules and may be either rounded or angular at the base. The latter is more finely sculptured than the sides. The umbilicus is nacreous, lirate and edged with five blunt nodules. It is grey or white marbled with blue, green or brown. It is very variable in shape and colour, as the two illustrations show.

Cardinalia conus Gmelin 1790 Indo-Pacific, 70mm high. A heavy solid shell, the lower whorls slightly convex, and with a rounded edge to the base. Spiral rows of nodules on the early whorls disappear on the later whorls, which have strong, radial growth lines. It is white with red, flare-like marbling which becomes long, narrow dashes on the base.

Tectus dentatus Forskäl 1775 north-west Indian Ocean, 80mm high. Finely striated radially, and wrinkled. The sutures have prominent, blunt, solid knobs, up to ten on the last whorl. The base is flat and largely polished with obsolete concentric striae near the centre. The base is white with a blue-green area near the columella which is twisted forwards; rest of the shell is a dull grey.

T. pyramis Born 1778 Indo-Pacific, up to 70mm high. This is rather shorter in relation to its height than the preceding species. The early whorls have radial lines of small nodules becoming obsolete towards the base. These whorls have small blunt nodules at the sutures, again becoming obsolete towards the base. The base is flat and spirally striated with the columella twisted forwards. It is a drab grey-brown marbled with mauve, generally axially streaked.

T. niloticus L. 1767 Indo-Pacific, up to 150mm high. It is the largest of the genus, almost smooth on the later whorls, except for fine radial striae. Above the sutures are angular, tubular nodules on a raised band on the early whorls, both of which disappear on the later whorls. In large adults the last whorl is sometimes very swollen, the otherwise straight side becoming very concave at the bottom. The base has fine growth lines and axial striae; umbilicate with a smooth columella. It is white with red-brown axial stripes. A common shell, it has been fished commercially for making 'pearl' buttons and the like.

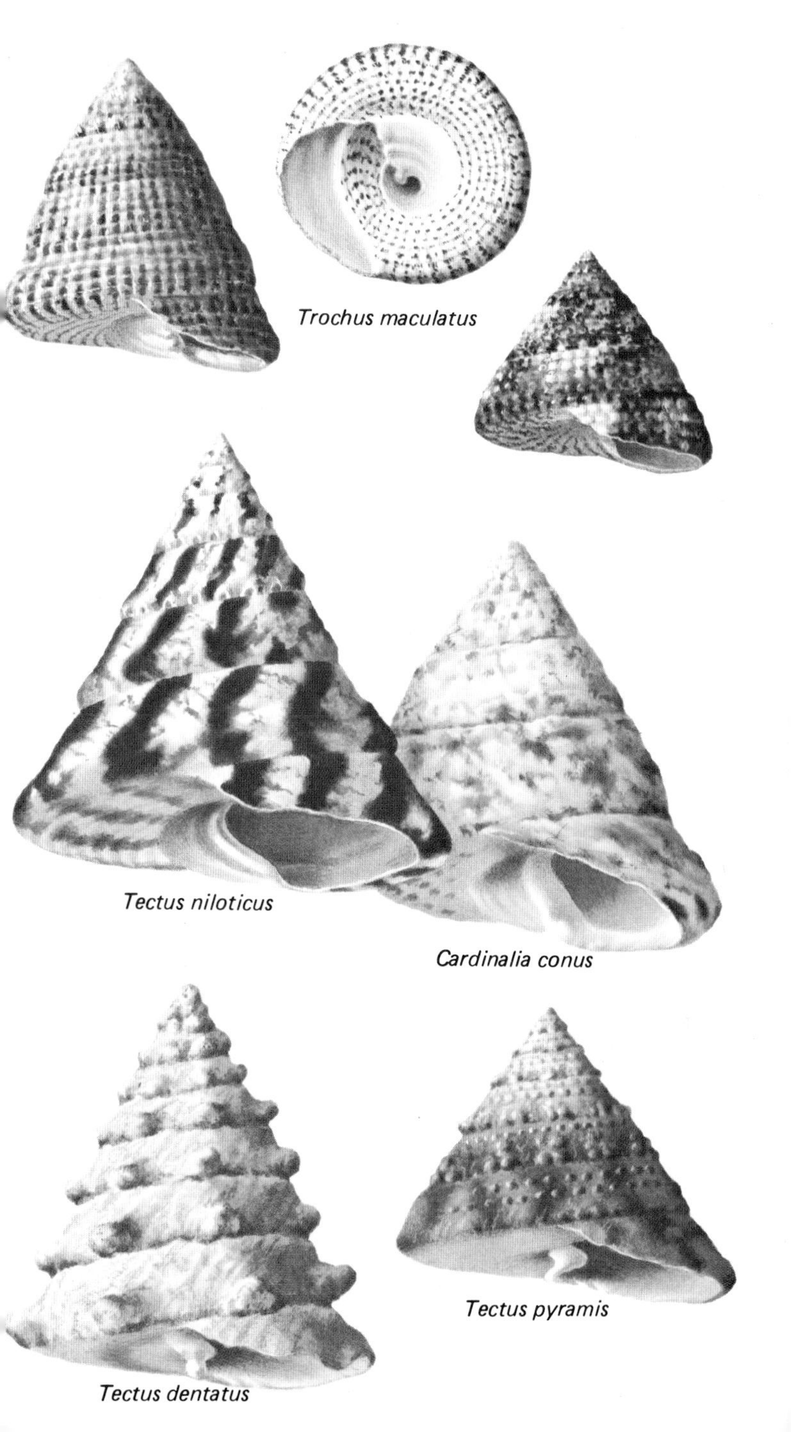
Trochus maculatus
Tectus niloticus
Cardinalia conus
Tectus pyramis
Tectus dentatus

Maurea punctulata Martyn 1784 New Zealand, 40mm. It is turbinate with spiral rows of tiny nodules, a rounded base and a smooth columella. Illustrated is *M. p. stewartiana* Oliver 1926 which is tan coloured.

M. cunninghami Griffiths and Pidgeon 1833 New Zealand, 50mm. It has a somewhat angular edge and spiral rows of minutely noduled striae. Early whorls are slightly concave and later ones convex. Slightly convex base has spiral lirae, minutely nodulose near the centre. Smooth columella. Cream background with light brown striae; base white.

Tristichotrochus formosensis Smith Taiwan, 40mm. Slightly concave sides and spiral rows of tiny nodules, the two nearest the suture being bigger. Base is sharply angulate and spirally striate, striae being finely nodulose near smooth columella. Flesh-coloured, red-brown clouding and spots on sutural beading; base, the same colour with red-brown clouding near columella and spotted at the edge.

Monodonta canalifera Lamarck 1816 west Pacific, 30mm. Turbinate with flat spiral cords and a curved tooth projecting from bottom of columella. Inside of lip white and lirate; edge of lip shows colours of exterior, dark green, green, and cream spots.

Tegula regina Stearns 1892 south California, 45mm. Six or seven whorls, axially unevenly ridged, with the lower third angled outwards. Dark purple-brown; base concave with close, narrow, black, radial ribs, cream in the interstices; interior and columellar area gold; columella and the area between the gold and the outer lip is silver-white.

Calliostoma monile Reeve 1863 Australia, 25mm. Rounded keel, flat sides and pointed spire. Whorls are finely striate and ridged above suture. Convex base; smooth columella with an obsolete tooth at lower end. Transparent creamy white; the sutural ridge a pure white with mauve spots almost as big as the white areas between them.

Clanculus cruciatus L. 1758 Mediterranean to Cape Verde, 10mm. Smooth with obsolete spiral ribs, final whorl inflated. Deeply umbilicate; columella has a tooth at each end, that at the top larger. Lip is strongly ridged within. Dark red-brown except on base around umbilicus where ribs become stronger and are coloured with alternate pink and dark red-brown spots.

C. pharonium L. 1758 the Strawberry Top, Indian Ocean, 25mm. Columella ends in a tooth, and has one large and three small folds. Tooth at top of lip projects over the umbilicus. Lip is finely dentate. Spiral rows of fine rounded nodules, mostly rich pink-red, but some are pearl-coloured; two rows of dark purple dots on each whorl; base similar, but there is an area round columella and inside lip which is white and ribbed.

Oxystele sinensis Gmelin 1790 South Africa, 30mm. Rather flattened with large, inflated body whorl and rather rough surface. No umbilicus. Body whorl deep blue-black, earlier whorls green-yellow; base grey; rose shading to white on smooth columellar edge.

Chrysostoma paradoxum Born 1780 Indo-Pacific, 25mm. Low spire and large body whorl. Smooth with a shallow groove below the suture. Columella and its callus, aperture and interior all rich orange-gold; rose-brown irregularly flecked with cream.

Calliostoma monile

Monodonta canalifera

Maurea punctulata stewartiana

Clanculus cruciatus

Tegula regina

Maurea cunninghami

Tristichotrochus formosensis

Clanculus pharonium

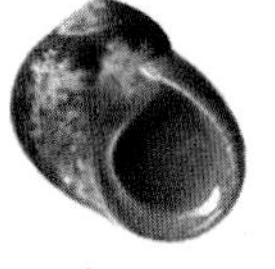

Chrysostoma paradoxum

Oxystele sinensis

Monodonta lineata da Costa 1778 Great Britain to north-west Africa, 25mm high and slightly narrower. Heavy and coarse, with rough grained surface. Body whorl swollen and penultimate whorl to a lesser extent. Rounded keel; umbilicus sealed with a callus. Apex often worn away in adults, the nacreous interior showing. Columella has a blunt tooth. Dark brown with very small tan flecks; columella white; inside lip dark brown backed by white; interior nacreous.

Umbonium giganteum Lesson 1831 Japan, 45mm across. Flat with rounded keel and smooth bar faint striae near bottom of last two whorls. Convex base has no umbilicus but has an umbilical callus. Pale blue-green with row of purple dashes below suture, becoming paler on body whorl; keel white with oblique, radial, purple stripes showing as a white and purple band at bottom of early whorls.

U. moniliferum Lamarck 1804 Japan, 20mm. Shaped like preceding species. Cream, heavily patterned with red-brown dots and flecks: base and columellar callus brown. Illustration shows a view of the base.

U. vestiarium L. 1758 Indian Ocean and East Indies, 12mm. Shaped like preceding species. Pattern and colour very variable—greens, browns, greys, purples, in dots, spiral lines, tiny axial flames, etc.

Cittarium pica L. 1758 West Indies, 80mm high and wide. Solid, heavy with rough uneven surface. It has a shallow channel below suture, rounded keel and is deeply umbilicate. Smooth columella with heavy callous deposit surrounding half the umbilicus and a deep notch in the middle. Black with white marbling; base around umbilicus white; interior silvery-white.

Angaria delphinula L. 1758 Indo-Pacific, 70mm wide. The genus *Angaria* have the thin horny opercula of Trochidae but are included by some with the Turbinidae. Solid, heavy with depressed spire. The encircling, coarse spiral cords have hollow spines on shoulder row, two rows of small ones on periphery, and on base around and inside deep umbilicus. Grey; spines sometimes darker; aperture pearly.

A. distorta L. 1758 South China Sea, 65mm wide, 50mm high. Heavy, solid with large body whorl. Angulate at shoulders with almost flat area from suture to shoulder which carries blunt knobs. Coarse, uneven spiral ridges, some nodulose, on body whorl. Deep umbilicus, opening to which is rough, almost spinous. Various shades of grey from white to black, over pale pink base colour.

A. melanacantha Reeve 1842 the Imperial Delphinula, Philippines, 40mm high, 55mm wide including spines. Sunken spire; suture disappears near aperture when body whorl no longer touches earlier whorl. Small, nodulose spiral ridges and a row of long upward and inward curving spines on shoulder. Six other rows of spines vary in length from row to row and line the umbilicus. Various shades of brown; nacreous, round aperture.

Family: Stomatellidae

Mostly small, they sometimes resemble the Haliotidae in having a depressed spire, large body whorl and aperture, but without perforations. Interior is nacreous. There are about fifty species.

Pseudostomatella decolorata Gould 1848 Philippines, 35mm long, 30mm high. Four whorls, the last greatly expanded, with very fine ridges showing more strongly on penultimate whorl. Wide aperture and smooth columella. Pale brown to green; ridges carrying small arrowhead-shaped

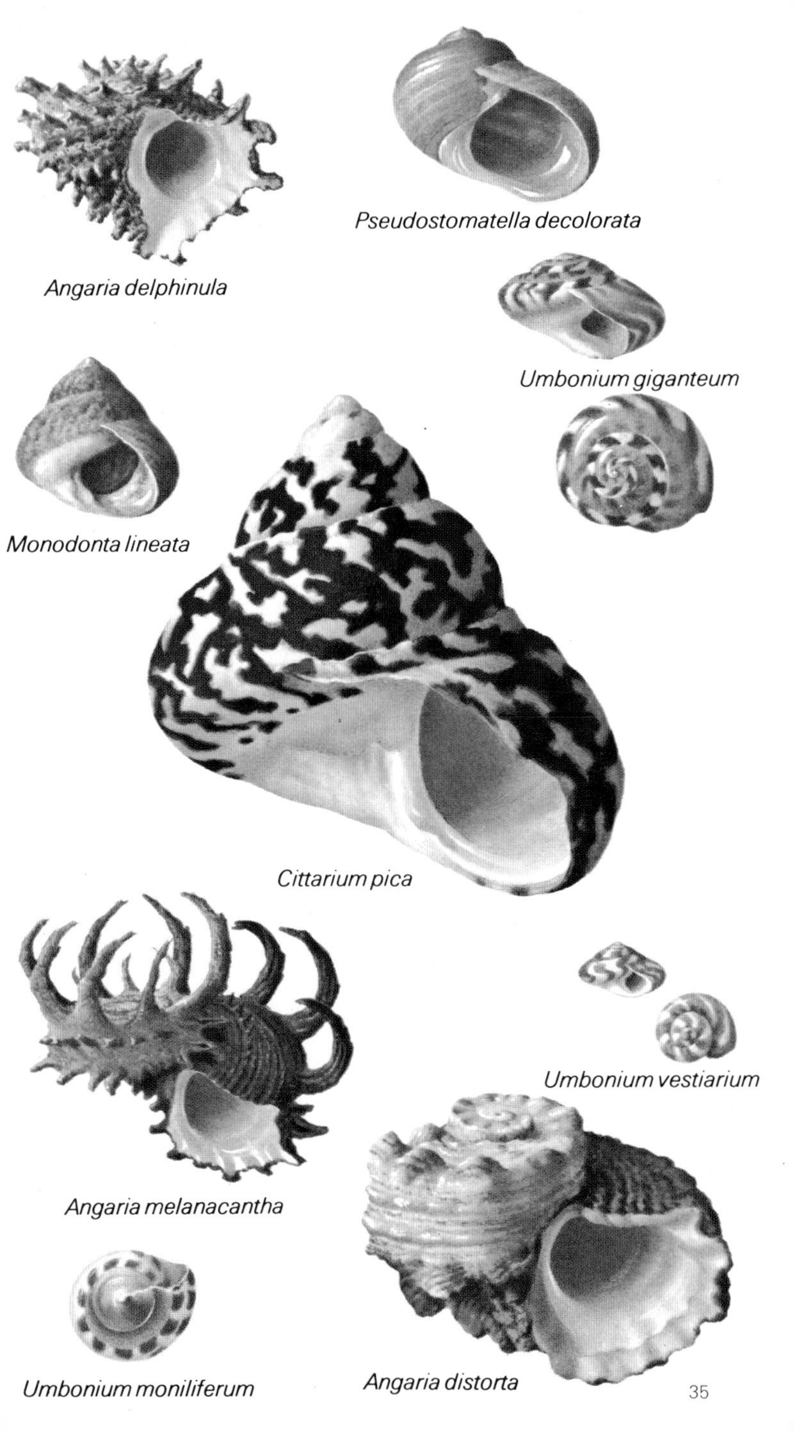
Angaria delphinula
Pseudostomatella decolorata
Umbonium giganteum
Monodonta lineata
Cittarium pica
Umbonium vestiarium
Angaria melanacantha
Umbonium moniliferum
Angaria distorta

white marks; alternate dark and light blotches below suture; white crescent-shaped callous area beside columella.

Family: Turbinidae

Mostly fairly large and medium-sized shells, and turbinate as the name suggests. They can be distinguished from the Trochidae to which they are somewhat similar by the calcareous operculum which is flat on the side attached to the animal, and usually spherical, but sometimes with sculpture, on the exterior side.

Turbo marmoratus L. 1758 East Indies and Australia, up to 200mm. The largest of the family, it is solid and heavy. The body whorl is very expanded and has three ridges carrying blunt knobs; that nearest the apex very angular and raised, a small one on the middle of the whorl, and one near the base, also angular. The columella is smooth and in adults has a large, rough, wavy fasciole. It is green or brown-green with spiral lines of light green and tan, and on the early whorls lines of dark green and white blotches of varying widths; a band of the same colours below the suture; interior is nacreous.

T. cornutus Lightfoot 1790 Japan, 90mm high. It has four large ridges and about three small ones on each whorl, all becoming obsolete on the body whorl on which the outer two of the large ridges develop hollow horns, growing larger nearer the aperture. It also has coarse growth lines, a wide simple mouth and smooth columella. It is green and brown, the colour changing along the growth lines; interior and columella are nacreous with a white area bordering the columella.

T. rugosa L. 1758 Mediterranean, Portugal, Azores and Canaries, 35mm high, 45mm wide. It bears nodulose spiral ribs and a row of rectangular knobs below the suture, arranged radially. Light brown; the simple columella which is silvery like the interior, is surrounded by a scarlet, callous area.

T. sarmaticus L. 1758 South Africa, 65mm high, 80mm wide. The largest South African turbo, it has a large body whorl, two spiral rows of nodules on the shoulder and other uneven ridges. It is dull brown, covered by a green periostracum with one black band above the suture, a second on the inside of the outer lip, and a third bordering the top of the columellar area dividing the nacreous interior from the cream colour of the shiny columella; columella bordered by a rich red callous area and pustulate operculum (illustrated) is creamy white.

T. (Ninella) torquata Gmelin 1791 Australia, 110mm wide, 70mm high. Solid, heavy shell. Whorls have a carina at their widest part which disappears in the body whorl of specimens from east Australia. It has other indistinct ridges, a row of nodules on the shoulder, and is sculptured with fine close axial lamellae. Umbilicus is deep and operculum (illustrated) has somewhat ear-shaped ridges. Pale red-brown and white with some green splashes; columella, umbilicus and operculum white; interior nacreous.

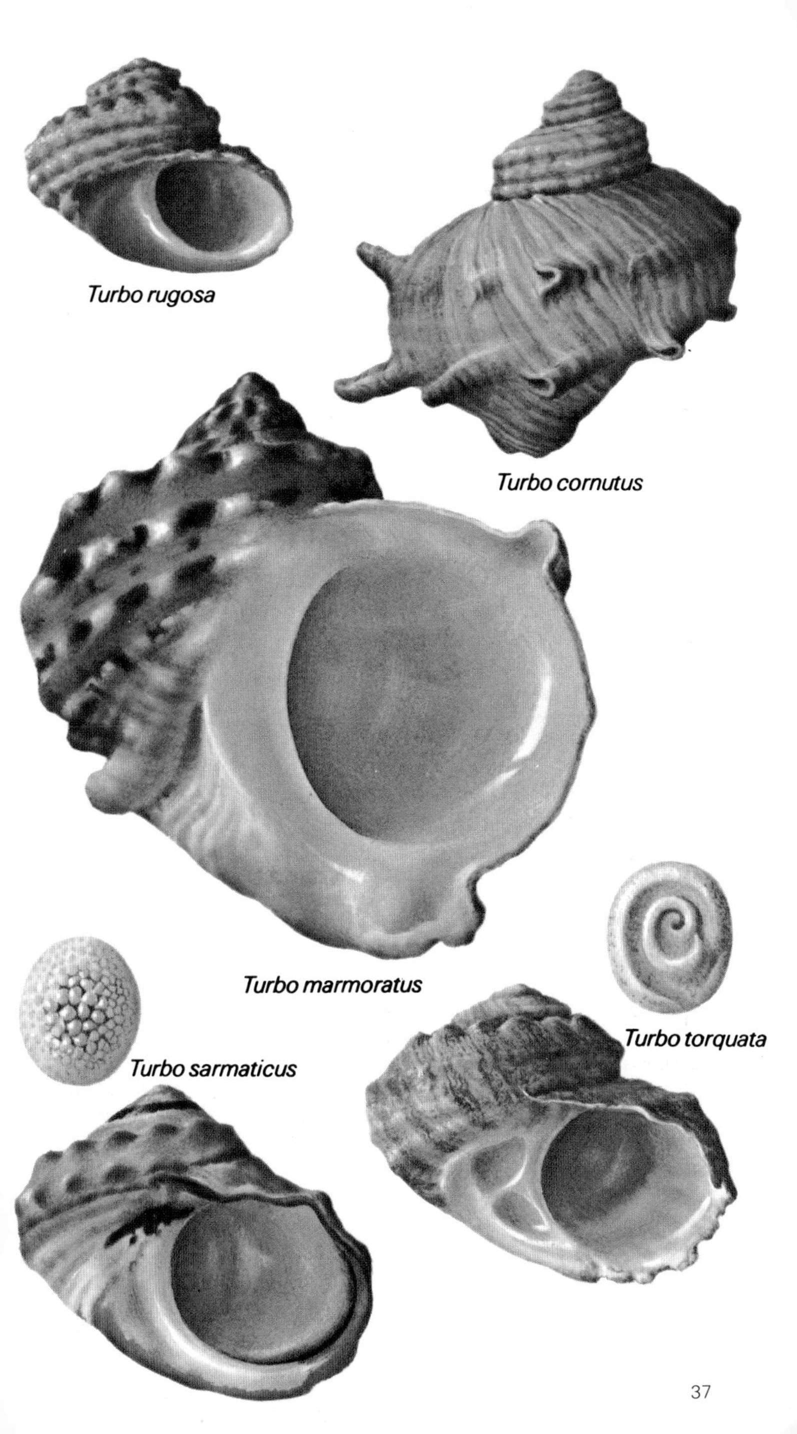
Turbo rugosa
Turbo cornutus
Turbo marmoratus
Turbo torquata
Turbo sarmaticus

T. (Ocana) cidaris Gmelin 1790 east coast of South Africa, 40mm high, 45mm wide. Rather depressed, smooth with fine growth lines. Smooth columella; operculum with white nodules rather like *T. sarmaticus.* Colour very variable, greens and browns with white markings especially below suture; spiral bands and dotted lines.

T. petholatus L. 1758 Philippines, 75mm. Highly polished, variable in shape, pattern and colour. Spire more or less high; smooth columella. Various shades of green, brown, fawn, cream and white, usually with flame-shaped light markings below suture. Spiral rows of dark spots and white arrowhead marks, especially evident on lower half of body whorl; interior nacreous. Operculum is well-known 'cat's-eye' used in jewellery. Two colour varieties are illustrated. Orange variety shows the 'cat's-eye'.

T. (Lunella) cinereus Born 1778 Indo-Pacific, 30mm high, 40mm wide. Very depressed spire; fine spiral striae becoming fainter on body whorl; fine growth lines. Deep umbilicus and somewhat produced siphonal canal giving a peg-top shape. Cream, heavily mottled with dark brown with green tinge, though less mottling round the periphery; aperture and columella nacreous; operculum finely granulated, white shading to dark at outer edge.

T. (Marmarostoma) bruneus Röding 1798 Indo-Pacific, 50mm long, about 47mm wide. Scabrous spiral cords of uneven sizes, largest on the shoulder giving a sub-angular look. Very narrow umbilicus. Cream with almost black, slightly wavy, axial flames; columella and inside of lip nacreous bordered with white, as is the interior; finely granulated operculum dark purple on columellar side, becoming pale, almost white, at outer edge.

T. (M) chrysostomus L. 1758 Indo-Pacific, 70mm long, 60mm wide. Sculptured with coarsely scabrous spiral cords, which on the angular shoulder carry short, broad, open spines; cord on the periphery and another a little lower have much smaller, open spines. Nearly circular aperture; very narrow umbilicus. Creamy fawn with dark green-brown flames from suture to shoulder, and spots and dashes below; interior and columella rich golden orange, bordered with yellow and then white; pale fawn at lip. Operculum almost smooth at centre becoming granulate and then striate at outer edge; dark brown in middle becoming golden orange with a white edge towards columella, and flesh-coloured with a golden orange edge towards outer margin.

T. (M) argyrostomus L. 1758 Indo-Pacific, 90mm long, 80mm wide. Bears unequal, scabrous spiral cords, one on the shoulder being especially large and one a little smaller, as far below the periphery as the large one is above, giving a flat-sided effect. May or may not be umbilicate. Pale green-cream with dark red-brown axial flame marks; aperture finely edged with green; columella and lip nacreous, silver, bordered with white; interior white; operculum green and white, granulate.

Phasianella australis Gmelin 1791 southern Australia. There are some forty species of *Phasianella* or pheasant shells, which are also turbans. Most are small and all are delicate and richly coloured. *P. australis* is the largest, 100mm. Elongate with pointed spire, convex whorls and no umbilicus. Smooth, highly polished and rather fragile. Operculum is polished, white and pointed at one end. Almost limitless variety of colours and patterns, but generally spirally marked with bands of plain colour, arrowhead marks, wavy lines, dots and dashes, greens, browns, reds, pinks, yellows and white. The two illustrations show a 'banded' variety and a young shell with a sub-axial flame pattern.

Turbo cidaris

Turbo petholatus

Turbo bruneus

Turbo cinereus

Phasianella australis

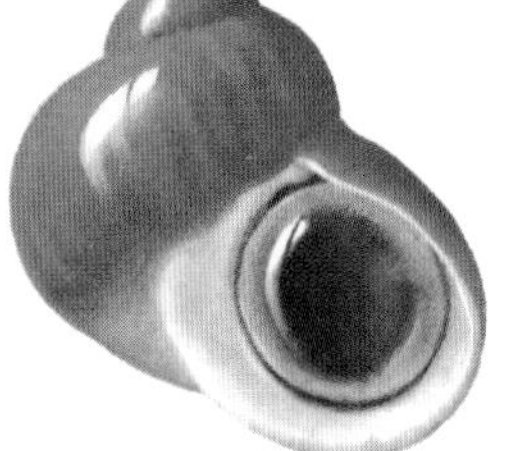
Turbo petholatus

Phasianella australis

Turbo argyrostomus

Turbo chrysostomus

Astraea tuber L. 1758 Florida and West Indies, 50mm high and wide. Low pointed spire with squat, rounded axial ribs running from suture to shoulder of each whorl; knobs on periphery. Overall diagonal rows of small nodules. Columella has shallow channel, and fold outside it. No umbilicus. Operculum has thick comma-shaped ridge. Green-brown; nodules white; interior nacreous; columella white.

A. kesteveni Iredale 1924 Western Australia, 30mm high and wide. Very fine spiral striae and growth lines; base of each whorl unevenly overlaps suture giving skirted effect. Base has fine concentric ridges. White; silver-white interior and columella, latter surrounded by pale blue; smooth, uneven operculum is purple, shading to green-white.

A. phoebia Röding 1798 West Indies and Florida, 55mm wide, 30mm high. Spiral rows of small nodules becoming hollow and subspinous towards aperture. Edge of each whorl bears saw-tooth spines curving slightly upwards. Base with spiral rows of nodules, may or may not be umbilicate. White; interior silvery; pale green tinge around umbilicate area.

A. calcar L. 1758 Philippines and Malaysia, 50mm wide, 20mm high. Depressed spire; low, blunt axial riblets; periphery of each whorl has long, blunt, dark-tipped spines; suture of last whorl may be deeply indented. Base with fine, spiral, scabrous ridges and no umbilicus. White; white columella surrounded by yellow-green; interior yellow-green; aperture has orange tinge; band of white between.

A. tuberosa Philippi Indonesia and Malaysia, 25mm high and wide. Low axial riblets; constricted a little below suture. Periphery has two rows of small spines. Base with spiral rows of fine lamellae, no umbilicus. White; interior and columella nacreous, pale lavender area around columella; faintly pustulate operculum, dull purple, white tinge at edge.

A. stellare Gmelin 1791 Australia, 30mm high, 45mm wide. Fine spiral ridges; low, small axial riblets. Base of each whorl has blunt hollow spines open from below. Base has spiral ridges. White; interior and columella silver-white, latter bordered by opalescent blue area.

A. heliotropium Martyn 1784 New Zealand in deep water, 70mm high, 120mm wide. Convex whorls with small, uneven, wavy, scabrous, spiral ridges; edge of each whorl with flat, triangular, hollow spines curving up at tips. Base has five spiral rows of small, lamellated spines, within which are axial lamellae running into deep open umbilicus. Grey; base light tan; interior silvery; columella white; operculum ear-shaped, white, with yellow-green tinge at wide end.

Cookia aureola Hedley 1907 Queensland, Australia, 75mm wide, 50mm high. Convex whorls bear oblique rows of small nodules and short, axial, low, blunt ribs below suture; final whorl and some of penultimate whorl carry short, triangular, hollow spines. Base has spiral, scabrous ridges. Dull brick red; columella white surrounded by rich golden orange; interior nacreous.

Guildfordia yoka Jousseaume Japan, 100mm wide including spines, 30mm high. Slightly pustulate; growth lines; upper edge of lip 'S'-shaped. Periphery of body whorl bears nine, long, backward-curved, hollow spines. Convex base; shallow umbilicus; white callus. Light brown; base paler; columella, interior nacreous; operculum ear-shaped, white, finely striate from an eccentric growth point.

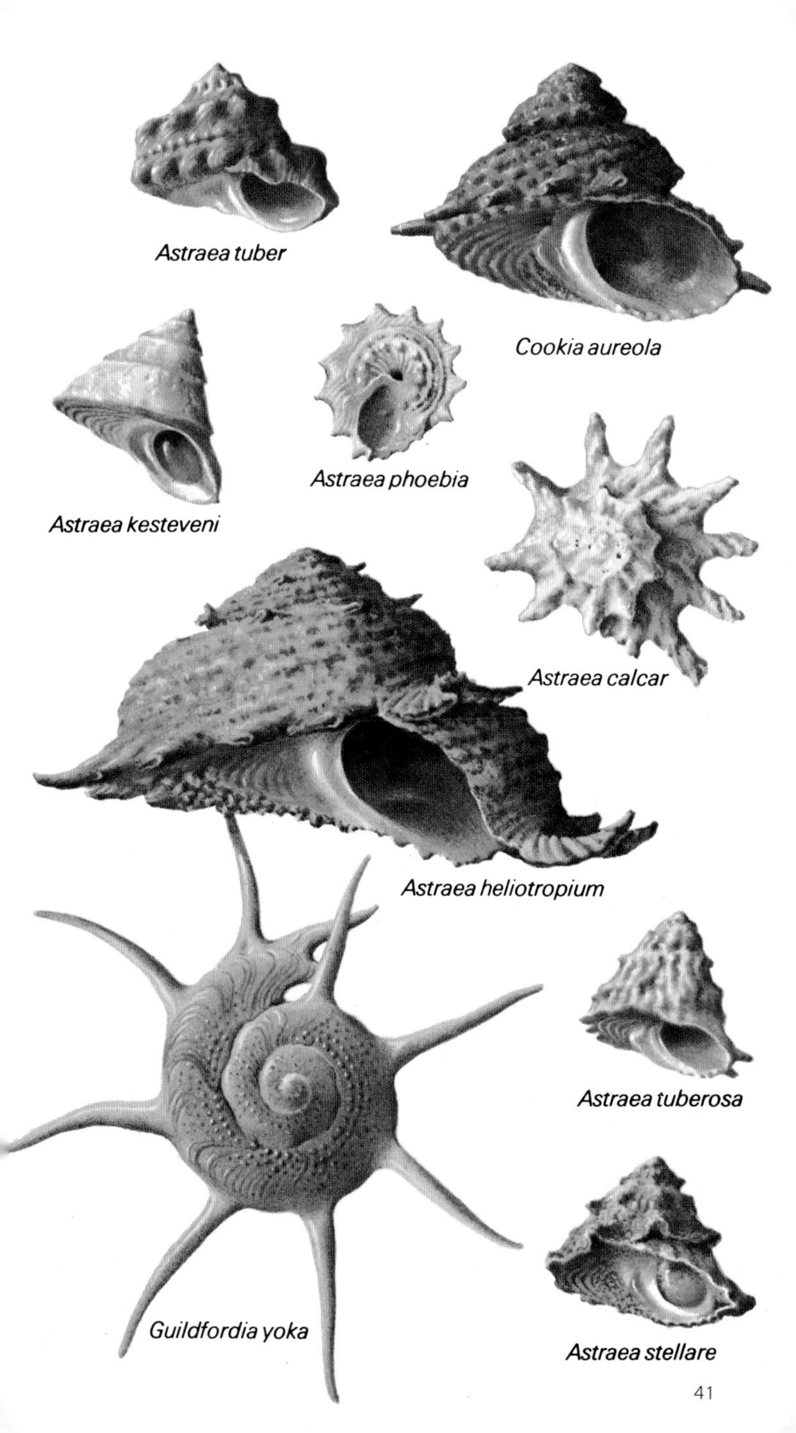
Astraea tuber
Cookia aureola
Astraea phoebia
Astraea kesteveni
Astraea calcar
Astraea heliotropium
Astraea tuberosa
Guildfordia yoka
Astraea stellare

Superfamily: Neritacea

Family: Neritidae

Generally small with large body whorl, depressed spire, semicircular aperture. Columella has two or more teeth; callous parietal area is smooth, pustulate or ribbed. Operculum is close-fitting with small projection, locking behind columella, holding it in place. They live in tropical and sub-tropical areas, in deep or brackish water in mangrove swamps, rivers and seas; on rock, coral and seaweed. Several hundred species. Vegetarians.

Smaragdia souverbiana Montrouzier South Africa and Mauritius, 22mm. Smooth columella with nine teeth, two outer ones large. Finely striate. White or yellow; dark zigzags; parietal area, interior white.

Nerita tessellata Gmelin 1791 West Indies, Florida, 20mm. Broad, spiral cords; two columellar teeth. Black, spotted with dark blue and white; parietal area white, pustulate; operculum black.

N. lineata Gmelin 1791 Indo-Pacific, 40mm. Depressed; spiral cords. Shiny parietal area, underlying plicae. Indented columella has four teeth; lip has twenty, fine teeth. Spiral cords black; interstices dull pink; parietal area yellow; interior cream; operculum with fine granules, purple-green.

N. undata L. 1758 Indo-Pacific, 40mm. Flat, spiral cords; narrow interstices. Plicate parietal area; columella has three teeth and lip has twenty short ridges. Variable colour, usually cream, dark green or brown, flames or spiral rows of dashes; sometimes uniformly dark purple-black; parietal area, columella, lip, interior, white.

N. textilis Gmelin 1791 east Africa, 40mm. Flat spire; coarse, spiral cords; growth lines. Pustulate parietal area; two columellar teeth; eighteen, long teeth on lip. White; dark blue dashes on cords; parietal area white, green-yellow stain; blue-black, pustulate operculum.

N. exuvia L. 1758 Indo-Pacific, 25mm. Fifteen, flat, broad-topped ridges overlapping deep channels. Pustulate parietal area; three columellar teeth; twenty long teeth on lip. Ridges beige; channels black; wavy, cream, axial lines; parietal area, interior, lip, dull cream-beige.

N. albicilla L. 1758 Indo-Pacific, 30mm. Elongate; rough growth lines; flat, spiral ribs. Pustulate parietal area; four columellar teeth; dentate lip. Cream or green marbled with dark brown, black or orange; parietal area, lip, pale green; columella, interior, white; pustulate, green operculum.

N. planospira Anton 1839 Indo-Pacific, 35mm. Flat spire; shouldered with spiral cords, deep interstices. Lip starts beyond apex and behind parietal area, which is pustulate and plicate; four columellar teeth. Grey-black maculated with red-brown; lip, interior, columella, parietal area, white; latter with purple spot posteriorly.

N. polita L. 1758 Indo-Pacific, 40mm. Elongate; fine growth lines, smooth parietal area and four or five columellar teeth. Cream, white, green, marbled or lined with brown, green or orange; smooth, green operculum.

Neritina paralella Röding 1798 Japan, China, 20mm. Thin, inflated; fine growth lines; ten columellar teeth; smooth lip. Variable, generally black, axial, wavy lines over olive-brown; smooth, grey operculum.

N. communis L. 1758 Philippines, 20mm. Similar to preceding species, but shorter spire. Colour and markings in almost limitless variations.

Neritina communis

Nerita tessellata

Neritina paralella

Nerita lineata

Nerita undata

Nerita textilis

Nerita exuvia

Nerita albicilla

Nerita planospira

maragdia souverbiana

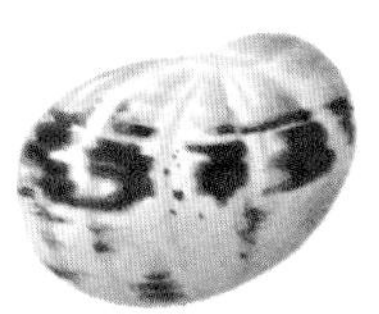
Nerita polita

Order: Mesogastropoda

Superfamily: Littorinacea

Family: Littorinidae

The periwinkles are small to medium, occasionally umbilicate and have a smooth columella and horny operculum. They live on rocks, mangroves and seaweed in intertidal areas. Tropical to temperate regions. Vegetarians.

Littorina neritoides **L. 1758** Mediterranean, UK to Madeira, 7mm. Pointed spire. Generally olive-grey; purple-brown spiral band; dark aperture.

L. littoralis **L. 1758** north Atlantic from the Mediterranean to New England, 15mm. Fine growth lines. Variable colour, yellows, reds, browns and greens; plain, banded or hatched; columella white; interior and lip brown.

L. littorea **L. 1758** same area as above, 25mm. Fine, spiral striae and more or less coarse growth lines. Generally dark brown to black, but also green, dark yellow or red, usually in spiral bands; columella white; interior chocolate. The common European Winkle which has spread across the Atlantic.

L. melanostoma **Gray 1839** Indian Ocean to South China Sea, 30mm. Thin; angulate keel; spiral cords. Pale yellow; brown dashes; purple-brown banding; lip and interior pale yellow; upper columella dark purple-brown.

L. scabra **L. 1758** Indo-Pacific and round to West Africa, 35mm. Spiral cords; carinated keel showing on spire. Cream, pale grey or brown, maculated with darker brown flames on lighter background below suture; columella white; interior as exterior but white band within lip.

L. intermedia **Philippi 1845** north Indo-Pacific, 15mm. Spiral, tessellated cords and angulate periphery. Tessellations are blue-white on pale chestnut; purple-brown spire; may have darker streaks along some growth lines.

L. saxatilis **Olivi 1792** north Europe and north-east America, 18mm. Coarse growth lines; faint, spiral cords anteriorly. Yellows, reds, browns, to purple-black; plain, tessellated or banded; columella, inside lip flesh-coloured.

L. undulata **Gray 1839** north Indian Ocean to Japan, 25mm. Spiral cords. Dark or light brown dashes on cream, which may form axial flame marks; violet columella.

Tectarius pyramidalis **Quoy and Gaimard** Japan and China, 18mm. The genus *Tectarius* is non-umbilicate. Spiral cords; row of nodules on shoulder, periphery of body whorl and on spire whorls. Blue-grey with white nodules and chocolate aperture.

T. pagodus **L. 1758** Indian Ocean, 65mm. Coarse, uneven, flat, spiral cords; oblique, axial cords ending in row of blunt spines; row of smaller spines on body whorl on which suture is formed in earlier whorls. Convex base with spiral rows of blunt nodules and ridges; smooth columella; spirally ridged inside lip. Creamy white with brown-grey shading; columella and interior cream with row of brown dashes level with large spines.

T. rugosus **Wood 1828** Philippines, 25mm. Four, nodulose, spiral cords per whorl; spiral rows of nodules on base; interior ridged. Apex colourless; lower three rows of nodules through white, straw yellow to orange at lip; lower row pale pink to dark purple; base, aperture white; columella pale brown.

Echininus cumingii **Philippi 1847** Pacific, 25mm. This genus is umbilicate and has nodulose, spiral cords bearing short spines. Blue-grey; umbilicate area white; interior brown.

Littorina neritoides

Littorina intermedia

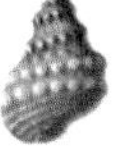

Tectarius pyramidalis

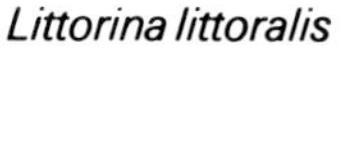

Littorina littoralis

Littorina littorea

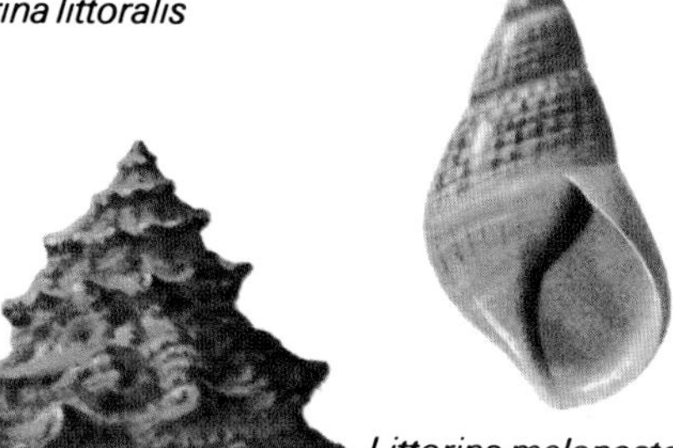

Littorina melanostoma

Tectarius pagodus

Tectarius rugosus

Littorina saxatilis

Littorina scabra

Littorina undulata

Echininus cumingii

Superfamily: Cerithiacea

Family: Turritellidae

There are over fifty species in this family which live in the tropics in shallow, muddy places. They have a long, pointed spire with many whorls, no umbilicus and a horny operculum.

Mesalia brevialis Lamarck Portugal to Senegal and the Atlantic Islands, 37mm. Solid with only slightly flattened, convex whorls and up to five ridges below the suture, the strongest anteriorly, the weakest posteriorly. It is minutely spirally striated with a curved and smooth columella and a sub-circular aperture. Cream or white with light brown axial flames, generally darker below the suture.

Turritella crocea Kiener China Seas, 90mm. Twenty or more convex but rather flattened whorls and about seven, spiral ridges with fine threads in between. Brown whorls with a pale area below the indented suture.

T. communis Risso 1826 Europe, 60mm. Convex but flattened, with eighteen whorls and about ten, spiral ribs of varying sizes; convex base. Light brown to almost white.

T. rosea Quoy and Gaimard 1834 New Zealand, 50mm. Flat-sided whorls and fine, spiral ridges overall, two ridges near the base of each whorl being twice as big as the others. Keel sharply angulate and base also finely ridged. Brown; lighter below suture.

T. bicingulata Lamarck 1822 Canary and Cape Verde Islands and west Africa, 65mm. Eighteen, finely striated whorls. Constricted at the suture, then sloping out to two flattened ridges with a shallow channel between them. Cream with fine, axial, pale red-brown lines overall, but tending to disappear in the channel, and with dark purple-brown flame marks from the suture to the first ridge, and on the base which has three or four, flattened ridges.

T. duplicata Lamarck 1816 Indo-Pacific, 150mm. About eighteen whorls, the early ones being convex with many, fine, spiral ridges. After the first six whorls a central ridge becomes elevated into a strong keel, most of the others tending to disappear. After about ten whorls a second elevated ridge begins to appear, but less prominently, and over the last two or three whorls both of these gradually become less conspicuous. The whorls at each end are therefore rounded and those in the middle are sharply angular. The upper half of each whorl is a medium-dark brown and the lower a pale cream-brown.

T. terebra L. 1758 Indo-Pacific, 125mm. About twenty-five rounded whorls with a very thin, sharp spire. It has six sharply developed spiral ridges and fine cords in the interstices. Deep sutures; four spiral ridges on the base. Dark or light brown.

T. leucostoma Valenciennes 1832 Gulf of California, 115mm. About twenty, flattened whorls, each with five, spiral ridges expanding from the constricted suture to the lowest ridge. Cream with many axial red-brown flames; earliest whorls dark brown.

T. gonostoma Valenciennes 1832 Gulf of California, 115mm. About sixteen whorls; flat-sided, but bevelled at each side to a constricted suture. Whorls may have fine, spiral cords or may be almost smooth. Cream with heavy, dark purple-brown mottling; on earlier whorls cream only shows at suture.

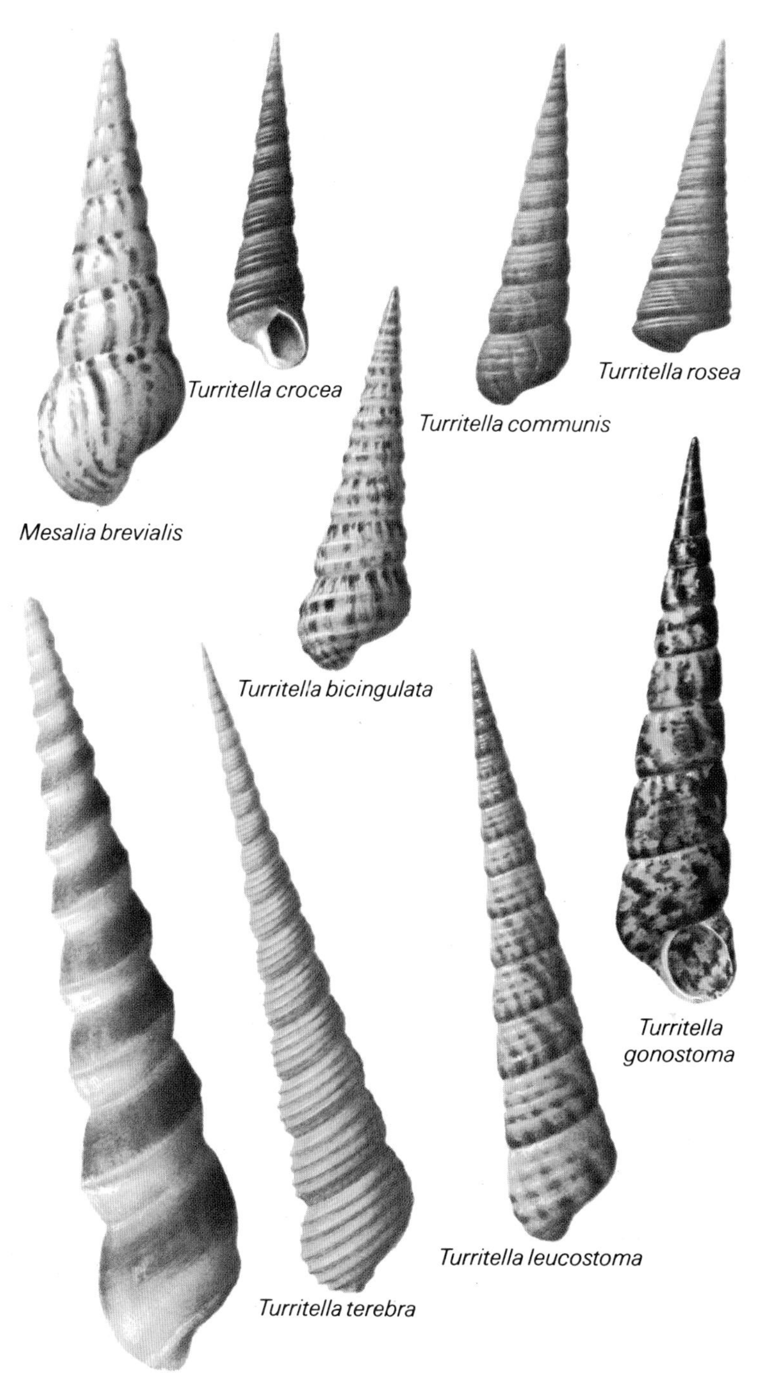
Turritella crocea
Turritella rosea
Turritella communis
Mesalia brevialis
Turritella bicingulata
Turritella gonostoma
Turritella leucostoma
Turritella terebra
Turritella duplicata

Family: Architectonicidae

The sundial shells, about forty species, live in tropical or subtropical, mostly shallow, waters. Depressed and circular, with a flat or slightly convex base. Deep, open, usually crenulated, umbilicus; small aperture and horny operculum.

Architectonica maxima **Philippi 1849** Indo-west Pacific, 60mm wide. Convex spire and a narrow, incised suture below which is a flat ridge, a narrow groove, a second ridge and a slightly wider groove. Both ridges have numerous axial striae, deeper on early whorls, making a tessellated pattern, and are fawn or flesh-coloured, with red-brown marks below the suture. These are followed by two ridges divided by a wide channel which have alternate dark red-brown and white blotches. Base from periphery inward: ridge; channel; smaller ridge; somewhat plicate, wide area; channel; finely nodulose cord; and finally a denticulate ridge lining the umbilicus. Apex mauve; base fawn, with brown radial mottling.

A. perspectiva **L. 1758** Indo-Pacific, 60mm. From deep, very narrow suture outward: narrow ridge with brown line; white band; deep very narrow groove; broad convex area with red-brown band; broad grey-brown band becoming fawn on last whorl; narrow, dotted, brown and white band; deep groove in which lower whorls form suture; small ridge white with pale-brown spots, hidden except on last whorl. Overall fine, axial growth lines. Base flat or slightly convex. From periphery: ridge; channel; cord; plicate area; channel; row of nodules; groove; denticulate ridge lining umbilicus. Base fawn, with brown dots on outer ridge and inner edge of plicate area; nodules white and denticles stained brown.

A. nobilis **Röding 1798** Central America, 45mm. Like *A. maxima* except that first two ridges below suture are not divided by a groove. Where *A. maxima* is fawn, *A. nobilis* has a pink tinge.

A. perdix **Hinds 1844** north Australia to Malaya and Sri Lanka, 30mm. From suture outward: minutely spirally striated band; narrow groove; broad striate convex band; narrow groove; broad ridge; channel; narrow ridge hidden under suture except on last whorl. Overall fine growth lines. Broad band pinky flesh; other bands and ridges white with chestnut spots; apex mauve. Base has two ridges divided by channel at periphery; broad, smooth (excepting growth lines) area; groove; row of nodules; groove; then denticulate ridge lining umbilicus. Outer ridges flecked with brown; opaque, blue-white, broad band; white umbilical ridge.

A. laevigatum **Lamarck 1816** India, 30mm. Five, roughly equal, flat ridges and axial growth lines overall. Top band is fawn with a few brown blotches; second and middle bands mauve towards aperture, becoming pale blue like apex, and indistinct pale brown dots; bottom bands fawn with large pale brown marks. Base sculpture as *A. perdix*, fawn with a blue tinge on broad middle area, brown radial markings near periphery.

Philippia radiata **Röding 1798** Indo-Pacific, 23mm. Relatively high and smooth with obsolete, spiral striations and two, very small ridges above suture. A brown band below suture sometimes has oblique axial rays spreading from it over the white background. Rather convex base with groove at periphery and white umbilical area. Fawn-flesh with brown specks on rim.

Heliacus stramineus **Gmelin 1822** India, Philippines, New Guinea, 25mm. Whorls becoming rounded to an almost circular mouth; fine spiral grooves between deep narrow suture and periphery; three ridges between periphery and base; small cord between lower two ridges; spirally ridged

Architectonica maxima

Architectonica perspectiva

Architectonica nobilis

Architectonica perdix

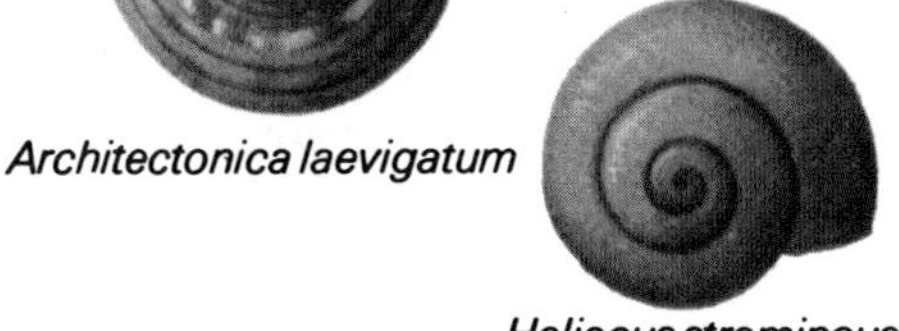

Architectonica laevigatum

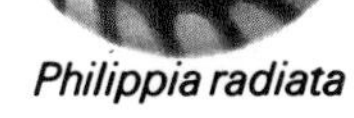

Philippia radiata

Heliacus stramineus

base, rather narrow, minutely denticulate umbilicus; oblique growth lines overall. Colour tan; darker at apex.

Family: Planaxidae

Small, solid, conical shells rather like the Littorinidae, and living in the same habitat. There are marine and freshwater members of the family and they are found in warm waters worldwide.

Planaxis sulcatus **Born 1780** Indo-Pacific, 30mm. Solid, heavy, high-spired and sculptured with spiral cords, about ten on the body whorl. Finely denticulate or smooth columella. It may have a narrow umbilicus. Lirate inside lip. Purple to brown with light grey or cream spots on the cords, profuse or sparse; inside of lip has brown stains.

Family: Modulidae

Somewhat similar in shape to the genus *Angaria* of the Trochidae family, they have a single tooth near the base of the columella and are narrowly umbilicate. They inhabit sandy, weedy areas.

Modulus tectum **Gmelin 1791** Indo-Pacific, 25 mm. Solid, depressed spire, angulate shoulder with strong axial ribs. Body whorl expands rapidly to a wide mouth and has spiral ridges, some nodulose. Columella smooth, prominent tooth at bottom. Lirate interior. Dirty white, some ridges with black specks; outer edge of columella and tooth deep purple-brown; a line of the same colour may run into interior at top of columella.

Family: Vermetidae

Unusual shells which are long tubes, evenly spirally coiled at the apex and then loosely and irregularly so. The genus *Siliquaria* has a narrow opening along the top of the tube.

Vermetus tulipa **Chenu 1843** Panama, west Central America. Early whorls small, 2–3mm diameter, growing to a length, if stretched, of some 60mm. The whorls are acutely angulate where they are joined to the substrate. Surface of tube is wrinkled. Cream or brown with grey or brown shading.

V. arenaria **L. 1758** Mediterranean. Generally fairly even spiral coils. Somewhat flattened tube with a slight ridge on the upper surface; coarsely wrinkled. Shades of grey.

V. cereus **Carpenter 1856** Philippines. Rather flattened tube, smooth apart from growth lines and closely coiled. Light brown. The specimen illustrated looks remarkably like a trilobite.

V. tokyoense **Pilsbury 1895** Japan. Slightly flattened tube with the end becoming free of the substrate and turning up. Rough surface. Light brown. The illustration shows the shell attached to a piece of dead coral.

Family: Siliquaridae

Vermicularia fargoi **Olsson** Florida and Caribbean. Tightly coiled spire; coils then open and become haphazard. Has three coils on tube, conspicuous on early coiled part but tending to disappear towards mouth. Brown, early coils paler. Sometimes found crawling about on mud.

Siliquaria ponderosa **Mörch 1860** Australia. Early whorls closely coiled then growing haphazardly. Whorls rounded except at the top where flattened. At the outer edge of this area is a narrow slit running the length of the shell except at the flattened apex. Pale tan or dirty white. The slit, or a series of holes, is common to the genus *Siliquaria*.

Vermetus tulipa

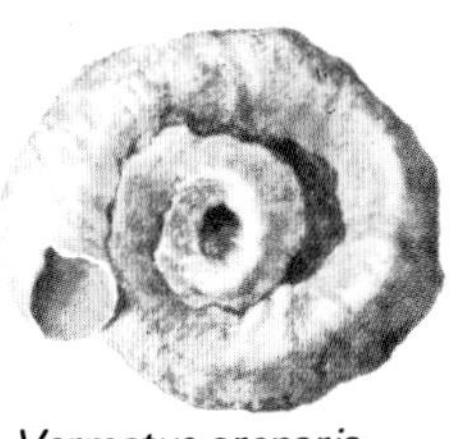

Vermetus arenaria

Planaxis sulcatus

Vermetus cereus

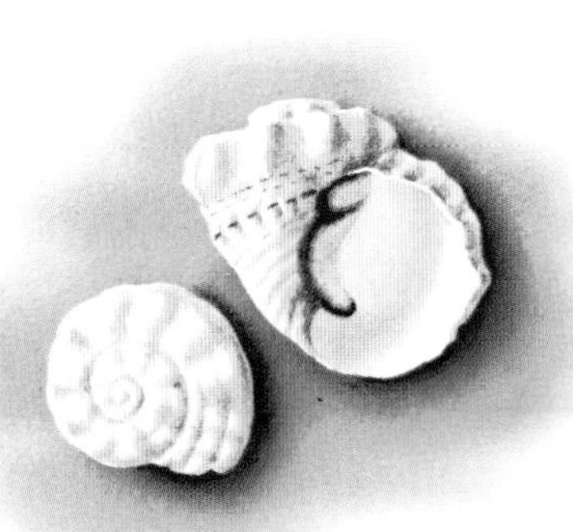

Modulus tectum

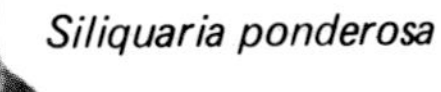

Siliquaria ponderosa

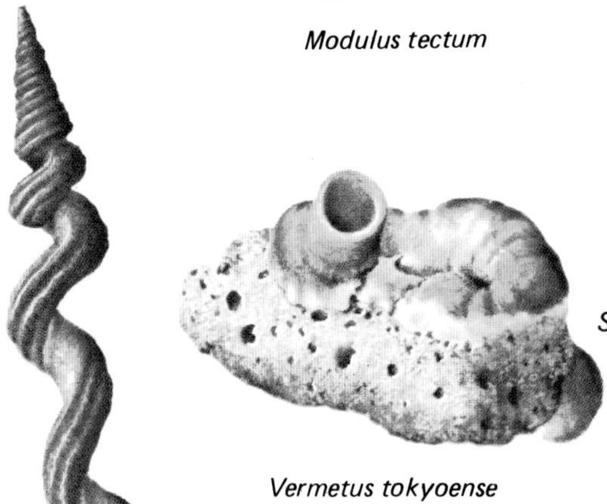

Vermetus tokyoense

Vermicularia fargoi

Family: Potamididae

One of two families of horn shells, they are long, tapering and have many whorls, usually large apertures and horny opercula. They inhabit muddy places, mangrove swamps and estuaries, and are vegetarians.

Terebralia sulcata **Born 1778** Indo-west Pacific, 60mm. Axial ribs; spiral grooves. Expanded lip joins body over short, tubular siphonal canal. Blunt, axial rib above siphonal canal. Grey or grey-brown; tessellations darker; obsoletely lirate interior; columella and inside of lip callous, shiny and grey-brown.

T. palustris **L. 1767** Indo-Pacific, 120mm. Coarse axial ribs and three, deep, spiral grooves. Flared lip curved over short siphonal canal. Obsoletely grooved aperture; columella, parietal area and lip callous. Dark brown; interior purple-brown; lip fawn, anterior purple; parietal area and columella chocolate, latter white centrally.

Cerithidea cingulata **Gmelin 1791** Indo-west Pacific, 50mm. Oblique, axial ribs and two, deep grooves form three rows of blunt nodules per whorl. Lip very expanded at each end. Blunt rib above siphonal canal. Nodules dirty white, interstices brown; interior white.

C. obtusa **Lamarck 1822** Indo-west Pacific, 50mm. Small, coarse, spiral cords; axial ribs. Expanded, recurved lip; short canal; apex usually eroded. Light brown or dirty white; brown markings; interior, lip, columella white.

Telescopium telescopium **L. 1758** Indo-west Pacific, 100mm. Straight-sided with four, unequal, flat, spiral ridges per whorl; last whorl rounded at base which has one large, many small cords and a wide, shallow channel around the strongly twisted columella; lip extended anteriorly. Dark brown or black; interior shiny blue-black; columella mauve-brown.

Family: Cerithidae

The other family of horn shells, they are similar but generally more colourful, and have simpler apertures. More or less distinct varices. The opercula are not circular as in Potamididae and nucleus is off-centre. They generally live in clearer water among coral, and prefer sand to mud.

Cerithium nodulosum **Bruguière 1792** Indo-Pacific, 120mm. Coarse, spiral cords; blunt, heavy nodules, eight on last whorl, one nearest aperture very swollen; three, large, nodulose cords on base. Expanded lip curves in front of short siphonal canal; lirate aperture. Columella ridges, bordering anal and siphonal canals, run into interior. White; blue-black, spiral dashes on nodules and base; interior, columella, lip, canals, white; interior sometimes with blue-black shading.

C. aluco **L. 1758** Indo-Pacific, 80mm. Smooth or obsolete fine ridges and a row of six, short, blunt spines on upper part of whorls. Aperture like *C. nodulosum* except siphonal canal bent back 90° and aperture lirae obsolete. White; profuse purple flecks and blotches; lip and columella opalescent, purple flecks anteriorly.

Rhinoclavis nobilis **Reeve** Philippines, 130mm. Straight-sided; first eight or nine whorls spirally ridged, last three of these axially ribbed, and prominent ridge at base. Expanded lip; siphonal canal recurved 90°; smooth columella; reasonably developed anal canal. Cream; profuse, small, pale tan flame marks; aperture opalescent.

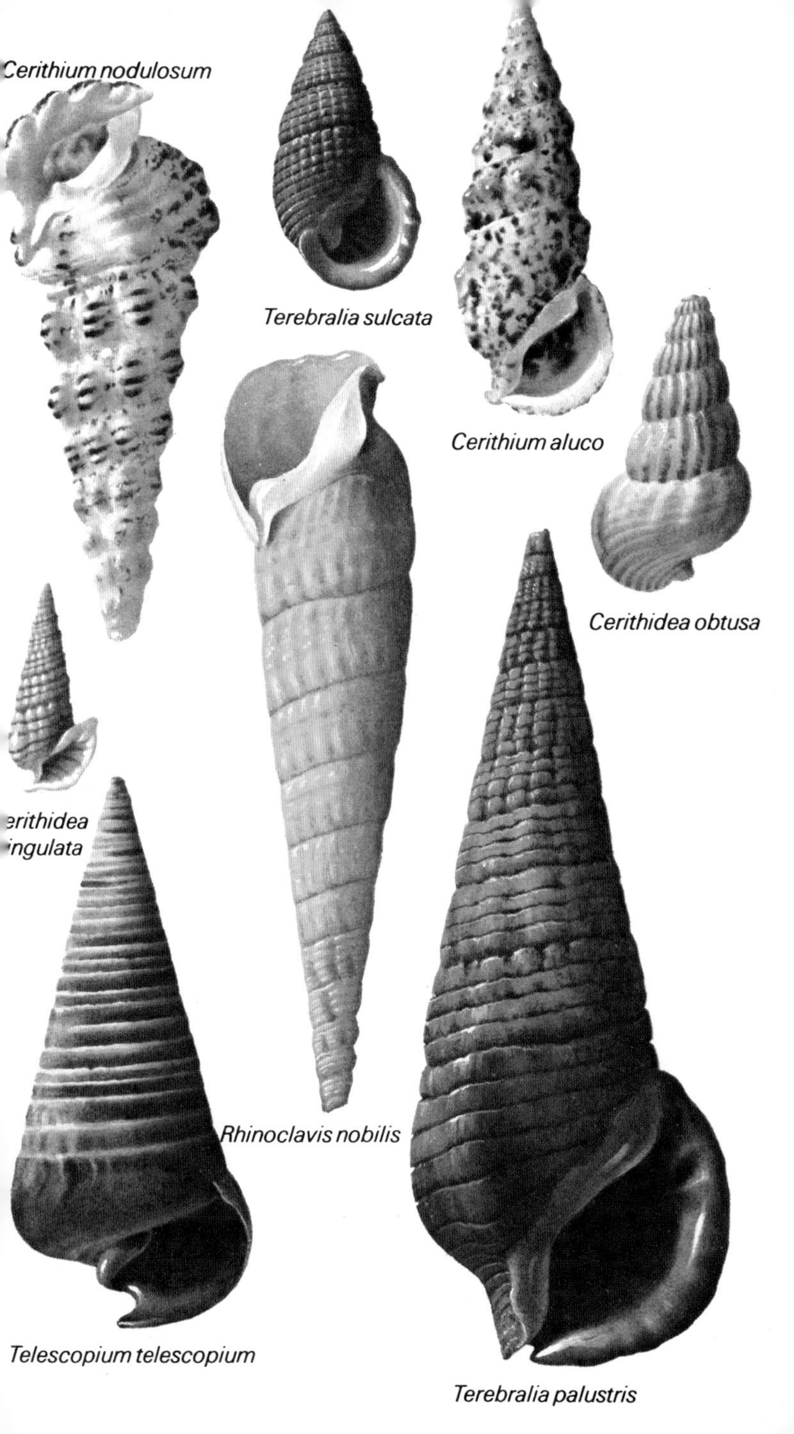
Cerithium nodulosum
Terebralia sulcata
Cerithium aluco
Cerithidea obtusa
erithidea
ingulata
Rhinoclavis nobilis
Telescopium telescopium
Terebralia palustris

R. vertegus **L. 1767** throughout Indo-Pacific, 65mm. Solid, narrowing anteriorly; uneven, deep sutures. Spiral ribs, axial cords, obsolete on last two whorls. Few early varices; slightly expanded lip; siphonal canal at 90° to axis. Small anal canal; callous columella, central fold; shallow groove around callus. Light brown; apex, early whorls, aperture white.

R. sinensis **Gmelin 1791** Indo-Pacific, 65mm. Row of blunt nodules below suture, then spiral rows of small nodules. Fine axial and spiral threads between rows and between shoulder nodules. Axial swelling between siphonal canal and suture. Faint early varices. Expanded lip; columella with central fold; siphonal canal at 90° to axis. Cream; faint clouding, spots of purple-brown especially on large nodules; varices whitish; aperture off-white.

R. bituberculatum **Sowerby 1865** Western Australia, 45mm. Four rows of blunt nodules on each whorl; fine axial and spiral striae between. Expanded lip; siphonal canal at 90°; columella with callus and one tooth. Flesh-coloured; dark blue flecks especially on last whorl, where they form spiral lines of dashes which show on interior; aperture white.

R. asper **L. 1758** Indo-Pacific, 60mm. Three, spiral cords (four on body whorl); oblique axial ribs; sharp nodules at intersections. Lip only slightly expanded; columella with central fold; siphonal canal at 90° and almost opposite lip. Cream or white; sometimes brown stains; aperture white.

R. fasciatus **Bruguière 1792** Indo-Pacific, 80mm. Deep suture; fine, axial riblets and faint, spiral striae, both obsolete on last whorls. Slightly expanded lip. Siphonal canal bent back 90°. Callous columella has one fold. White or fawn; two or more light brown or chestnut bands obscure white; bands showing on interior and lip; columella, callus, interior, lip white.

Clypeomorus traillii **Sowerby 1855** Indo-west Pacific, 30mm. Stubby; fine, spiral striae; fine ridges, some nodulose. Early varices; expanded, thickened lip; short siphonal canal at 45°; distinct anal canal. Creamy brown; some nodules dark brown; aperture white; interior may be spotted brown.

Cerithium erythraeonense **Lamarck 1822** Red Sea, 70mm. Spiral, fine ridges; angulate whorls; ten ribs form short, pointed nodes at periphery; very large rib above siphonal canal. Expanded, thickened, crenulated lip curved to cross strong siphonal canal at almost 90°. Smooth columella; distinct anal canal. Fawn; grey-brown markings; aperture white; lip fawn.

C. caeruleum **Sowerby 1865** Indian Ocean, 40mm. Fine spiral striae; rows of nodules on body whorl, two on earlier whorls; one row with larger nodules than others forming angulate shoulder. Blue-grey; apex pink; nodules, lip and short, backward-curved siphonal canal, dark purple-brown; rest of aperture white.

C. vulgatum **Bruguière 1792** Mediterranean, west Africa, 75mm. Fine, deep, spiral striae; axial ribs; restricted below suture, then angular; ribs form short spines at the angle. Slightly flared lip; distinct anal canal; short, slightly recurved siphonal canal. Variable. Grey or brown; wavy, axial, darker lines; interior similar; columella white, colour showing through.

C. ruppellii **Philippi 1848** Red Sea, 45mm. Narrow. Three spiral rows of nodules, five on body, striae between. Early, white varices; expanded lip; long, straight canal. Pale grey-brown; aperture white; interior grey-brown.

Tymphanotonus aurita **Müller** west Africa, 55mm. Finely spirally ridged with rows of rectangular, oblique nodules in middle of whorls, and rectangular aperture. Chocolate-brown; nodules often worn to white.

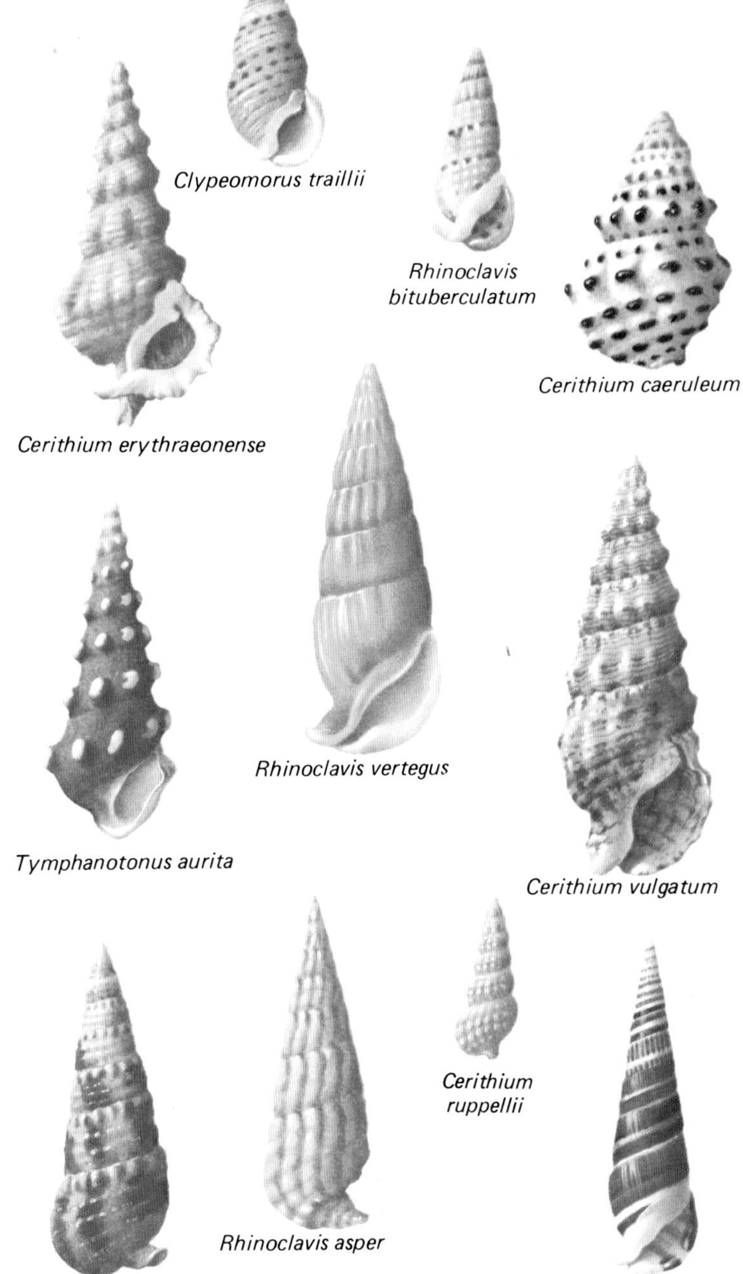
Clypeomorus traillii
Rhinoclavis
bituberculatum
Cerithium caeruleum
Cerithium erythraeonense
Rhinoclavis vertegus
Tymphanotonus aurita
Cerithium vulgatum
Cerithium
ruppellii
Rhinoclavis asper
Rhinoclavis sinensis
Rhinoclavis fasciatus

Superfamily: Epitoniacea

Family: Epitoniidae

The wentletraps or staircase shells, are found worldwide, often among soft coral or sea anemones, in shallow water. There are some 200 species. They are rather delicate, conical, with many, often loosely coiled, whorls, on which are slightly oblique, axial lamellae or varices which were formed as the reflected outer lips at earlier pauses in the growth of the animal. Some are umbilicate. They are usually white, but a few have brown tinges or markings. Their opercula are horny, with few whorls, and the nucleus is nearly central.

Cirsotrema zelebori **Dunker 1866** New Zealand, 22mm. Thin with about nine whorls which have oblique, rather closely set lamellae, fifteen per whorl. There are low, spiral ridges, nine on the last whorl, which form blunt nodules on the lamellae. White.

Epitonium multistriatum **Say 1829** south-east USA, 12mm. Narrow and very pointed; about fourteen, axial lamellae per whorl. Shining white.

E. lamellosum **Lamarck 1822** West Indies, 20mm. Eight whorls, twelve lamellae on last whorl. A pale fawn, darker at the suture; lamellae, apex and aperture white.

E. pallasi **Kiener 1838** Australia, Philippines, China Seas and Mauritius, 25mm. Rather broad at the base compared with the preceding species. About eight whorls often joined only by the lamellae, of which there are about ten per whorl. It has a deep umbilicus. Shiny pale orange-brown apex; lamellae and aperture white; interior pale orange-brown.

E. scalare **L. 1758** east Asia and Australia, 65mm. The Precious Wentletrap, for long prized by collectors, is the largest of the genus, about half as wide as it is high. Whorls joined only by the lamellae, eight per whorl. A deep umbilicus. Very pale flesh colour; lamellae white. This species was once very rare and fetched high prices. Copies were made in China of rice paste, which are reported to have deceived collectors. If they were on the market today they would probably fetch a higher price for their interest's sake than the real shell.

E. perplexa **Pease 1860** west Pacific and Indian Ocean, 40mm. About seven whorls with twelve lamellae per whorl. White or pale fawn; may have a purple band below the suture after the early whorls; lamellae, apex and aperture white.

E. dubia **Sowerby 1844** Australia and East Indies, 40mm. Very thin and delicate. Deep suture; expanded whorls; minutely spirally striate; small lamellae on the early whorls become thin, small ridges on later whorls. White.

Clathrus clathrus **L. 1758** the Common Wentletrap, Mediterranean to the North Sea, 35mm. Rather narrow; about fifteen whorls with nine lamellae per whorl. Cream or fawn. Often there are two or three, spiral, pale purple-brown bands which show most clearly where they cross the lamellae. Both colour forms are illustrated.

Amaea raricostata **Lamarck 1822** Mauritius and Sri Lanka, 25mm high and over 20mm wide. Stubby; cancellated by fine axial ribs and spiral cords. It has occasional, heavy, thick varices especially on the last whorl. White.

A. magnifica **Sowerby 1847** China Seas, up to 100mm. Very thin, delicate, and high-spired. Spirally finely ribbed; low, thin and irregularly spaced lamellae. White; apex mauve. This is a rather rare shell and a collectors' item.

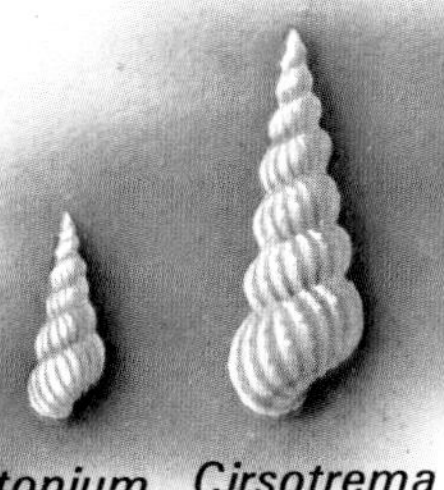

Epitonium multistriatum

Cirsotrema zelebori

Epitonium lamellosum

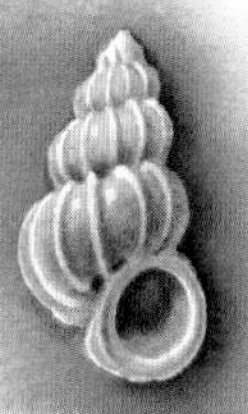

Epitonium pallasi

Clathrus clathrus

Epitonium scalare

Epitonium dubia

Clathrus clathrus

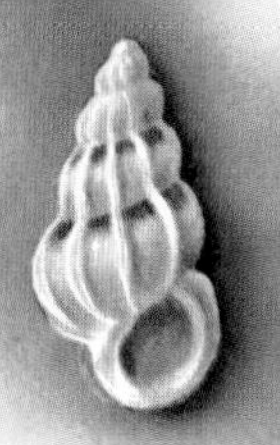

Epitonium perplexa

Amaea raricostata

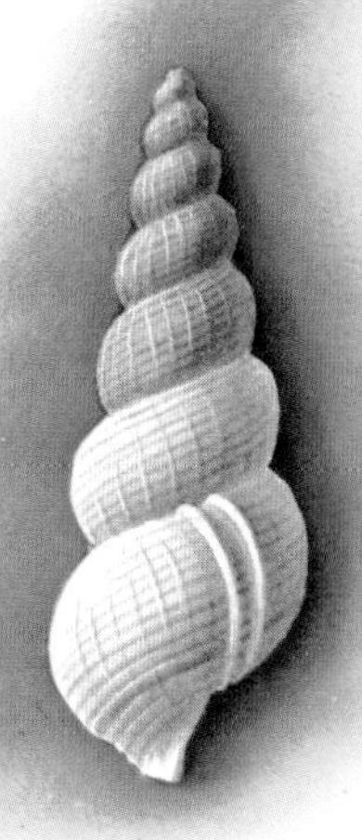

Amaea magnifica

Family: Janthinidae

The violet snails live in all warm-water seas. Very thin-shelled, they produce a float of jelly-like bubbles to which they remain attached and move where the winds and currents take them. They live on plankton and emit a violet fluid when attacked. About thirty species.

Janthina globosa **Swainson 1822** 40mm. Globose with rounded whorls, deep suture, depressed spire and extended columella. Wide lip with a restriction in the middle which shows in the V-shaped growth lines; fine spiral striae. Violet spire, paler on last half of body whorl, very pale just below suture.

J. exigua **Lamarck 1816** 20mm. Relatively high spire; whorls with V-shaped striae, the 'V' lining up with an indentation on lip; long columella; lip produced anteriorly. Purple; white band below suture.

J. janthina **L. 1758** 40mm. Globose; angulate whorls; depressed spire; fine growth lines. Flat base, spirally striate; columella produced and twisted anteriorly. Early whorls pale blue, body whorl white; base, columella violet.

Superfamily: Hipponicacea

Family: Capulidae

Cup-shaped, they are parasites, preying largely on other molluscs.

Capulus ungaricus **L. 1758** Europe, 50mm wide, 30mm high, but variable. Hooked apex points downward and inward; radial striae; coarse, uneven growth lines. Thick, dark brown periostracum; dirty white; interior pink.

Superfamily: Calyptraecea

Family: Calyptraeidae

The slipper limpets often have a shelf or cup-like projection internally. Very variable in shape, depending on their substrate.

Crepidula fornicata **L. 1758** the American Slipper Limpet, North America and accidentally introduced to Europe around 1900, 50mm. Recurved apex; strong growth lines; about half the aperture covered by 'shelf'. Dirty white, fawn and cream. Very variable. Often grouped one on top of another.

Crucibulum lignarium **Broderip 1834** Indonesia, China Seas and west tropical America, 25mm. High and cup-shaped with coarse growth lines and radial striae. Interior cup attached at apex and on one side. Light brown.

C. auriculatum **Gmelin 1791** Central America, 30mm. Round and depressed with pointed, recurved apex and spiral threads covered with short, sharp spines. Cup attached at one side; ear-shaped. Creamy white with pale purple-brown axial rays more clearly seen in the shiny interior.

C. scutellatum **Gray 1828** Gulf of California and Panama, 65mm. Coarse large radial ribs. Cup attached at apex. Brown; pink interior.

C. extinctorium **Lamarck 1822** Malaysia and China Seas, 40mm. Very variable, flat or elevated; sharp apex, usually recurved; coarse spiral striae. White with very pale pink rays. Illustrated attached to a *Natica*.

Trochita trochiformis **Born 1778** west Central America to Chile, 60mm wide and 30mm high. Depressed and conical. Whorls carry coarse, oblique, radial ribs. Thin, wavy-edged aperture runs from centre to near edge of shell, almost opposite end of last whorl. Pale brown flesh; base lighter and shiny.

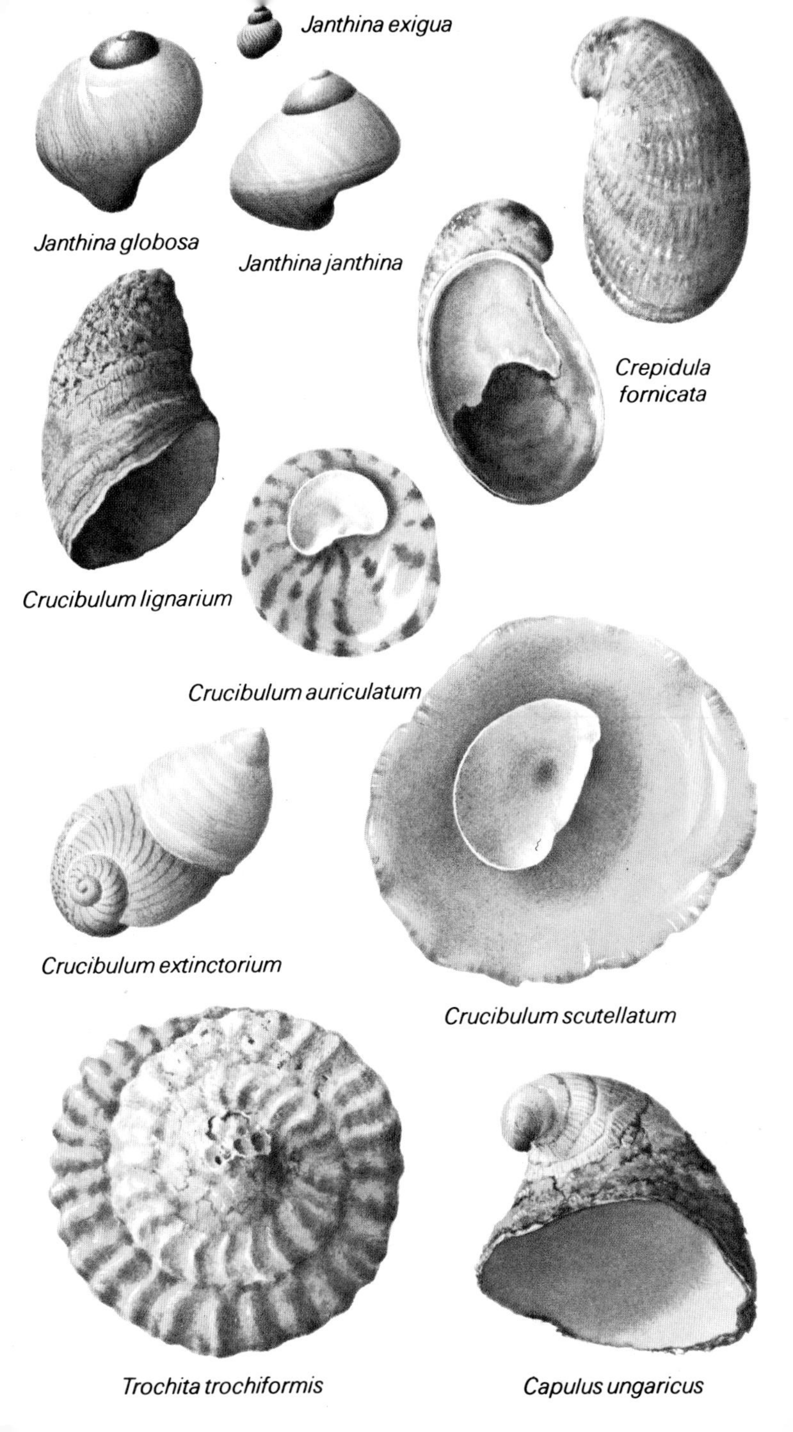
Janthina exigua
Janthina globosa
Janthina janthina
Crepidula fornicata
Crucibulum lignarium
Crucibulum auriculatum
Crucibulum extinctorium
Crucibulum scutellatum
Trochita trochiformis
Capulus ungaricus

Family: Xenophoridae

The carrier shells owe their popular name to the animals' habit of attaching empty shells – if bivalves, inner side upwards – pieces of coral, stones or grains of sand, to their shells. In most species this only occurs during their young state and the attachments are only on early whorls. In others such as *X. pallidula* and *X. neozelanica* it continues throughout growth. They live in tropical seas, in shallow and deep water; those from shallow water are the most enthusiastic 'collectors'. They are active and move quite quickly.

Xenophora pallidula Reeve 1843 Japan, Philippines and South China Sea, 100mm. Conical; spiral striae; coarse growth lines. Slightly concave base, free of any attachments, has coarse growth lines – almost folds – and narrow umbilicus which may be covered by callus on inside lip. White or light brown; base light brown and cream; aperture callus rich brown fading to white at edge.

X. crispa Koenig 1831 Mediterranean, 45mm. Somewhat depressed and narrowly umbilicate, with oblique growth lines and generally axial, small, wavy riblets. Wavy, uneven periphery. Base has oblique radiating rows of small nodules; striated around umbilicus. Shells, stones and so on attached to suture and periphery. Cream; base cream to fawn, darker around umbilicus.

X. corrugata Reeve 1843 Indo-west Pacific, 40mm. Depressed with deep suture and uneven surface like rough hewn marble. Base has small, oblique, radiating ridges with coarse growth lines, some dark brown. Usually well hidden under empty shells, stones and so on.

X. calculifera Reeve 1842 Japan to Philippines, 70mm. Rather thin with uneven surface, fine oblique striae. Suture barely perceptible where not obscured by attachments which are small and only on the suture. Base cancellate with small, radial, curved ribs and spiral striae. Wide, deep umbilicus. Cream-grey, yellow staining; base the same.

Tugurium indicus Gmelin 1791 Australia, 85mm. Very thin, delicate, rather depressed, with suture only a faint ridge. Outer edge of shell wafer thin. Surface is sculptured with close-set, fine, radial riblets which are crossed by oblique, radial, uneven striations. Small pieces of stone debris attached only to first two or three whorls. Concave, deeply umbilicate base. Outer edge of aperture forms a flat ridge, about 2mm wide, with fine growth lines/folds running from ridge into umbilicus. Outside the ridge is a thin, smooth, shiny, irregular skirt, very jagged at edge. Suture, visible through skirt, is further from apex than ridge of lip. Cream to pale fawn; base ridge to umbilicus white; skirt fawn.

T. exutus Reeve 1842 west Pacific, 46mm. Depressed and similar to *O. helvacea* but sculptured with fine, oblique, axial ridges; broad, wavy ribs run at 90° to these. Ridge on base has four rows of minute nodules. Open umbilicus is carinated with striae running into it; ridge to carina smooth with a few faint concentric threads and growth lines. No attachments, or a few grains of sand near apex. Fawn, darker at apex.

Stellaria solaris L. 1767 Philippines, 75mm. Low spire; somewhat convex whorls with hollow, protruding spines at periphery of whorls, hiding suture. Surface uneven with fine, generally spiral threads; oval aperture. Base concave and umbilicate with strong, wavy, uneven, nodulose, oblique, radial cords. Ends of spines curve along edge of base, forming continuous strong ridge. Light brown; aperture smooth, shiny, darker brown.

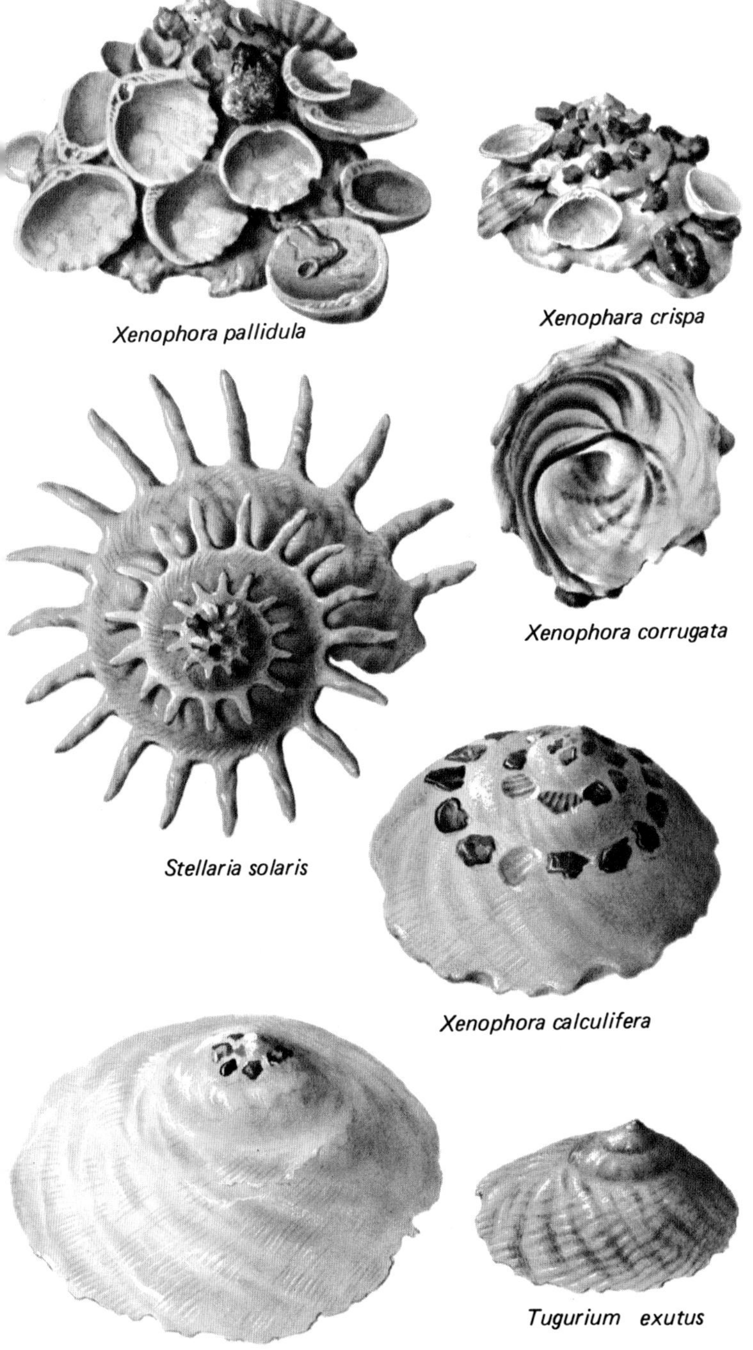

Xenophora pallidula

Xenophara crispa

Xenophora corrugata

Stellaria solaris

Xenophora calculifera

Tugurium indicus

Tugurium exutus

Superfamily: Strombacea
Family: Strombidae

The family includes the genera *Strombus, Lambis, Tibia* and *Terebellum*. *Tibia* contains only six species and two in the subgenus *Rimella*. All are from the Taiwan to Red Sea area. None are common, though they may not be rare locally. Most are collectors' items. *Tibia serrata* Perry 1811 is a 'lost' shell – its locality is unknown. They are high-spired, long, have an extended siphonal canal which is generally smooth, except in *Rimella*. Sculpture is confined to lip and early whorls, and colour to shades of brown. They generally inhabit deep waters.

Tibia fusus L. 1758 Philippines, 200mm long and 35mm broad. Very high-spired with eighteen whorls. Early whorls have narrow radial ribs and spiral ridges. Ridges disappear first, about six whorls from the base, the ribs three whorls later. Last three whorls are smooth apart from fine growth striae and spiral grooves on base of body whorl. Very long, narrow, delicate siphonal canal. Lip has five, finger-like projections outlined by a low ridge which forms a half circle on penultimate whorl and disappears at other end on siphonal canal. Columella with tooth posteriorly and callus extending under curved end of the ridge, together forming anal canal. Horny, leaf-shaped operculum. Shiny tan-fawn, lighter at apex; ridge on the fingers red-brown. Rather rare shell and highly prized by collectors.

T. martinii Marrat 1877 Philippines and Taiwan, 150mm. Rare. Lightest of the genus; microscopic axial and spiral striae; slightly constricted at suture. Five, short, stubby projections – two or three obsolete ones – on lip, outlined by small ridge which continues on siphonal canal. Light brown; some darker shading, especially below suture; fingers white; ridge red-brown; columella and callous inside lip white; interior grey-brown. A form, *melanocheilus* A. Adams 1854, has a broad purple-brown band inside lip and other minor differences.

T. insulae-chorab Röding 1798 Indian Ocean from Red Sea to Philippines, 160mm long, 45mm wide. Early whorls axially ribbed with microscopic spiral striae between, later whorls smooth. Short siphonal canal often curves up slightly towards aperture; five, short, stubby fingers and a ridge from siphonal canal outlining lip, fingers, siphonal and anal canals. In the form *curta* Sowerby 1842 this ridge ends on penultimate whorl instead of on antepenultimate whorl (see illustration). Both forms have strong columellar callus, with blunt tooth posteriorly, continuing towards apex forming anal canal with ridge. Shiny light brown; form *curta* with darker band below suture; axially rather streaky.

T. powisi Petit 1842 Japan and Philippines, 50mm. Smallest of the genus with ten whorls. Spiral ridges with interstices pitted. It has the usual five fingers, short siphonal canal and very short anal canal. Lip has deep furrow behind fingers. Columella with small callus, inner edge of lip has thick, callous ridge with lirae which continue into aperture. Light brown, some purple areas; columella, interior, ridge inside lip, white.

Rimella (Varicospira) cancellata L. 1758 Indian Ocean, Philippines, north Australia, 35mm. Solid with axial ribs and spiral grooves giving it a cancellate sculpture. Short siphonal canal. Thickened crenulate lip, backed by deep furrow, is produced towards apex over two or three whorls. Columellar callus follows lip to form long anal canal. Irregularly placed varices. Lirate lip and interior. Cream, exterior of lip, varices red-brown; columella, interior of lip white.

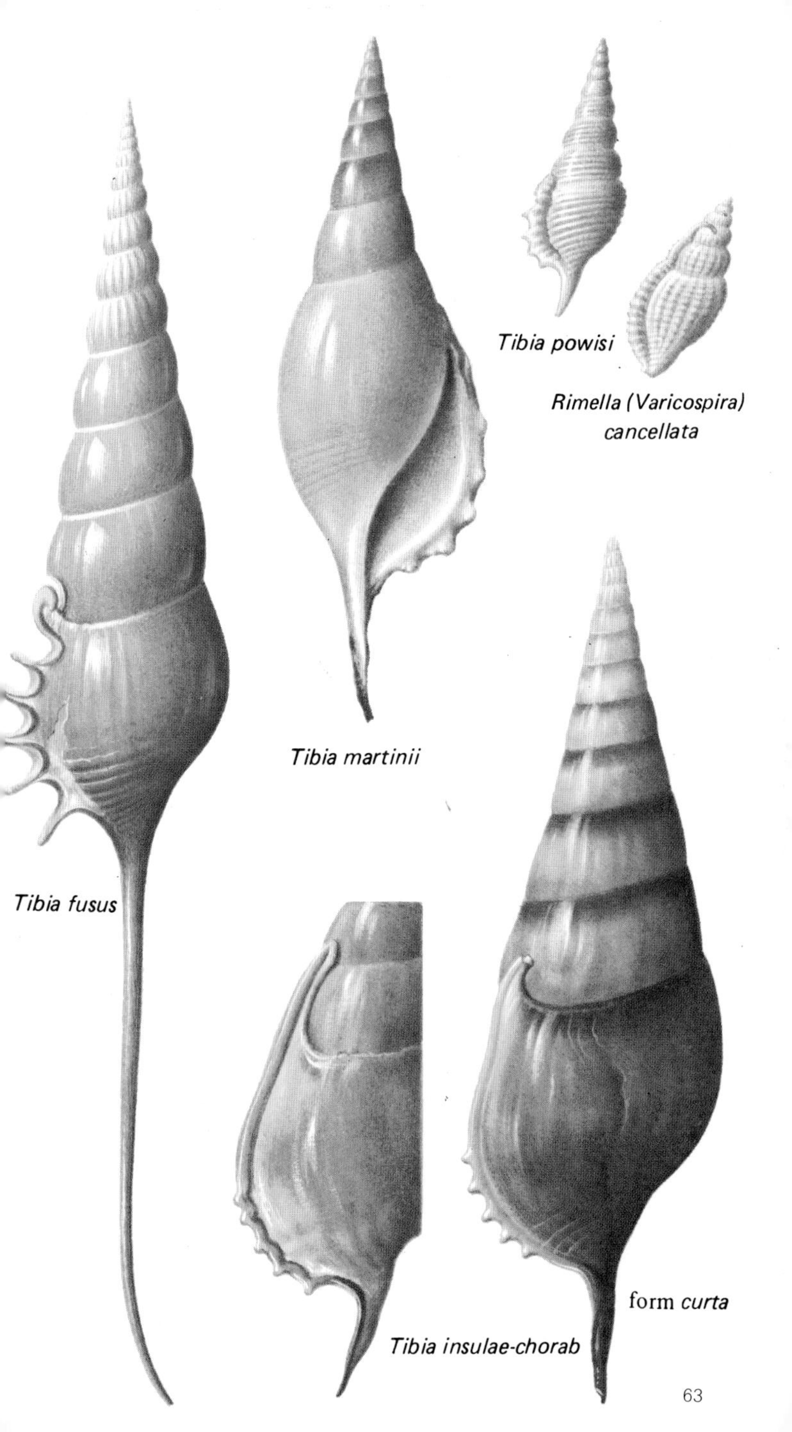
Tibia powisi
Rimella (Varicospira) cancellata
Tibia martinii
Tibia fusus
form curta
Tibia insulae-chorab

T. (R.) crispata Sowerby not illustrated. From the same area as and very similar to *R. cancellata* but extension of lip turns back sharply.

Genus: *Strombus*

The conch shells are found throughout the tropics. They are large and small, generally solid, with an expanded body whorl, narrow aperture, thickened outer lip and short siphonal canal. A characteristic of these shells is the 'stromboid notch', an indentation near the bottom of the outer lip through which one of a pair of tentacles, each carrying a well-developed eye, protrudes. The operculum is horny, long and narrow, with the nucleus at the anterior end. The foot is very muscular and can turn them over or move them along remarkably quickly. They live in shallow water on sand or sandy mud and are vegetarians. Some are edible.

Strombus pugilis **L. 1758** the West Indian Fighting Conch, Caribbean, 80mm. It has about eight whorls with about nine short spines on the shoulder of the body whorl projecting from just above the suture in earlier whorls; those on the penultimate are longest. All but last two whorls have spiral cords and blunt ribs which end in the spines. Spire concave and about a quarter of the length of the shell. There are a few spiral cords from the suture to the spines and at the base of the body whorl. Outer lip thickened and projected at posterior end; strong stromboid notch; short siphonal canal; columella, fasciole and inside lip callous. Light tan, almost white at apex; red edge to siphonal canal; callus golden brown; interior white.

S. alatus **Gmelin 1791** the Florida Fighting Conch, Florida and adjoining coasts. Same size as *S. pugilis* and rather similar in shape but the outer lip is not projected, and the spines are reduced to five short ones on the latter half of body whorl and elsewhere to blunt knobs. The spiral cords which extend from the apex to the suture of the final whorl are stronger. The colour, however, is a rich chocolate streaked with white on the spire; white at the apex; fasciole and inner edge of lip orange-red; parietal callus and columella dark red-brown; interior brown-mauve.

S. canarium **L. 1758** India to north-east Australia, the Philippines and Japan. Variable in size up to 100mm but generally nearer 50mm. Spire may be short and smooth or high with angular whorls, cancellate sculpture and varices. Body whorl striated at the base and either smoothly rounded or with faint ridges and shouldered; both varieties illustrated. Swollen outer lip; stromboid notch not strong; short siphonal canal; callous and very shiny columella. Colour either plain white with pale fawn stains and axial streaks, or dark brown on intermediate colours; it may have an overlying fine network of darker brown; interior and columella white; parietal wall sometimes metallic looking. The plain colour form has been named *turturella* Röding 1798 and *isabella* Lamarck 1822.

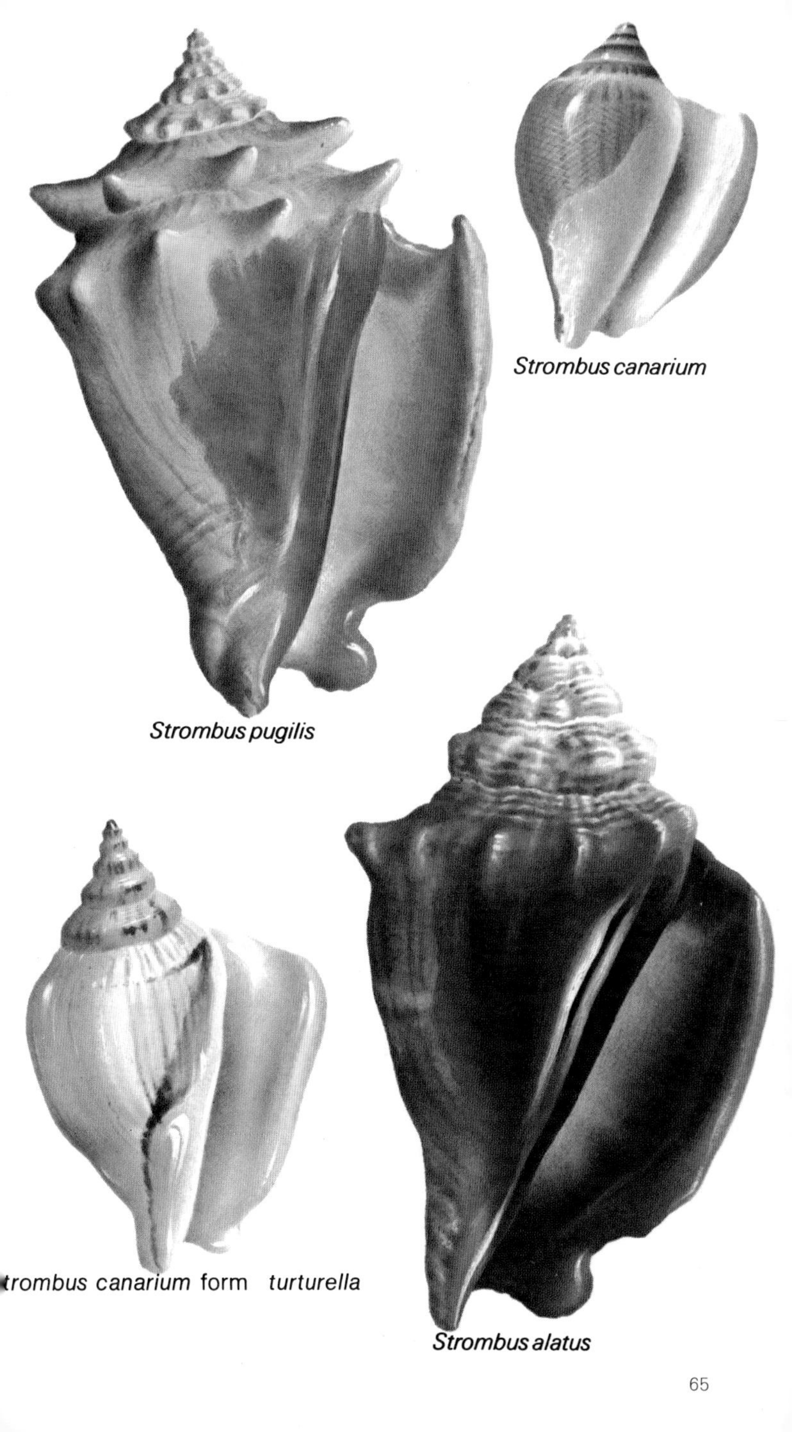

Strombus canarium

Strombus pugilis

trombus canarium form turturella

Strombus alatus

S. gigas L. 1758 the Queen Conch, Florida and the West Indies, 300mm. Moderate spire. Short, strong spines at the shoulders of the whorls. Coarse, flat, spiral ridges; spiral cords; axial striae; coarse axial ribs and folds especially on the parietal wall. Flared lip is produced posteriorly to almost level with the apex, and lip edge is wavy and simple. Narrow but deep stromboid notch. Heavily callous, smooth columella. Wide and glazed parietal area. Short, open and recurved siphonal canal. Cream and white; aperture a rich, shining rose pink with yellow shading. This species is taken in huge quantities for food and sometimes produces a pink pearl.

S. latissimus L. 1758 west Pacific, up to 200mm. A remarkably heavy solid shell. The carinate early whorls have small blunt knobs, which on the penultimate and body whorls become fewer in number, larger and very rounded, one on the shoulder farthest from the lip particularly so. The thick lip flares out beyond the apex, and opposite the columella is thickened even more into a ridge, sometimes forming an overhang. The stromboid notch is well towards the anterior and is narrow and deep. Heavily callous columella. It is cream heavily mottled with brown, generally axially. Interior and inside of lip dirty white, columellar callus light brown as is the area of the lip on and outside the ridge.

S. goliath Schröter 1805 (not illustrated) Brazil, up to 320mm. The largest of the strombs. Heavy with rather short, concave, pointed spire. Adpressed suture. Inflated body whorl with rather rounded shoulders carrying five, heavy, blunt knobs on the dorsal side, and five much smaller ones on the ventral side under the parietal glaze. This glaze covers the ventral side of the body whorl. The dorsal surface is somewhat malleate with coarse, broad, flat, spiral ridges having shallow, fairly narrow grooves between. The lip flares very widely, the edge making a smooth, sweeping curve, and the posterior end forming a nearly perfect half-circle, and extending well beyond the spire to hide it completely from the ventral side; it ends posteriorly on the suture above the body whorl. The aperture is relatively narrow. The cross-section of the inside of the lip is almost 'L' shaped, there being a strong ridge parallel to the straight, smooth columella. A narrow groove runs from the posterior end of the columella across the top of the parietal area where the lip joins the body whorl. Stromboid notch is almost non-existent. Short, open and slightly twisted siphonal canal. Cream with axial, wavy, purple-brown lines on early whorls; aperture and callus white to pink-white; interior darker. Periostracum is tan and shows faintly through the parietal glaze.

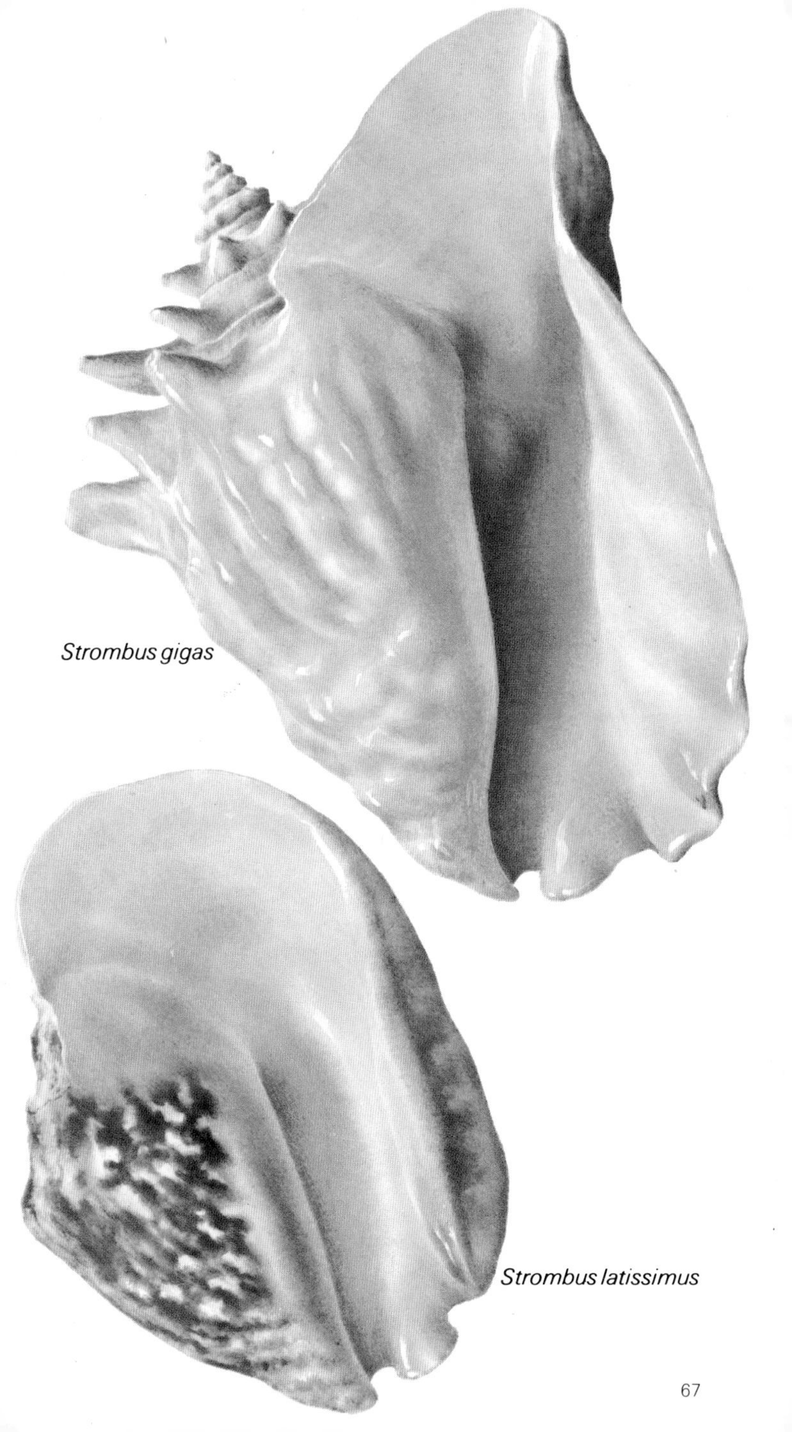
Strombus gigas

Strombus latissimus

S. galeatus **Swainson 1823** Gulf of California to north-west South America, 200mm. It has a pointed, concave, rather depressed spire, with spiral striae and axial riblets. These disappear on the last whorls giving way to broad, rather flat, spiral ribs. The lip is expanded and the stromboid notch not particularly well developed. The body whorl is white, the earlier whorls orange or brown; interior brown and orange. Young specimens have brown or orange mottling or bands and the interior is white. It has a thick, brown periostracum. The animal is popular as food and fished in large numbers in the Gulf of California. It is closely allied to the Brazilian species, *S. goliath* (see page 66).

S. costatus **Gmelin 1791** the Milk Conch, West Indies and Florida, 175mm. Very solid and heavy. Crenellated spire with shouldered whorls and two spiral cords above the suture and below the shoulder knobs. The solid outer lip is thick and ridged at the edge of the aperture. Columella, parietal area and outer lip heavily callous and glazed. Creamy white; parietal area and outer lip metallic looking; base of columella with a brown tinge.

S. oldi **Emerson 1965** (not illustrated) Somali coast of east Africa, 110mm. Due to its restricted range, a rare shell. Rather similar to *S. tricornis* (see page 70) from a little farther north on the other side of the Horn of Africa. The body whorl is narrower, the angle of the spire is less and the posterior extension of the lip is flatter, more open and less pointed. The knobs on the shoulder are smaller and more numerous. The body whorl has about ten, coarse, rounded, spiral ridges which extend on to and become bigger on the flared lip. There are small, spiral threads between the ridges. Narrow and deep stromboid notch. Straight columella; glazed parietal area. Short, open and slightly twisted siphonal canal. Cream, heavily mottled with dark brown; knobs generally white; inside of lip white becoming dark brown within the aperture.

S. listeri **Gray 1852** (not illustrated) Bay of Bengal and north-west Indian Ocean, 150mm. Until very recently a rare shell, but found in some numbers in the last few years. Light but strong. High, pointed spire; long, narrow body whorl. Early whorls with many axial ribs, which become obsolete on the last three whorls. The latter with strong but rounded shoulders. Constricted suture. Expanded lip, the edge roughly parallel with the columella and extending posteriorly in a round-ended projection about level with the suture above the penultimate whorl on the ventral side. Thickened outer third of lip. Very wide and deep stromboid notch. Smooth columella and narrow aperture. Extended, open siphonal canal curves away from the lip side. White profusely covered with brown, axial streaks; body whorl mostly brown on the dorsal side, lighter on the ventral, with four spiral rows of small, white, chevron-shaped marks; aperture white; thickened part of inside lip and interior stained with brown.

Strombus galeatus

Strombus costatus

S. tricornis Humphrey 1786 Red Sea and Gulf of Aden, 125mm. Solid with fairly high spire and knobbed above the suture. Body whorl is shouldered with blunt knobs, the one on the dorsum being large and elongated axially. Flared outer lip is produced anteriorly to about level with the apex, thickened and slightly inverted close to the aperture. Strong stromboid notch; heavily callous columella, parietal wall and outer lip. Cream with more or less heavy mottling of light or dark brown; callus a dirty yellow-brown, with rather a metallic look. It lives in shallow, sometimes very shallow, water.

S. gallus L. 1758 the Rooster-tail Conch, West Indies, 150mm. Fairly high spire and small, blunt knobs above the deep suture. Spiral cords on the early whorls become low, flat ridges below the knobs on the shoulder of the body whorl. There are about seven knobs becoming progressively larger to the penultimate knob; final knob small. The outer lip is flared and very produced posteriorly, slightly outward and backward, to give the 'rooster-tail' which curves in on itself to become channelled. It also flares out to obscure in part the spire. The outer lip is not thickened but is slightly frilled and has a deep stromboid notch. The siphonal canal is produced and curved backward and then forward and to the right. White with orange-tan mottling; interior white; inside of the lip, columella, parietal wall and siphonal canal shades of pale orange-brown; inner edge of outer lip is slightly metallic looking. This species is rather uncommon.

S. peruvianus Swainson 1823 Peru to Mexico, 150mm. Rather similar to *S. tricornis*, but bigger, and the top of the posteriorly extended lip is more recurved. Rather concave spire and spiral ridges with small nodules above the suture. Five blunt knobs on the shoulder of the body whorl, the penultimate being by far the largest. The body whorl has low flat, spiral ridges, about fifteen in number, and a second row of very rounded knobs or swellings below those on the shoulder. Outer lip is thickened and produced with the outer edge curled back. Stromboid notch is wide and lirate below the notch; short, sharply recurved siphonal canal. Columella with about eight plicae, the anterior ones bifurcated, and a well-defined channel running from where the lip joins the shoulder to the last whorl and down into the aperture. There is also a channel running from the produced top of the lip into the interior; this opens up into a deep 'well' which is the interior of the last of the knobs on the shoulder. Creamy white to light brown with mottling showing through the blue-white glaze on the parietal wall; interior and columella fleshy pink; the 'well' white; the thickened lip has a metallic look. It has a thick brown periostracum which, like those on other strombs, flakes off when dried.

Strombus tricornis

Strombus gallus

Strombus peruvianus

S. dentatus L. 1758 east Africa to Polynesia excluding north Indian Ocean, Japanese mainland and Australia, 55mm. Smooth, glossy, inflated, and bluntly axially ridged below suture on last three whorls; early whorls have small varices. Lip is slightly expanded with six blunt teeth; interior of lip strongly lirate. Columella shiny and smooth except for four long folds at posterior end and five short ones anteriorly. Thick callus; short, recurved siphonal canal. White maculated with brown, generally axial flame-like marks; siphonal canal with purple stain; teeth on lip, anal canal and columella white except for anterior folds; lirate area on inside lip rich dark purple, showing through on body whorl. Generally found in shallow water in or near coral reefs.

S. fusiformis Sowerby 1842 Red Sea, Gulf of Aden, east Africa and Madagascar, 45mm. Spiral band below suture and varices on early whorls. Two, low, blunt knobs on shoulder of body whorl, which is faintly spirally ridged except close to lip where ridges become prominent, and on lower third of body whorl where they are prominent throughout. Lip slightly expanded and strongly lirate within; shiny columella, ridged at each end; fairly shallow stromboid notch; short siphonal canal. Cream; sparse brown dots and blotches of light red-brown; columella, interior white. Generally found on sandy coral bottoms in deeper water.

S. erythrinus Dillwyn 1817 Indo-Pacific from east Africa and Red Sea to Hawaii, Samoa and Tonga, 50mm long and 17mm wide. Angular, ribbed and with some early varices. Spiral cords become bigger anteriorly and near lip; ribs become six, blunt knobs on dorsal half of body whorl. Constricted at suture. Lip slightly expanded, thickened, backed with a groove, lirate internally. Stromboid notch not strong; short siphonal canal; columella with plaits at each end. White, banded on body whorl and clouded with pale or dark yellow-brown; inside of lip, inner half and bottom end of columella, rich dark purple-brown. It is variable, and two varieties are illustrated. The variety *elegans* Sowerby 1842 is from Noumea in New Caledonia. The variety from Fiji, Tonga, Samoa and the Ellice Islands is given subspecific rank as *S.e. rugosus* Sowerby 1825 and has a coarser sculpture, the columella and aperture pure white.

S. sinuatus Humphrey 1786 west Pacific, 120mm long and 70mm wide. High-spired with body whorl two-thirds of total length. Whorls somewhat constricted below suture and spirally corded. Angular shoulder just above suture has row of blunt knobs, biggest on middle of dorsum. Lip wide, flaring and thickened, with sharp ridge running parallel to columella; lip joins spire of shell partly obscuring it; posterior edge of lip has four or five spatulate projections. Deep stromboid notch; short, recurved siphonal canal. White or cream mottled with yellow-brown; inside of lip pink; columella light golden brown; interior rich dark purple-brown. Lives in rather shallow water.

S. marginatus L. 1758 is divided into four subspecies:
S. marginatus marginatus L. 1758 north and east Bay of Bengal to north part of Strait of Malacca, 55mm. High-spired with varices, spiral cords, axial ribs and sharply angular shoulder. Ribs disappear on last two whorls leaving nodulose keel on shoulder. Base of body whorl striate. Thickened, sharp-edged lip ending posteriorly on keel of penultimate whorl. Shallow stromboid notch; short siphonal canal. White maculated with dark or light brown; four white spiral bands on body whorl; inside of lip lirate, white; columella with faint plicae anteriorly; interior white.
S. marginatus robustus Sowerby 1874 Strait of Malacca, around South China Sea to south Japan, 65mm. Much shorter spire than above and rounded shoulders without keel. Strong axial fold at outer side of parietal area. Posteriorly, lip ends on or above suture of second whorl.

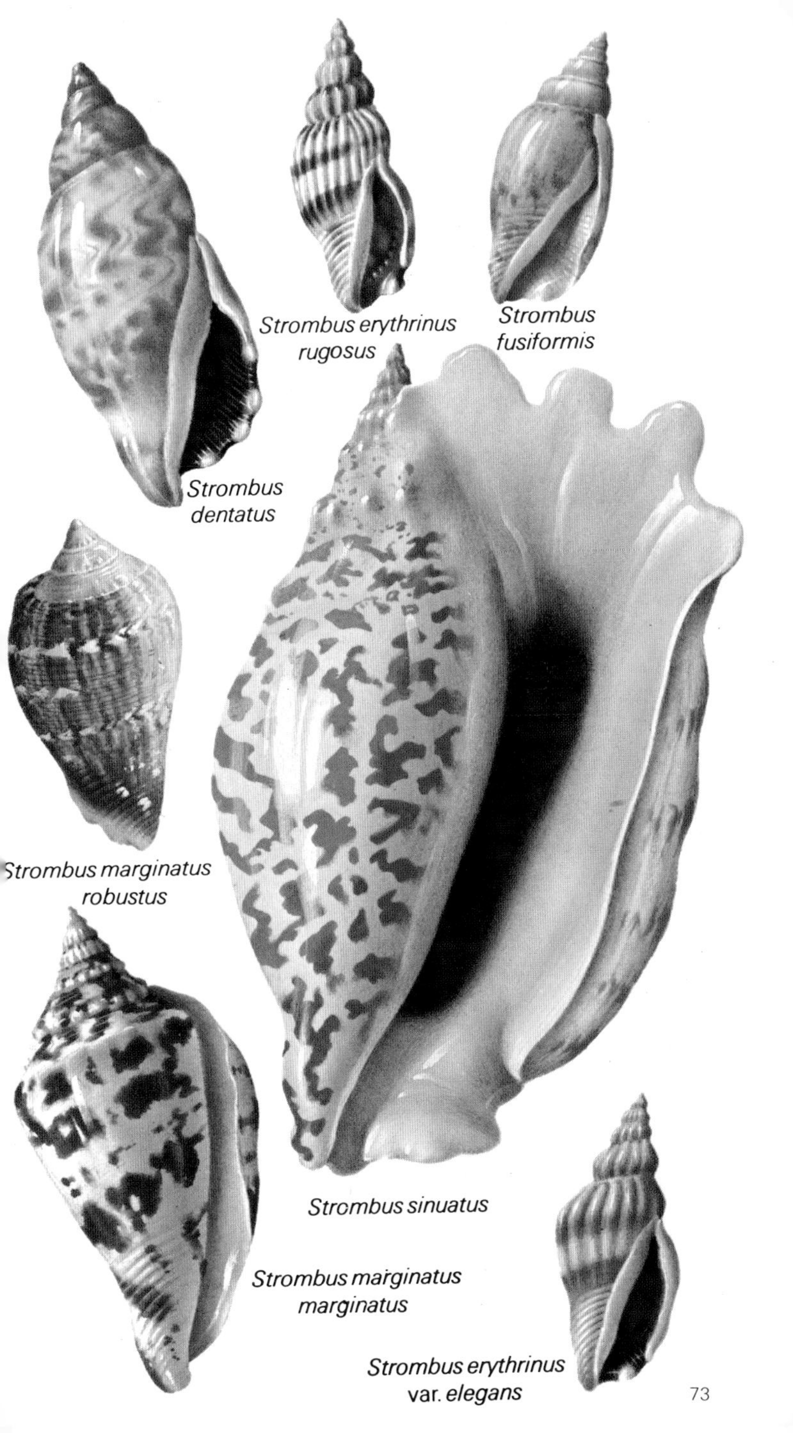
Strombus erythrinus rugosus
Strombus fusiformis
Strombus dentatus
Strombus marginatus robustus
Strombus sinuatus
Strombus marginatus marginatus
Strombus erythrinus var. elegans

S. marginatus succinctus L. 1767 (not illustrated) south-east India and north-west Sri Lanka, 50mm. High-spired and rather narrow. Large blunt knob on dorsum of body whorl. Colour paler.
S. marginatus septimus Duclos 1844 (not illustrated) Ryukyu Islands, Philippines, New Guinea, Solomons and New Caledonia, 48mm. High spire. Like *S.m. succinctus* but without knob on dorsum; sometimes with two or three small ones instead. Lip a little more flared. Dark brown; five, narrow, spotted, white bands.

S. maculatus Sowerby 1842 east and north Pacific, 35mm. Moderate spire with former varices and rounded, faintly knobbed shoulders. Narrow constricting groove below suture which disappears on body whorl. Shoulders on dorsal half of body whorl with about five, very blunt knobs, corded on bottom half of body whorl and near outer lip; latter swollen and turned in. Well-developed stromboid notch; short, almost straight siphonal canal; strongly lirate aperture; columella usually lirate with a smooth, central area, but may be lirate its entire length. Cream or white with usually very pale yellow-brown banding and mottling, but variable; aperture with pink tinge.

S. variabilis Swainson 1820 central Indo-Pacific from Indonesia and Malay Peninsula to the Marshalls and Samoa, 60mm. High-spired; finely spirally ridged on early whorls and base of body whorl; former varices on early whorls, which are also axially ribbed and angular, ribs becoming knobs on shoulder of body whorl. Flaring lip is recurved anteriorly and extends posteriorly to end on the penultimate whorl. Siphonal notch is not very deep; slightly recurved siphonal canal. White with brown mottling, mainly wavy, axial streaks, and areas of cross-hatching; about five, more or less indistinct, spiral, white bands show on lip; interior and columella smooth and white. Those specimens from the west of its range generally with a dark brown rectangle halfway along edge of columella; those from east without.

S. epidromis L. 1758 New Caledonia to New Guinea, the Philippines, Indonesia and Singapore and north to Taiwan, 90mm. Angulate and axial ribs which disappear on the last two or three whorls and give way to small knobs; spiral striae and varices on early whorls. Flared lip is thickened and recurved at edge, posterior end joining shell above suture of penultimate whorl and forming anal canal. Moderately deep and wide stromboid notch; slightly recurved siphonal canal; smooth and heavily callous columella and broad parietal area. White with faint, pale brown, axial mottling; interior and columella shiny white; lip edge and columella with metallic look. Lives in rather shallow water.

S. mutabilis Swainson 1821 Indo-Pacific except Hawaii, 40mm. Very variable. Rather short spire. Rounded, blunt, axial ribs, becoming heavy knobs on dorsal half of body whorl. Faint, spiral cords becoming much sharper on base and near lip. Deep, spiral, narrow, constricting groove below suture. Lip flared, thickened, with a short, blunt projection at shoulder forming outer side of anal canal. Shallow stromboid notch; short, only slightly recurved siphonal canal; lirate aperture and columella. White or cream mottled and spotted with light and/or dark browns generally in axial streaks and spiral bands. Often has a central light band on body whorl and sometimes up to seven, narrow, light bands. Columella and lip white or pink; deep interior sometimes mottled as outside. Colour form *zebriolatus* Adam and Leloup 1938 from Kenya has dark brown, wavy, axial streaks on body whorl.

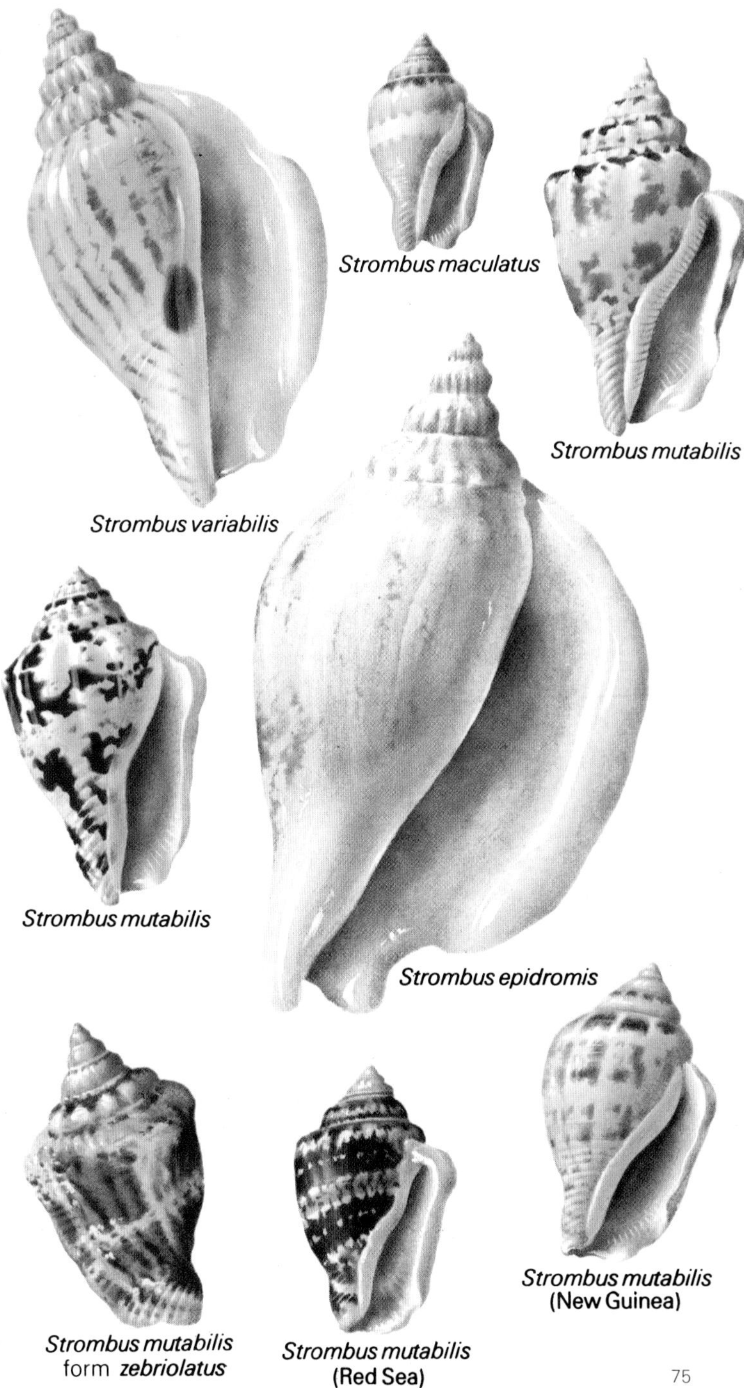

Strombus maculatus

Strombus mutabilis

Strombus variabilis

Strombus mutabilis

Strombus epidromis

Strombus mutabilis (New Guinea)

Strombus mutabilis form *zebriolatus*

Strombus mutabilis (Red Sea)

S. plicatus Röding 1798 is divided into four subspecies:
S. plicatus pulchellus Reeve 1851 west Pacific, 35mm. High, concave spire; varices, spiral cords, axial ribs; cord below suture more prominent especially on earlier whorls. Axial ribs become nodules on shoulder of last two whorls. Three or four blunt knobs on dorsum of body whorl; penultimate knob biggest. Corded on bottom of last whorl and near lip. Flared, slightly recurved lip, extending to suture of penultimate whorl. Wide, moderately deep stromboid notch; short, reflexed siphonal canal; lip lirate and columella also, though almost smooth centrally. White or cream; light brown maculations; four, narrow, white spiral bands with brown spots on body whorl; both sides of lip white; columella white or cream; interior with purple-brown lirae; end of siphonal canal mauve.
S. plicatus plicatus Röding 1798 (not illustrated) Red Sea, 60mm. Colourless aperture and lip; some of columellar lirae brown.
S. plicatus sibbaldi Sowerby 1842 (not illustrated) Gulf of Aden to south India and Sri Lanka, 35mm. Similar to *S.p. plicatus* but stubbier body whorl. Columellar lirae purple-brown; interior and lip white.
S. plicatus columba Lamarck 1822 east Africa, Madagascar and Seychelles, 45mm. Purple-brown or brown blotch posteriorly on columella, latter strongly lirate; one or two blotches well inside lip.

S. microurceus Kira 1959 Java to south Japan, Samoa and New Caledonia, 25mm. Similar shape to *S. mutabilis* but columella yellow-orange on its outer half and dark purple-brown with yellow striae on the inner; interior white; band of purple-brown with white spiral lirae within lip; siphonal canal has purple mark.

S. terebellatus Sowerby 1842 east Africa and west Pacific, 50mm. Elongate, narrow, rather fragile, inflated and smooth. Narrow band below suture on early whorls. Oblique groove on bottom of body whorl. Lip slightly thickened behind edge. Wide, very shallow stromboid notch; smooth, thinly callous columella; siphonal canal short but extending beyond end of lip. Cream maculated with light or dark brown; darker below suture; brown lirae inside aperture; columella white, brown marks showing through callus.

S. urceus L. 1758 west Pacific from the Solomons to Malay Peninsula and Sumatra, 60mm. Rather narrow with high spire of angular whorls, spirally striate and ribbed. The ribs become more or less heavy knobs on shoulder of dorsal side of body whorl; spiral ridges on base of body whorl. Thickened lip; stromboid notch sometimes deep, sometimes hardly apparent; columella smooth in the middle and lirate at each end; aperture usually strongly lirate. It is, however, a very variable species. White, cream or brown, with spots, flecks, bands or axial streaks of light or dark brown; columella and inside of lip may be dark or pale pink; interior white or purple-brown or deep rich purple; the whole columella, lip and interior may be almost black. Illustrated are: a brown, orange-mouthed specimen from the Philippines; a deep-water specimen, light in weight and colour and with more rounded whorls from the south Philippines; and *ustulatus* Schumacher 1817, the white with black mouth variety, often bigger than average, from Singapore. The latter form is found less commonly elsewhere around the South China Sea.

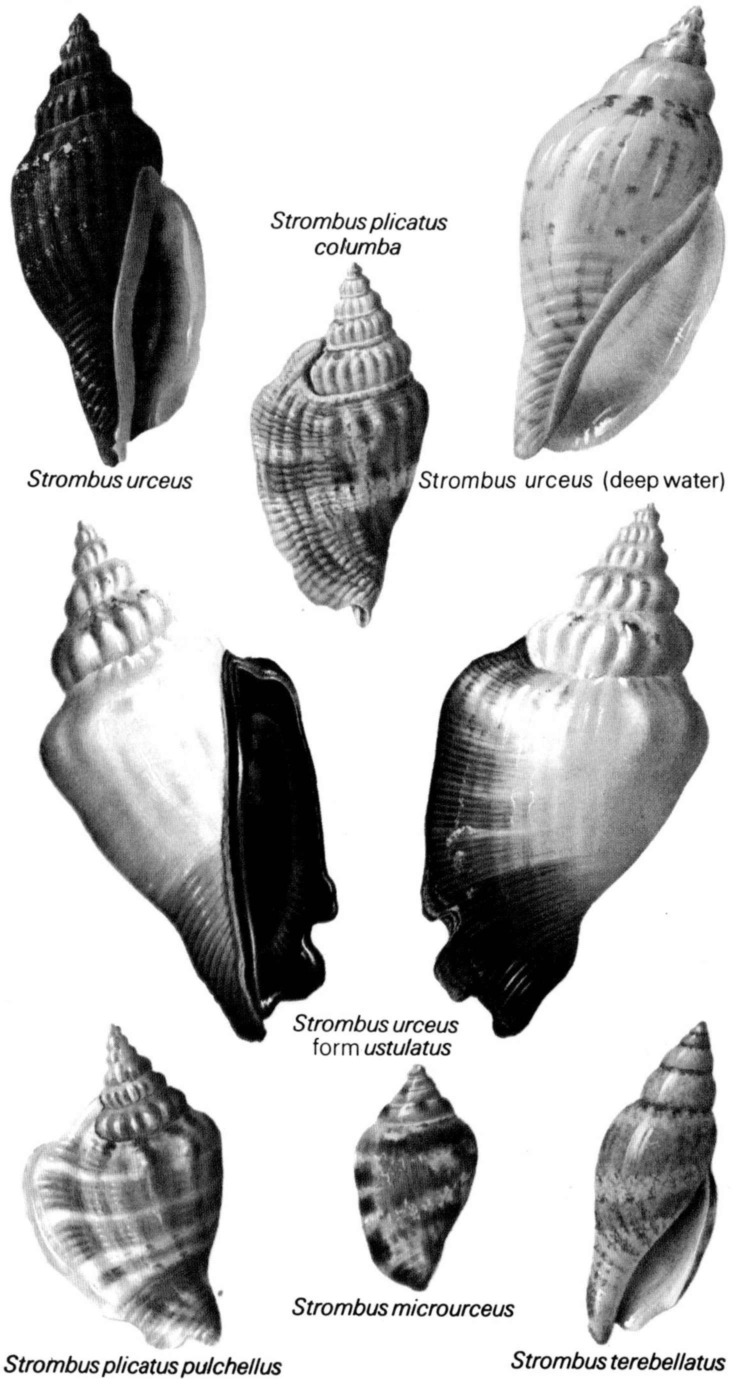

Strombus plicatus columba

Strombus urceus

Strombus urceus (deep water)

Strombus urceus form ustulatus

Strombus microurceus

Strombus plicatus pulchellus

Strombus terebellatus

S. minimus L. 1771 west Pacific, 40mm. High spire; spirally ridged and axially ribbed, ribs replaced by knobs on body whorl. Narrow band below suture. Flared, thickened lip is lirate within, projecting posteriorly to top of antepenultimate whorl. Broad, shallow stromboid notch; callous columella, callus extending posteriorly forming, with lip, the elongate anal canal; short, almost straight siphonal canal. Light or dark brown; tiny cream spots; sometimes spiral row of white dots on body whorl; aperture yellow; columella white or pinkish.

S. fasciatus Born 1778 Red Sea, 50mm. Solid with low spire and sometimes narrow band below suture which disappears on last whorls. Knobs, increasing in size, developing on penultimate whorl, about ten on body whorl, largest on middle of dorsum. Slightly flared lip; wide, moderately deep stromboid notch; slightly curved siphonal canal; smooth columella; small fasciole. White; five to nine, narrow, irregular, broken, spiral lines of very dark brown-black; columella yellow; interior yellow-orange; ends of spiral lines show just within lip and lines themselves show faintly through lip and sometimes under columellar callus.

S. vittatus L. 1758 is divided into three subspecies:
S. vittatus vittatus L. 1758 South China Sea to New Guinea and Fiji, 85mm. Stout, pointed spire with axial ribs. Groove below suture forming band on which ribs form nodules; below groove, ribs become obsolete on middle whorls; also two, fine, spiral cords on this band. On body whorl only cords remain. Area below suture concave. Spire may have former varices. Shoulder of body whorl has one or two slight 'lumps' and one blunt nodule in middle of dorsum. Base with up to twenty, spiral grooves. Expanded, thickened, lip curling in a little; broad but not deep stromboid notch; inside of lip finely lirate, lip ending on suture above penultimate whorl. Columella smooth centrally, pustulate posteriorly, with up to five weak lirae anteriorly, and bounding short, straight siphonal canal. White; mottled with light or dark brown, especially on subsutural band, lip and last part of body whorl; in latter area fine, axial, minutely zigzag, brown lines, and about five, spiral bands; columella, interior and inside of lip white. Illustrated is high-spired form named *australis* Schröter 1805.
S. vittatus campbelli Griffith and Pidgeon 1834 (not illustrated) north and east Australia, 65mm. Similar to above but knob on dorsum bigger, and band below suture has strong nodules.
S. vittatus japonicus Reeve 1851 south Japan, 65mm. Lower spire; fine, spiral cords and small, axial ribs.

S. lentiginosus Jousseaume 1886 tropical Indo-Pacific, east Africa to Society Islands, not generally on Asian coast except south India and Malay Peninsula, 105mm. Solid and heavy with short, pointed spire and spiral striae. Axial ribs nodulose on shoulder of later whorls and bearing eight knobs, longer axially than broad, on dorsal side of body whorl, antepenultimate in middle of dorsum very large. Body whorl with five rows of much smaller nodules, three other coarse ridges without nodules, coarse striae at base. Slightly flared lip is greatly thickened opposite columella and extended and curved posteriorly, to join spire on suture above antepenultimate whorl; this part of lip has two notches, one quite deep. Deep stromboid notch; short, recurved siphonal canal; straight, smooth columella is heavily callous, as is parietal area which extends over two whorls above body whorl and, with end of lip, forms deep, long anal canal. White, heavily speckled with dark brown-grey; interior and columella pink-cream; lip edge with about seven, spiral, purple-brown stripes; parietal area somewhat metallic.

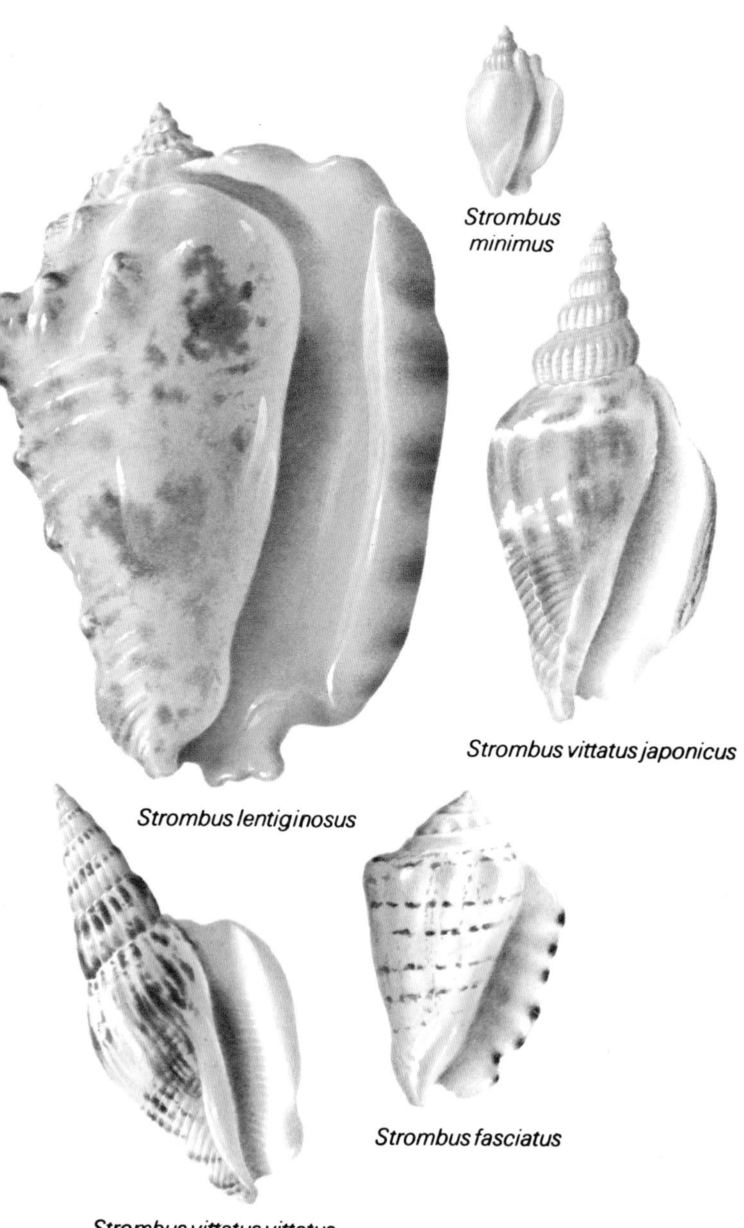

Strombus minimus

Strombus vittatus japonicus

Strombus lentiginosus

Strombus fasciatus

Strombus vittatus vittatus

S. gibberulus L. 1758 is divided into three subspecies:

S. gibberulus gibberulus L. 1758 Indian Ocean, Malaya and north-west South China Sea, 70mm. Moderately elevated spire; spirally corded; earlier swollen varices. Penultimate whorl is expanded and bulges over suture on dorsum of body whorl; latter also inflated at shoulder, and with cords disappearing except at base where they reappear on lip. Lip expanded, thickened before its sharp edge; posterior end on inflated part of body whorl, well below suture. Moderately deep stromboid notch; siphonal canal long for the genus and somewhat recurved. Smooth columella. White heavily marked with broken light brown axial lines; more or less faint, narrow, white spiral bands; columella pinky white, sometimes with mauve inner side at bottom; lip white with deep mauve lirate band about 3mm inside; interior white with mauve tinge; anal canal mauve; siphonal canal white.

S. gibberulus albus Mörch 1850 Red Sea to Kenya, 55mm. Smaller than *S. g. gibberulus*. White with very pale fawn markings and purplish broken band below suture; columella white; same purple band on inside lip.

S. gibberulus gibbosus Röding 1798 south-east South China Sea and outward to Society Islands. Smaller, 55mm, and markings usually darker browns and yellows. Columella white or yellow, sometimes rich chocolate brown. Spiral cords on lip rather weak.

S. aurisdianae aurisdianae L. 1758 east Africa to Ryukyus and the Solomons, 75mm. Solid; moderately high spire; spirally corded and axially ribbed. Somewhat angulate and with short blunt spines on shoulders. Below suture is a strongly beaded band. On body whorl the blunt spines become quite large. Two cords near middle of last whorl have short blunt knobs. Lip expanded, thickened, but coming to a sharp edge, with two or three uneven folds on outside before the thickening; a short finger-like projection from posterior end of lip and a deep stromboid notch at the other. Siphonal canal deep, slightly twisted to the right and recurved almost through 90°. Lirate anal canal; columella pustulate at top, otherwise smooth and callous; wide parietal wall, callus almost extending to apex. White or pale brown mottled with darker brown; lip with about eight pale red-brown spiral bands; columella and parietal area light brown at anterior end, creamy white at the other; inside lip pinky white; interior orange.

S. aurisdianae aratrum Röding 1798. This is a subspecies of the above from north-east Queensland, Australia and from the east coast of Malaya and perhaps areas between, 90mm. It is bigger, more elongate, has a sharper spire and the projection on the lip is about twice as long. Siphonal canal is also longer and a little less recurved. Spines and knobs bigger and fewer. Colour is tan, similarly mottled; stripes on lip, projection and edge of lip are deep purple; columella and interior are rich shiny orange-brown, and upper half of parietal area is dark brown; inside of lip fades to pink.

S. bulla Röding 1798 Indonesia, Philippines, Taiwan, the Ryukyus and New Guinea east to New Caledonia and Samoa, 70mm. Similar to *S. aurisdianae* but much smoother; spiral cords barely raised; subsutural cord obsolete; knobs smaller; projection from lip longer and narrower; no lirae on anal canal; callous area covers about half the shell and sometimes whole of spire. White, heavily mottled with light brown; callus white, pink-brown anteriorly; interior rich shiny orange-brown.

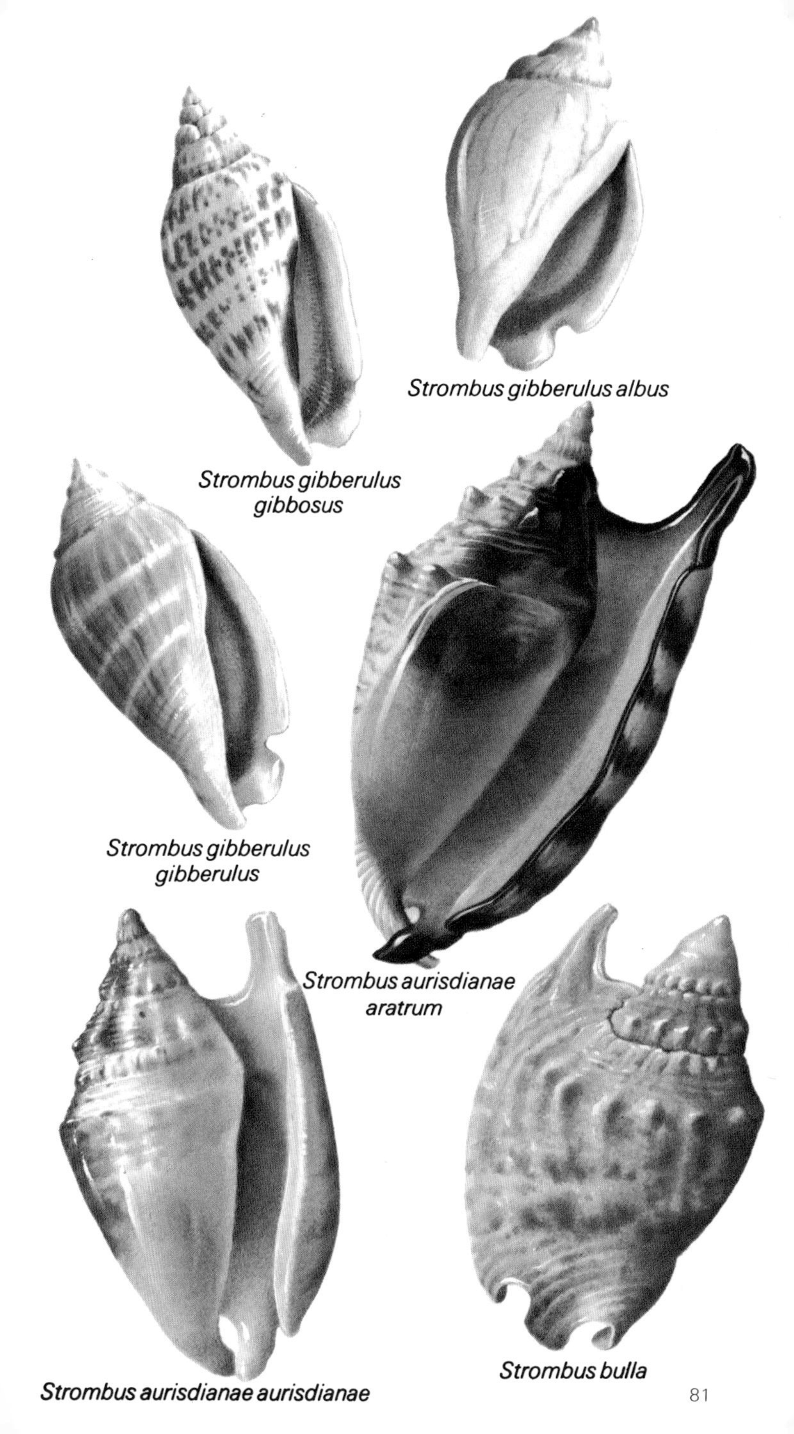
Strombus gibberulus albus
Strombus gibberulus gibbosus
Strombus gibberulus gibberulus
Strombus aurisdianae aratrum
Strombus bulla
Strombus aurisdianae aurisdianae

S. pipus Röding 1798 east Africa and south and east tropical Pacific, 70mm. Rather short spire, slightly angulate, spirally corded and axially ridged with early varices. Body whorl a little inflated with four spiral rows of small knobs on middle and one near base; each knob has a dark purple-brown spot on its lip side. Expanded lip is thickened before sharp edge which is curved inward and extends onto penultimate whorl, with two notches at posterior end; inside of lip faintly lirate. Deep stromboid notch; very short, recurved siphonal canal; smooth, callous columella; parietal wall has clear glaze. White with tan-brown mottling; inside of lip, anal and siphonal canals rich purple-brown; edge of lip white crossed by bands of the same colour; columella, lower end of parietal wall and deep interior white.

S. luhuanus L. 1758 Strait of Malacca to south Japan, the Line Islands, Fiji and east Australia, 70mm. Conical with low, smooth or ribbed, concave spire and last whorl sharply angulate. Lip curved inward, faintly lirate within. Deep stromboid notch; short, almost straight siphonal canal; smooth columella. White or cream; broken, brown, axial, zigzag streaks forming six, broad, spiral bands on body whorl; inside lip and aperture rose pink; columella with broad, straight band of very dark brown callus.

S. decorus Röding 1798 north and east Indian Ocean, 70mm. Similar in shape to *S. luhuanus*. Conical with depressed spire and pointed nipple-like apex. Shoulders bumpy rather than knobbly, one large 'bump' on dorsal side of body whorl. Expanded lip has thin edge. Deep stromboid notch; almost straight, short siphonal canal; smooth columella. White; broken, axial, wavy streaks of dark brown forming seven, spiral bands on body whorl; columella, inside lip, both canals white; interior apricot. Young specimens, before the lip develops into its final shape, can very easily be mistaken for a species of *Conus*.

S. latus Gmelin 1791 west African coast from Rio d'Oro to Angola and Cape Verde Islands, 150mm. Only known *Strombus* from the east Atlantic. Solid, heavy with moderate spire. Spirally striate on early whorls and axially ribbed; ribs forming knobs on shoulder. Seven large blunt knobs on body whorl, increasing in size near lip; second row of smaller, blunt knobs; and a third and fourth, smaller still, below these. Body whorl has rough, axial, uneven ridges. Expanded, thickened lip; wide, deep stromboid notch; short, almost straight siphonal canal; and smooth, heavily callous columella. White, heavily mottled with brown; inside lip and interior pink; thickened edge of lip metallic pink; columella and around notch with orange tinge.

S. granulatus Swainson 1822 Ecuador to the Gulf of California, 75mm. High spire; spirally striate with angular shoulders, ribbed on very early whorls then bluntly spined and restricted at suture. Body whorl has nine, quite large, spinous knobs and three other spiral rows of small knobs; axial cords begin to develop on penultimate whorl and grow stronger towards lip which has fine, axial growth striae, and thickens before curving inward to a sharp edge. Lip is pustulate within. Very deep stromboid notch and short siphonal canal, recurved through 45°. Columella has six faint ridges anteriorly. Parietal wall callous anteriorly. White heavily clouded with light and dark brown; inside lip, interior and columella white with pinky tinge in stromboid notch and siphonal canal; parietal wall pinky, a little metallic looking at base, clear glaze on upper half.

S. gracilior Sowerby 1825 (not illustrated) Peru to the Gulf of California, 75mm. Very similar to the above but sturdier spire, and less well-developed knobs; lacks the pustules inside lip.

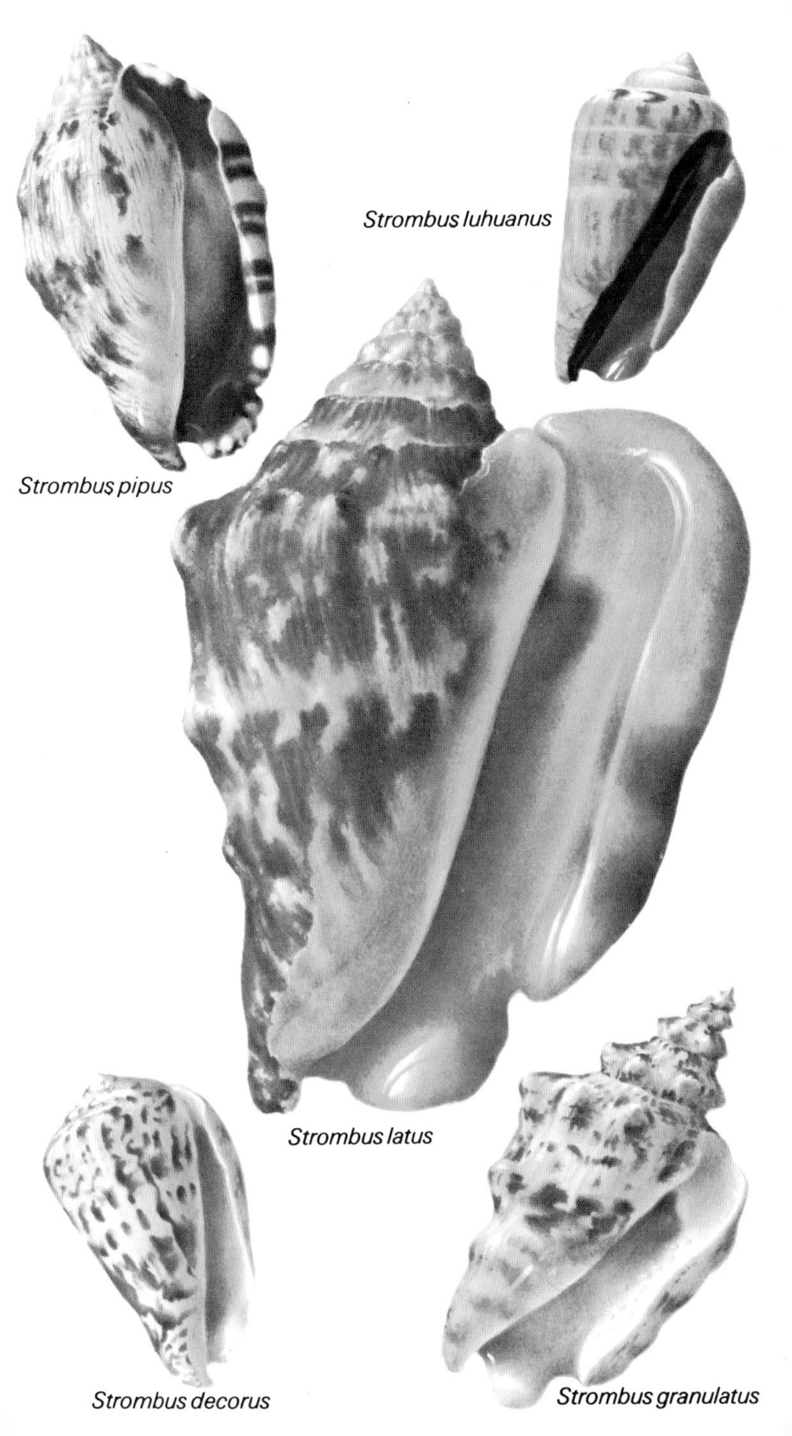
Strombus luhuanus
Strombus pipus
Strombus latus
Strombus decorus
Strombus granulatus

Genus: *Lambis*

The genus *Lambis*, known as the spider or scorpion shells, includes nine species all of which are found only in the tropical Indo-Pacific. None are small; they all develop long projections on the outer lip of the adult shell, have a long siphonal canal and the characteristic stromboid notch. Some are divided into subspecies. The female is often bigger than the male.

Lambis lambis L. 1758 east Africa and Red Sea to the Marshall Islands and Tonga, 200mm. Pointed spire of about eight whorls, spirally and axially striate, with sharp angular shoulders carrying small knobs. Body whorl has fine, blunt knobs; that nearest the parietal wall a mere lump; the next not much bigger; the third well-developed; the fourth, the largest, and the fifth are joined; a second row of much smaller knobs and a third on which they are barely apparent. Lip flares widely and is spirally and axially corded; edge of lip forms six long spines, posterior two being the longest, one lying against spire. In the male, which is about forty per cent smaller than the female, the anterior three spines are quite short and are only slightly recurved, whereas in the female they are at least twice the length and curved up at about 50° to 60°. Long siphonal canal is curved beyond plane of lip. Inner edge of lip is callous between base of spines, which are formed by the folding inward of the edge of what would otherwise be broad, spatulate projections. Very deep and wide stromboid notch and smooth columella. Parietal callus covers half body whorl and up to, but not contiguous with, the apex itself. Small ridge on columella and on inside of lip at end of anal canal; aperture itself is very narrow. Creamy white, heavily mottled with brown, or brown and purple-brown; spines retain the periostracum even when dried; columella, parietal area and interior flesh-pink.

L. crocata Link 1807 Indian Ocean to the Ryukyus, Marshalls, Samoa and northern tip of Australia, 150mm. Moderate spire of about twelve whorls; spirally and axially corded with a row of beads above suture. Body whorl has one spiral row of about four knobs, the last two large, and two other rows of more and smaller knobs. Lip has six spines which are longitudinally corded. Wide and deep stromboid notch; long siphonal canal is curved through 90° in two planes. Cream maculated with yellow-brown; inside lip, interior and columella pinky-orange. There is a subspecies from the Marquesas Islands, *L. c. pilsbryi* Abbott 1961 which is twice the size of *L. crocata*, with the smaller projections less curved.

L. millepeda L. 1758 east Indonesia and the Philippines, 145mm. Rather short spire with about eight whorls, spirally ribbed and angulate with blunt spines on shoulder, restricted at suture. Body whorl has a row of three large and one small knobs and two rows of small knobs. Narrow aperture; lip has three long projections posteriorly and six, short, hooked ones opposite columella. Very deep stromboid notch and only slightly curved siphonal canal. Inside lip strongly lirate except at edge and columella smooth in the middle, lirate at each end. Parietal callus wide and extending to first projection, ending just short of apex. Cream with brown maculations; suture stained purple; inside lip, columella and callous area mushroom-coloured; lirae white, interstices purple; two white teeth at base of anal canal.

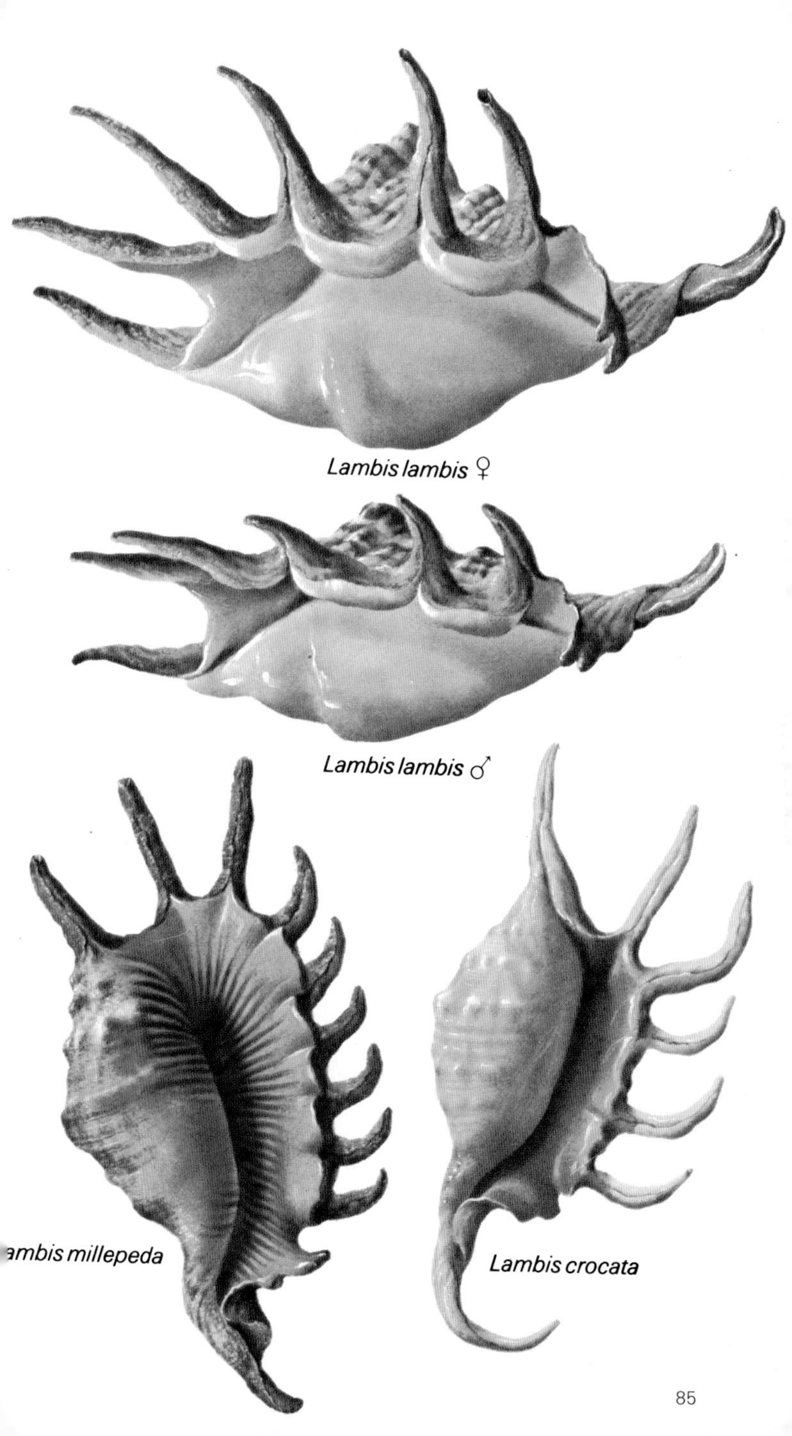

Lambis lambis ♀

Lambis lambis ♂

ambis millepeda

Lambis crocata

L. truncata Humphrey 1786 is divided into two subspecies:
L. truncata truncata Humphrey 1786 Indian Ocean, 400mm. All spire, except for last three whorls, is very depressed, almost flat, giving a truncated effect; spirally corded and fine, axial cords. Whorls with angular, coarse, large knobs on shoulder; body whorl has especially large hump on high point of dorsum and other smaller humps and knobs. Lip has six, rather open spines; deep, wide stromboid notch has crenulated edge. Siphonal canal is as long as average spine, a little reflected backward and toward aperture. Heavily callous, smooth columella; parietal area spreads over part of body whorl but is 'indented' at shoulder of whorl. Cream; brown maculations; aperture, inside of lip, columella and canals pink; darker on parietal area.
L. truncata sabae Kiener 1843 (not illustrated) Red Sea and Gulf of Oman and from Indonesia eastward to the Tuamotu Archipelago, up to 300mm. Differs from the above in size and in having a pointed, not truncated, spire.

L. chiragra L. 1758 is divided into two subspecies:
L. chiragra chiragra L. 1758 Sri Lanka eastward to Tuamotu Archipelago, north to Ryukyus and south to northern Australia. Female form much bigger, up to 250mm, and differently marked from the male. Rather low spire of nine whorls; spirally corded, becoming very coarse on body whorl. Whorls angulate with row of nodules on shoulder, bearing four, large knobs on body whorl, the last a double one and very long; three or four other rows of knobs. Lip is posteriorly curved through 90° across top of columella and behind apex but covering nearly half the spire. Five, long, strong spines on lip, tips more or less pointing posteriorly—one at end beyond top of columella; one at end of anal canal; three on lip, one of which is below the deep stromboid notch—and a sixth projection at base of body whorl. Columella lirate, inside of lip faintly so; wide parietal area. At top of columella and further into aperture than end of anal canal is a ledge or flange, 10mm to 15mm long. White mottled with brown; lip, columella and interior shades of pink; interstices of lirae and lip edge darker.

The male, once thought to be a separate species named *L. rugosa*, is up to 175mm, similar to female but smaller with five knobs on body whorl. Columella lirate throughout, strongly at each end, and lip also lirate. Aperture, interior, lip and columella pinky-red; lirae white with interstices; bottom of columella purple. It has the same sunken shelf at the top of the columella. Illustrated on page 89.
L. chiragra arthritica Röding 1798 western half of Indian Ocean, 200 mm. Female only slightly larger than the male, and the species is remarkably similar to male form, *rugosa*, of *L.c. chiragra*. Lip and columella strongly lirate, more so than in *rugosa*. Lirae white, interstices purple over a rich red-brown or pink background. The main structural difference is no 'shelf' within aperture at top of columella. Instead there is a rounded projection from top of columella (not so deep in aperture) over which the lirae from columella continue and which forms one side of very narrow anal canal.

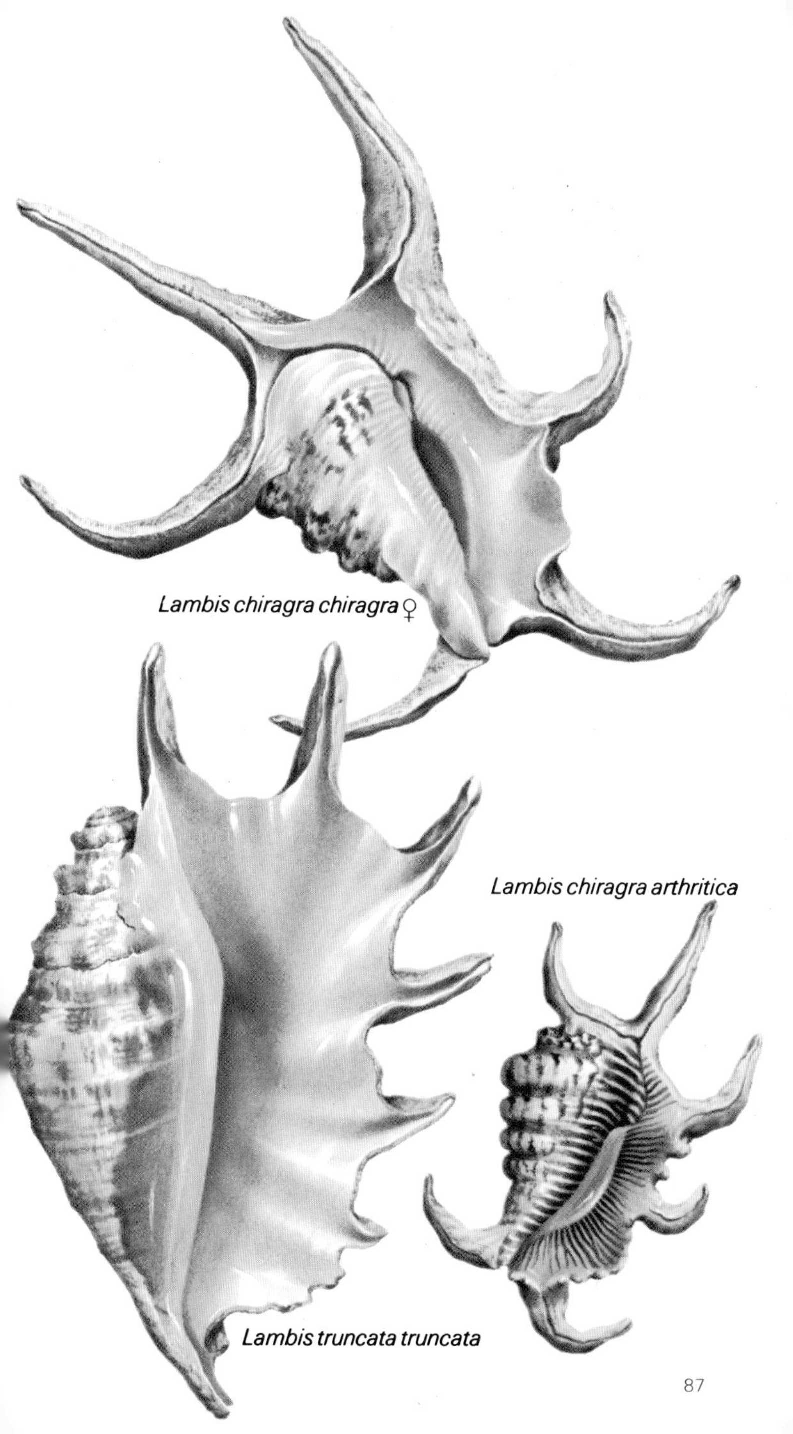

Lambis chiragra chiragra ♀

Lambis chiragra arthritica

Lambis truncata truncata

L. digitata **Perry 1811** east Africa to Samoa, 140mm. High spire; spirally striate; low ribs ending in sharp knobs on angular shoulder of whorls. Body whorl has spiral row of six knobs, biggest in middle of dorsum and three rows of small knobs. Lip, flattened at edge and extending as narrow ledge over inside of lip, has eight projections, two long ones posteriorly – one joined to and projecting beyond spire and tending to bifurcate – and six short ones opposite columella. Deep, wide stromboid notch with scalloped edge as is rest of thickened lip below. Moderately long, almost straight siphonal canal. Columella, narrow parietal area and inside of lip strongly lirate. Outer end of columellar lirae sharply defined and forming a ridge. Narrow ridge on each side of inner end of anal canal ends in deep well. White, heavily maculated with brown; inside of lip white or pinky white; lirae on lip and columella white, interstices dark purple-brown; interior yellow.

L. scorpius **L. 1758** Strait of Malacca to Ryukyus, north Australia and Samoa, 170mm. A subspecies *L.s. indomaris* occurs in the Indian Ocean. Short spire; spirally corded including lip; angular shoulders with small knobs; beaded band below suture. Body whorl has row of five large knobs, two rows with smaller knobs, and a fourth with still smaller ones anteriorly. Posterior end of lip curves round to engulf half of spire and forms rounded, triangular flange on far side of first projection; three posterior projections are long; three opposite columella short and hooked; all six point posteriorly. Deep stromboid notch with scalloped edge, as is remainder of lip. Quite long, curved siphonal canal is sealed at end. All projections are coarse, lumpy and of irregular width. Strongly lirate columella and aperture. White and pale grey; sparse brown spots and clouding; inside of lip pink; lirae white; interstices, interior purple-brown.

Family: Aporrhaidae

Pelican's foot shells. About six species all from north Atlantic and Mediterranean.

Aporrhais pespelecani **L. 1758** Mediterranean to north Norway, 50 mm. High spire; spirally corded and angular with axial ribs across last three whorls. Body whorl with second row of riblets, concave between the rows, and a beaded cord. Very expanded, thickened lip with coarse growth lines, bearing four projections – one short and squat partly attached to spire, a second longer, a third short, a fourth forming siphonal canal. Wide, shallow stromboid notch. Two biggest ridges on body whorl form 'backbone' of middle two projections. Smooth columella; callous parietal area. Dull grey-brown; lip lighter; aperture dirty white.

Family: Struthiolariidae

The ostrich foot shells. Six species: *S. papulosa gigas* Sowerby 1842; *S. papulosa papulosa* Martyn 1784; *S. pelicaria vermis* Martyn 1784; *S. pelicaria tricarinata* Lesson 1880 from New Zealand; *S. scutulata* Martyn 1784 from Australia and one species from the Indian Ocean.

Struthiolaria papulosa papulosa **Martyn 1784** 80mm. Solid; very angular and carinated at shoulder of last four whorls; spirally corded. Rather flat-sided body whorl has coarse growth lines. Expanded lip, almost semicircular columella, and heavily callous lip and parietal area, lower half swollen. Short, curved siphonal canal. Grey; axial, indistinct, brown streaks; aperture white; interior light brown.

S. pelicaria vermis **Martyn 1784** North Island, 35mm. Similar in shape to above but smooth except for faint, low nodules on shoulders and sometimes fine, beaded, spiral cords. Variable colour, purple through brown to yellow.

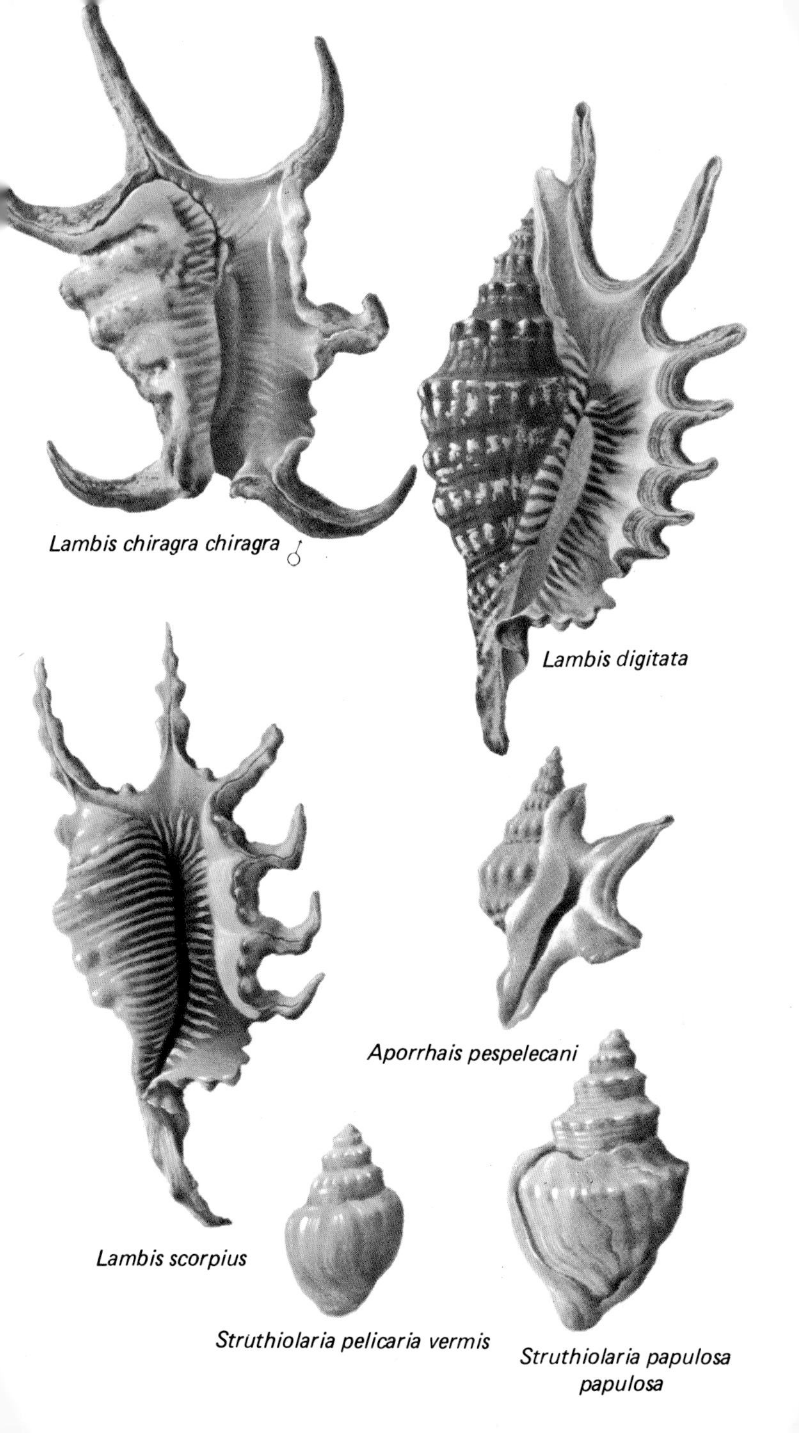

Lambis chiragra chiragra ♂

Lambis digitata

Lambis scorpius

Aporrhais pespelecani

Struthiolaria pelicaria vermis

Struthiolaria papulosa papulosa

Superfamily: Cypraeacea

Family: Cypraeidae

The cowries are probably the most popular shells with collectors because of their beautiful glossy state when taken and their gorgeous colours. They are found in tropical and semi-tropical waters, and are herbivorous, feeding on algae. They grow spirally like other gastropods until they reach adulthood when the outer lip turns in, thickens and teeth are formed on both the lip and columella. They are active during the night and hide during the day. They prefer hard coral reef and shallow water. There are, however, exceptions to nearly all these generalizations.

Cypraea isabella L. 1758 Indo-Pacific, 30mm. Rather cylindrical with short, fine teeth. Fawn-grey with more or less profuse, interrupted, black, longitudinal lines; orange-red tips and spire; white base. *C. isabella mexicana* Stearns 1893 (not illustrated) found in west Central America is very similar but has brown, callous sides and the tips are more produced and brighter.

C. hungerfordi Sowerby 1888 Japan, 40mm. The commonest of the endemic Japanese cowries and lives in deep water. Specimens believed to be of same species recently found in deep water off Cape Moreton, Queensland. Pyriform with strong teeth. Flesh-coloured base and slightly callous sides merging into whitish band with dark brown spots which edge a broad, longitudinal, dark band of close, grey-brown stippling; darker transverse bands showing through faintly. Other endemic Japanese species are *C. teramachii, C. hirasei* and *C. langfordi*.

C. cinerea Gmelin 1791 Caribbean and down east coast of South America to Bahia Blanca, 30mm. A little swollen; teeth quite strong. Grey-brown with two, light, transverse bands; base lighter with rosy tinge; margins and extremities marked with minute spots and dashes of black as if dipped in ash. I have a beautiful specimen taken from a fish's stomach off Brazil which is smaller than those from the Caribbean, with a lighter base, finer teeth and dashes of rich red-brown all over the dorsum.

C. tessellata Swainson 1822 Hawaii, 30mm. Rare, but one of the most beautiful. High dorsum and strongly callous margins. Fine teeth. Blue-grey background with three darker brown-grey bands; rectangular red-brown and white spots on the sides and extremities. Base brown-red with white spots spreading over to teeth on each side.

C. lurida L. 1758 Mediterranean and east Atlantic from Azores in the north to St Helena and Angola in the south, 45mm. Rather light, ovate. Fawn-brown; three darker bands with two dark brown spots at each end; pinky-red tinge at each end of aperture; base white; cf *C. pulchra*.

C. pulchra Gray 1824 Red Sea, Gulf of Aden and south coast of Arabian Peninsula to Gulf of Oman, 50mm. Superficially similar to *C. lurida*, but the latter is much coarser. *C. pulchra* is less swollen, smoother, glossier and lighter in colour, and lacks the red tinge at ends. Its teeth are very much finer, are red-brown and in the middle of the columella they extend about one-third of the way across the base. Straighter aperture than in *C. lurida*. It is also a much more solid shell and as one of the most beautiful cowries is aptly named. Rare.

Cypraea isabella *Cypraea hungerfordi*

Cypraea cinerea

Cypraea tessellata

Cypraea lurida

Cypraea pulchra

C. chinensis **Gmelin 1791** Indo-Pacific from Tahiti westward to the Strait of Malacca and from south Japan to the north coast of Australia, and also the east coast of Africa from the Cape of Good Hope to Aden, 30mm. The callous sides give the shell a somewhat angular look. Teeth strong especially on the outer lip. The dorsum is mottled light brown to green; sides, which are more or less heavilv callous, are cream dotted with mauve spots; base and teeth are creamy white with orange between the teeth. May be more elongate than the illustrated specimen.

C. coloba **Melville 1888** south-east India and Sri Lanka, the Bay of Bengal, and central Indian Ocean from the Andaman Islands to the Chagos Islands (i.e., mainly the area between the two in which *C. chinensis* occurs, to which it is very closely related), 25mm. It is more depressed than *C. chinensis* and has extremely callous margins and very heavy, coarse teeth especially on the outer lip. Background colour is fleshy pink; dorsum covered with minute green-brown spots and with larger dark brown spots where the dorsum and margins meet; base pink-brown.

C. onyx **L. 1758** South Africa up the coast of Africa to Kenya; Madagascar, Mauritius, north Indian Ocean, Indonesia, Philippines and northward to south Japan; north-west Pacific including the Marshalls, Gilberts, Solomons and north New Guinea; 40mm. Base and sides are very dark brown to black, contrasting strongly with the pale blue-white dorsum through which two yellow bands are conspicuous on a young shell and show through faintly on the adult; the dorsal line is dark brown, revealing the yellow bands. In the variety *adusta* Lamarck 1820 from east Africa the dorsum is a very deep rich brown with a faint lighter dorsal line. The variety *succincta* L. 1758 (not illustrated) from India and Iran has a rich red-brown on base and dorsum with two faint pale lines across the latter; teeth lighter.

C. pyrum **Gmelin 1791** Mediterranean and west African coast to Cape Frio, 45mm. Pyriform with fairly strong teeth. The wide margins, base and extremities are orange-red; teeth and interior white; dorsum brown; mottling nearly obscures cream background, but three narrow cream bands are more or less conspicuous.

C. mus **L. 1758** is not by any means a common cowrie as it appears to be restricted to the north coast of South America from the end of the Isthmus of Panama to the Gulf of Venezuela, 45mm. It is a broad, somewhat angular shell with a humped dorsum. Moderately strong teeth. Its light fawn background is overlaid with darker fawn-grey wavy lines running from the dark brown teeth on the outer lip over the wide margin; these lines show only at the edge of the margin on the columellar side, and the teeth, also dark brown on that side, are covered by a dark brown band running the length of the shell; dorsum is speckled with the same fawn-grey strippling and the conspicuous pale dorsal line is marked on both sides with very dark brown spots tending to run together at the posterior end of the dorsum.

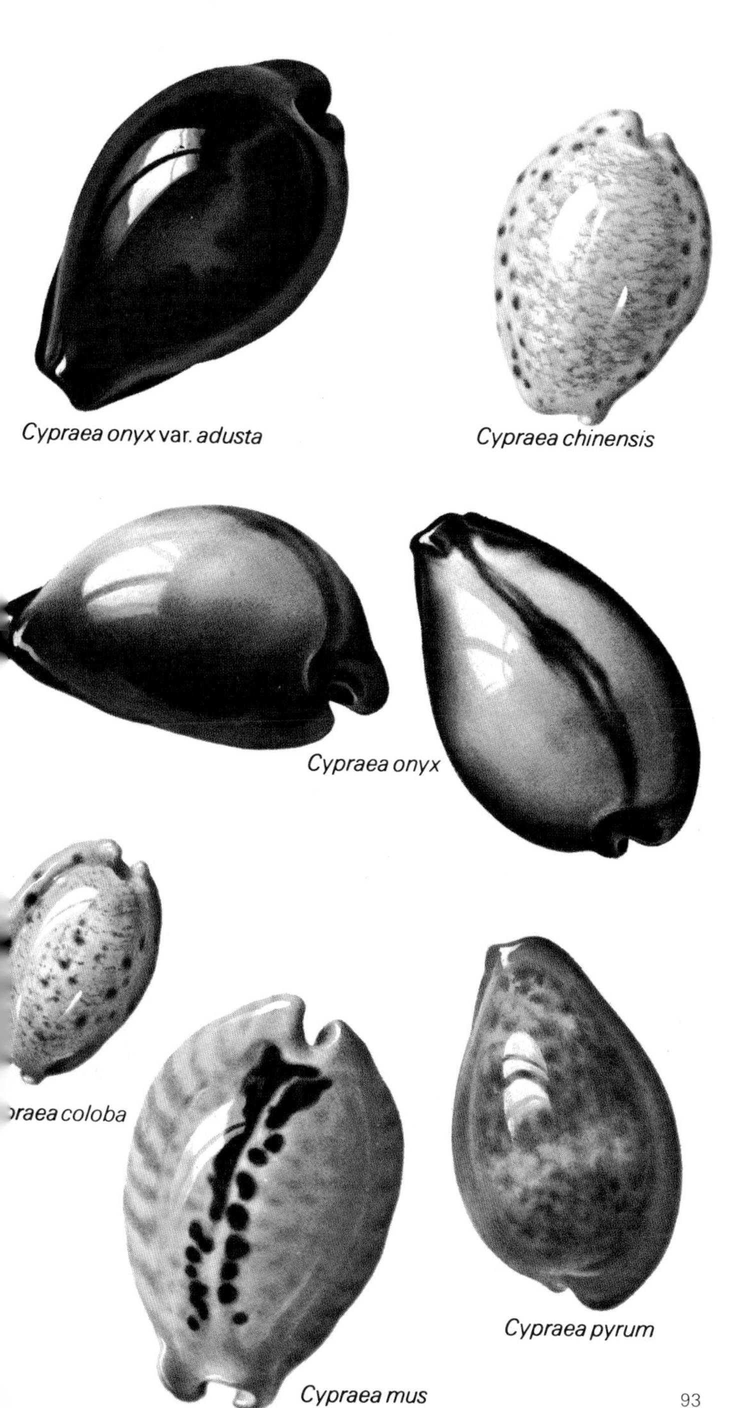
Cypraea onyx var. *adusta*

Cypraea chinensis

Cypraea onyx

[Cy]praea coloba

Cypraea pyrum

Cypraea mus

C. ziczac L. 1758 Indian Ocean from east Africa, east and north through the China Seas to south Japan, and south to north-west and north Australia and the Great Barrier Reef, then east across the central Pacific to Tahiti including New Guinea, the Solomons, New Hebrides, Fiji, Samoa and Cook Islands, 20mm. Pale brown; three bands of very light blue-white chevrons; base is light red-brown with black spots which spread on to the paler margins and encircle the extremities; small, dark spot on the sunken spire.

C. diluculum Reeve 1845 east Africa from Durban to Aden, the Seychelles, Mauritius and Reunion, 25mm. Somewhat similar in pattern to *C. ziczac* though larger. Ivory white background; three bands of dark grey-brown between which are chevrons of the same colour – the illustrated specimen has less colour than is typical. Margins and base are spotted with the same colour except in the variety *virginalis* Schilder and Schilder 1938 which has no spots and is found together with the more common variety. The sunken spire carries a small dark spot.

C. clandestina L. 1767 Indian Ocean and west Pacific including south Japan, Wake, Marshall, Gilbert, Cook and Lord Howe Islands and the Great Barrier Reef, 20mm. White, covered on dorsum with three, very pale, broad bands of grey; dorsum is also banded with red-brown hairlines which need a magnifying glass to be seen properly; extremities have very pale pinky tinge; base and teeth are white.

C. lutea Gmelin 1791 South China Sea including the Philippines and Singapore; Sulu and Celebes Seas, west New Guinea and western Australia; 20mm. Dorsum pale brown to green, sparsely dotted with small brown spots and having two, conspicuous, narrow, very pale blue-white bands; base and margins are orange-red and profusely dotted with dark brown spots; extremities are also dark brown.

C. saulae Gaskoin 1843 Philippines, Caroline Islands and Sulu Sea, and from the Torres Strait down the Great Barrier Reef to Brisbane, 25mm. This is a rare cowrie but I have included it as, to me, it is the most beautiful. There are three varieties and the illustration shows the variety *siasiensis* Cate 1960. It is a narrow shell with the dorsum somewhat inflated. It is white to light fawn, speckled with a few small chestnut spots which are larger on the margins and with a large almost square blotch on the dorsum. The extremities and teeth are orange-yellow or white.

C. coxeni Cox 1873 north New Guinea, New Britain and the Solomons, a small area and therefore not a common shell, 20mm. It is long, narrow and cylindrical. Variety *hesperina* Schilder and Summers 1963 (not illustrated) is smaller with larger, darker dorsal markings. Pale fawn-cream; dorsum heavily mottled with a rich brown, leaving broad, unmarked margins and base.

C. punctata L. 1771 Indian Ocean, China Seas and central Pacific including south Japan, Marshalls, Tahiti, Cook Islands, New Caledonia, and north-east, north and north-west Australia, 15mm. White, more or less heavily spotted on dorsum and sides with dark chocolate brown; teeth often lined with brown.

C. walkeri Sowerby 1832 Seychelles and Maldive Islands, the south tip of Malaya and Singapore, the north coast of Sumatra, the Sulu Sea and the north coast of Australia, 30mm. Found in two forms, one pyriform and inflated, the other, *surabajensis* Schilder 1937 from the Philippines, elongated and narrow. Fawn background is heavily mottled with purple-brown which forms a more solid band across the middle of the dorsum. Margins and extremities spotted

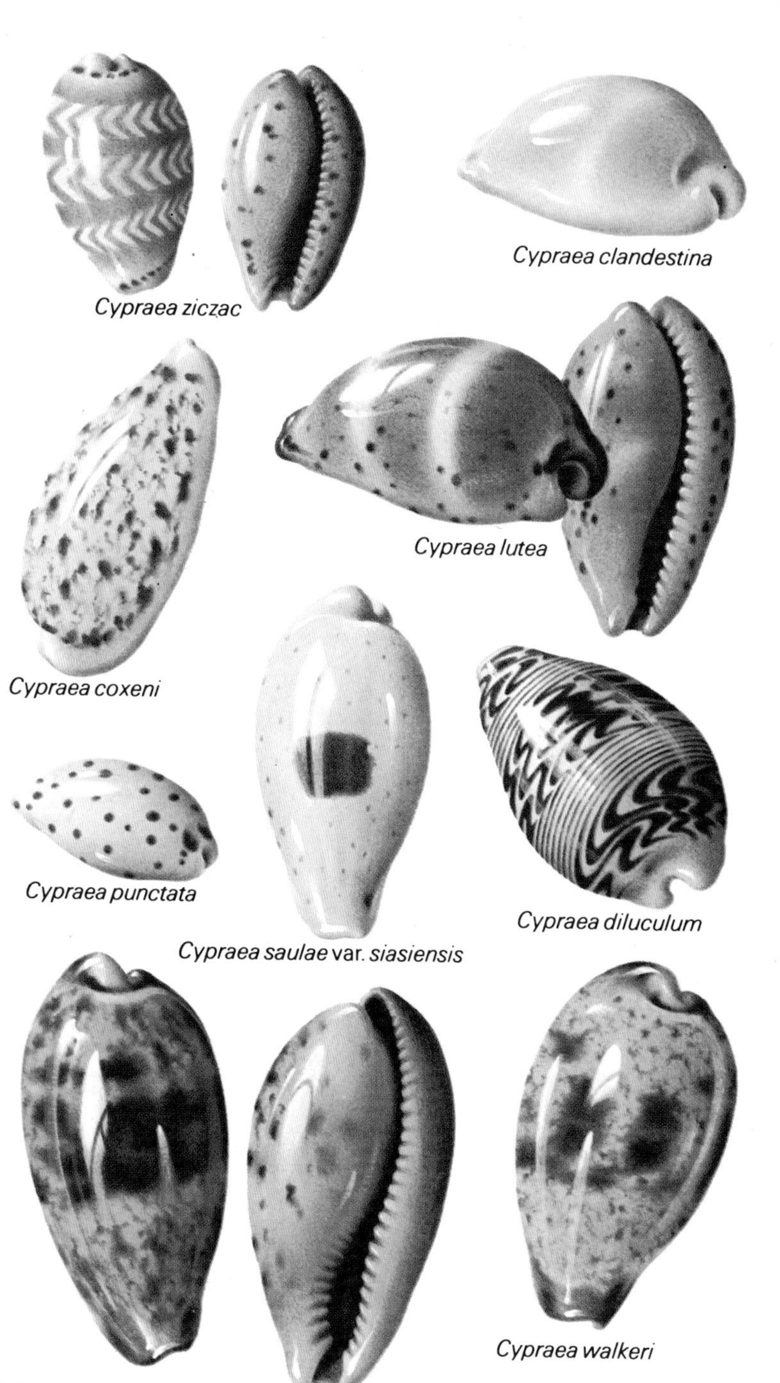

Cypraea ziczac

Cypraea clandestina

Cypraea lutea

Cypraea coxeni

Cypraea punctata

Cypraea diluculum

Cypraea saulae var. *siasiensis*

Cypraea walkeri

Cypraea walkeri form *surabajensis*

with same colour. In both forms the interstices of the teeth, and the labial side of the base are marked with purple.

C. pyriformis Gray 1824 Singapore to the south Philippines then south to north Australia, 35mm. Very pyriform and inflated. Blue-white heavily stippled with smudged, brown spots and three, interrupted bands of darker brown; base cream with quite long, dark brown teeth on the columella; labial teeth short and cream.

C. teres Gmelin 1791 Indo-Pacific from east Africa to the Galapagos, and from Japan to the northern half of Australia, 40mm. Found in three main forms, one cylindrical and narrow, one cylindrical but more inflated and one altogether rounder. The labial margin has a more or less developed callus which does not show on the illustrated specimen. Dorsum is blue-grey heavily mottled with green-brown in dots and larger marks, the latter in three broken bands; base white; callus white with dark brown, rather smudged dots.

C. asellus L. 1758 Indo-west Pacific to Samoa, 20mm. White with three broad bands of very dark brown-black edged with lighter brown.

C. quadrimaculata Gray 1824 northern half of Australia, Indonesia, east Malay Peninsula, Singapore, Philippines and New Guinea, 30mm. Rather square with strong teeth. White with a faint blue tinge; two, faint, narrow, darker bands on the dorsum and heavy stippling of light brown dots; two, large, dark brown spots at each end.

C. pallidula Gaskoin 1849 Japan, Philippines, north Australia and east to Samoa, 20mm. White or pale brown dorsum, more or less heavily speckled with green-brown dots and with four, equally spaced, narrow, blue-grey bands; base white. Closely related and very similar are *C. interrupta* Gray 1824 and *C. luchuana* Kuruda 1960.

C. irrorata Gray 1828 central Pacific, 15mm. Blue-white dotted with red-brown; dots darker on margins; may have a red-brown blotch on columellar margin as illustrated.

C. fimbriata Gmelin 1791 Indo-Pacific to the Tuamotu Islands and Tahiti, 20mm. Blue-white with pale brown speckling and interrupted bands; darker brown spots on the labial margin; two purple-brown spots on the anterior canal and one each on the spire and end of the posterior projection of the labial margin.

C. irrorata and *C. fimbriata* together with *C. gracilis* (see page 98) are three of a complex of closely related and very similar shells. Others in the group are *C. hammondae* Iredale 1939, *C. microdon* Gray 1828, *C. minoridens* Melville 1901, and *C. serrulifera* Schilder and Schilder 1938. *C. raysummersi* Schilder 1960 is a synonym of *C. hammondae*. All inhabit areas within the general limits of north Australia, east Indonesia, the Philippines and eastward across the Pacific.

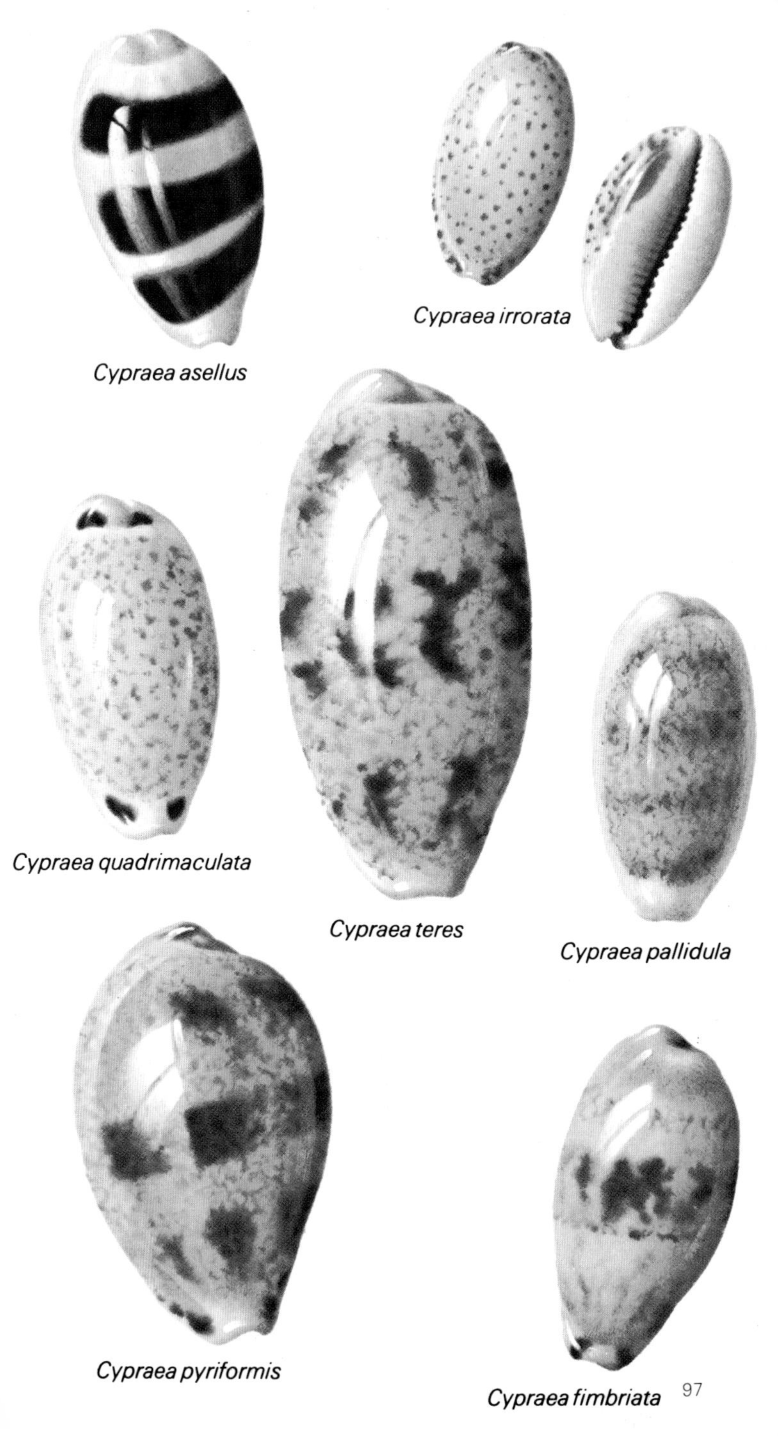

Cypraea asellus

Cypraea irrorata

Cypraea quadrimaculata

Cypraea teres

Cypraea pallidula

Cypraea pyriformis

Cypraea fimbriata

C. albuginosa Gray 1825 Gulf of California, the Bay of Panama, Ecuador and the Galapagos Islands, 20mm. Pyriform with rather fine teeth. Pale lilac base colour which forms a much darker line on the margins; dorsum is heavily ocellated with dark red-brown rings with white-lilac centres and also with small white-lilac spots; three faint bands show through; base is lilac at edges fading to white at aperture; teeth white.

C. poraria L. 1758 Indo-Pacific from east Africa to Tahiti including south Japan and north-east Australia, 15mm. Somewhat similar to *C. albuginosa* but heavier and rounder, and extremities less produced. Background rich purple-lilac, darker than in *C. albuginosa,* with white spots and ocellations, but rings are purple instead of red-brown. The colour tends to fade quickly after collection.

C. helvola L. 1758 Indo-Pacific from east coast of Africa to Tahiti, including south Japan and the northern half of Australia, 20mm. Beautifully coloured when fresh – like *C. poraria* it fades rapidly. Dorsum has background colour of pale blue heavily dotted with lighter spots, and is more or less sparsely blotched with large, dark red-brown spots which run together towards the margins to form solid uneven bands, one on each side, which do not join at the extremities. Margins, sharply divided from sides, are callous and a rich red spreading over the base and coarse teeth. Convex base bears darker red mark on centre of columellar side. This very handsome shell varies somewhat over the vast area in which it occurs.

C. gracilis Gaskoin 1849 Indian Ocean and west Pacific including south Japan, the Philippines, south coast of New Guinea, north Australia, the Great Barrier Reef and Fiji, 15mm to 25mm. This shell has a number of races and varies considerably in size and depth of colour. Background colour blue-grey, more or less heavily stippled with brown and one or more irregular brown spots across the centre of the dorsum; faintly banded; brown spots on margins tend to spread on to white base; two dark spots at each extremity which spread into the anterior and posterior canals.

C. spurca L. 1758 the only cowrie found throughout the Atlantic and Mediterranean area in which cowries occur, 25mm. The Mediterranean shell is long and narrow with slightly swollen base and very open anterior canal. The extremities are produced, especially anteriorly. Its dorsum is light tan with darker heavy stippling and larger dark brown spots around the margins which are slightly callous; base cream to light tan.

C. spurca acicularis Gmelin 1791 America, 20mm. Smaller, squatter and rounder than *C. spurca.* Heavily callous margins. Dorsum spots are orange; base and margins are white; dark spots on margins are more discrete. I have a beautiful specimen taken from the digestive tract of the fish *Amphyethys cryptocentum.*

C. boivinii Kiener 1843 south Japan, the Philippines, Malaya, Singapore and Sumatra, 25mm. Narrow shell with quite strong teeth. Very pale, smoky, blue-grey with indistinct brown spots, and a conspicuous dorsal line edged in pale brown, running from a spire blotch to the edge of the anterior canal; extremities, margins and base white.

C. labrolineata Gaskoin 1849 central Indo-Pacific from Sumatra to Fiji and Hawaii including Okinawa and north and north-east Australia, 20mm. Somewhat elongate; rather coarse teeth. Dark green dorsum heavily spotted with white; line of dark spots, often joining up around the margins, especially on either side of anterior and posterior canals; extremities, base white.

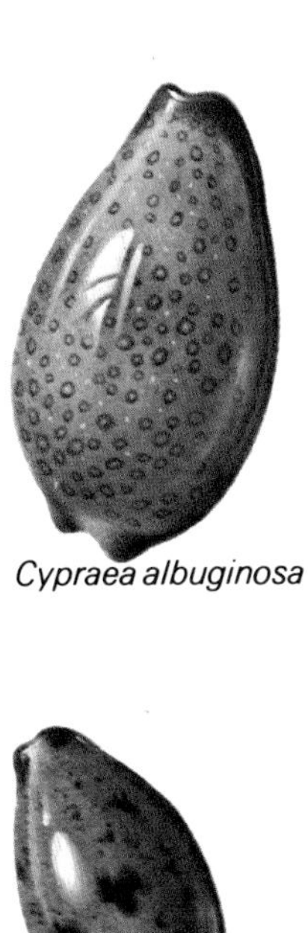

Cypraea albuginosa

Cypraea helvola

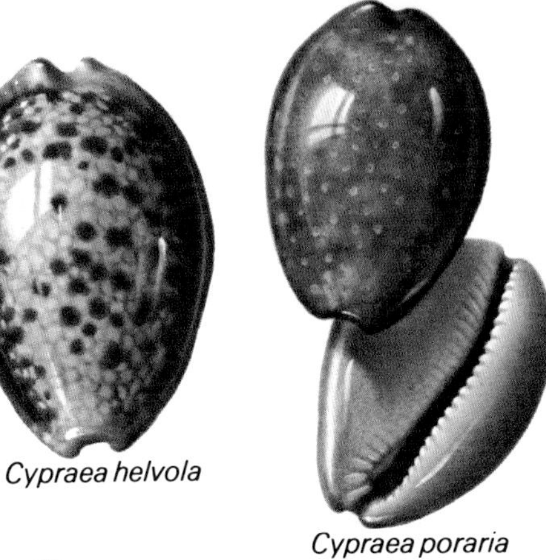

Cypraea poraria

Cypraea gracilis
(Philippines)

Cypraea spurca

Cypraea gracilis
(Singapore)

Cypraea boivinii

Cypraea labrolineata

Cypraea spurca acicularis

C. eburnea Barnes 1824 west and south New Guinea, Queensland and the Solomons, to Fiji and the Cook Islands, 50mm. Produced extremities; pitted labial margin; strong teeth. Pure white or with a very faint cream blush on the dorsum.

C. miliaris Gmelin 1791 north Australia, Philippines and Malay Peninsula to south Japan, 45mm. Shaped like *C. eburnea,* above. Dorsum pale brown or green-brown with small white spots of varying size and usually a conspicuous dorsal line without spots; base and margins white.

C. lamarckii Gray 1825 west Indian Ocean to the Strait of Malacca and north to Karachi, 50mm. Similar in shape to the two preceding shells and marked like *C. miliaris* but in addition some, at times many, of the dorsal spots are ocellated. It also has darker brown spots on the margins and axial lines of the same colour at the extremities; base white.

C. erosa L. 1758 Indo-Pacific from east Africa to Tahiti including south-east Africa, Gulfs of Aden and Oman, Queensland, Japan and Hawaii, 55mm. Variable within this large range but generally rather flattened with strongly callous and deeply pitted margins, and somewhat produced extremities. Base convex with very strong teeth. Dorsum brown or brown-green with profuse small white dots, a few sometimes ocellated, and some larger brown spots; margins cream with some brown dots and dashes; usually a rather square, more or less colourful, purple-brown blotch halfway along each margin which may spread well on to the base.

C. nebrites Melville 1888 Red Sea and Gulf of Aden to Zanzibar, the Gulf of Oman and the Persian Gulf, 35mm. Somewhat similar to *C. erosa* but more callous. The dorsum is darker as are the dots on the margin and the square blotches which do not spread on to the base; the latter has red-brown dots and dashes, especially on the outer ends of the labial teeth.

C. caputserpentis L. 1758 Indo-Pacific from east Africa to Tahiti, South Africa to Arabia and Japan, and from Queensland to Hawaii, 40mm. Rather flattened with broad, angular margins and a humped dorsum. A broad, dark brown band covers the wide margins and part of the base, fading to a white aperture; pale pink to cream, squarish blotches divide the band at each end of the shell; dorsum is also brown with small, white spots of varying size; extremities have a purple tinge.

Cypraea nebrites

Cypraea erosa

Cypraea lamarckii

Cypraea eburnea

Cypraea caputserpentis

Cypraea miliaris

C. caputdraconis Melville 1888, 30mm. An uncommon shell because it is found only on Easter Island in the east Pacific, and is the only cowrie found there. Superficially it resembles *C. caputserpentis*, but is less depressed, having a humped back and less expanded margins. Wide margins, base and interstices of teeth are very dark brown to black; spots on the lighter brown dorsum are smaller than in the average *C. caputserpentis* and are pale blue. Dorsal sulcus is red-brown.

C. sulcidentata Gray 1824 Hawaiian Islands, 45mm. A broad, hump-backed shell with callous margins, a convex base, and very close-set, deeply cut teeth. The margins and base are dark brown when fresh but very rapidly fade to a fawn colour (as illustrated), lightening to almost white at the teeth; dorsum has four, darker fawn bands against a blue-grey background.

C. schilderorum Iredale 1939 central Pacific from Guam to Society Islands, and from Midway and Hawaii to Fiji and Tonga, 30mm. Closely related to *C. sulcidentata* and *C. kuroharai* Hale 1961. It is smaller than *C. sulcidentata*, more depressed and has a pure white base and teeth, the latter being fine. The dorsal pattern and colour are very similar, except for a red-brown line at the upper edge of the margins.

C. ventriculus Lamarck 1810 south Pacific from Solomons and New Caledonia to the Marquesas, 50mm. It is a depressed, solid and heavy shell with large lateral calluses. The dorsum is banded as in the previous two species, and is rich red-brown fading to a grey-brown margin and then to white at the lips; uneven cream-tan band, showing the banding, runs axially across the dorsum.

C. carneola L. 1758 throughout the Indian and west and central Pacific Oceans, 50mm. Pale pink-brown with four darker bands; teeth bright purple. Very large specimens up to 100mm may be a separate species *C. laviathan* Schilder and Schilder 1937, which has a smaller range – north Australia and central Pacific.

C. talpa L. 1758 this beautiful cowrie is found throughout the Indian and west and central Pacific Oceans, 55mm. Its base, margins and extremities are very dark brown to black; dorsum banded with shades of golden brown on a creamy background. A very similar cowrie, *C. exusta* Sowerby 1832, is found in the Gulf of Aden; it has finer teeth and is more pyriform.

C. cribraria L. 1758 east Africa to the central Pacific, Japan to northern half of Australia, 25mm. The most common of a number of related and somewhat similar cowries including *C. cribellum* Gaskoin 1849, Mauritius and Reunion; *C. esontropia* Duclos 1833, Mauritius; *C. catholicorum* Schilder and Schilder 1938, New Britain and New Caledonia; *C. cumingii* Sowerby 1822, east Polynesia; and *C. haddnightae* Trenberth 1973, south Australia. It varies over its wide range and is divided into a number of races differing mainly in depth of colour and callosity of the margins. The base and margins are white and the dorsum red-brown, well-covered with largish, almost circular, white spots.

Cypraea caputdraconis

Cypraea sulcidentata

Cypraea schilderorum

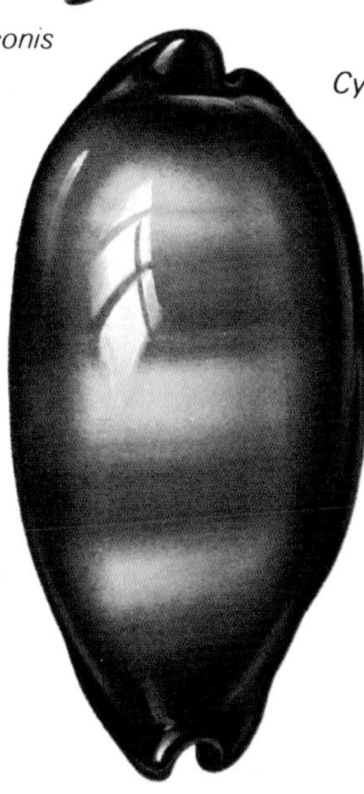

Cypraea talpa

Cypraea cribraria

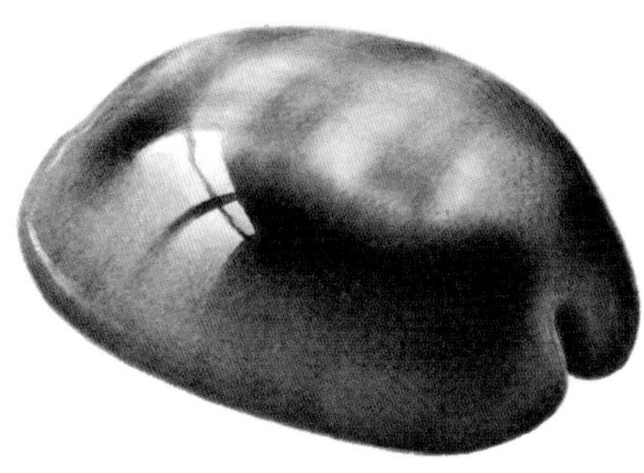

Cypraea ventriculus

Cypraea carneola

The *C. arabica* complex. The shells in this complex have long caused problems for the conchologist as they all have somewhat similar characteristics. However, it seems generally accepted today that there are seven species. While a typical specimen of one of the species is easy to classify, as in all species there occur atypical specimens in which one or more characteristics may resemble those of another species. Classification may then be very difficult even with a large series of shells for comparison.

C. arabica L. 1758 Indo-Pacific from east Africa to Tahiti and from Japan to north Australia, 80mm. Sub-cylindrical with strong callus on the margins. White, cream or pale brown; dorsum with overlay of fine, generally axial, brown lines interrupted by gaps or reticulations where the base colour shows through; callus is spotted with darker brown; the base white, cream or pale brown; no dark blotch on the columella; teeth, often longer in the middle of the columella, pale or dark brown.

C. eglantina Duclos 1833 central Indo-Pacific from Malay Peninsula to Samoa, including the northern half of Australia, Taiwan and the Marshalls, 70 mm. Cylindrical and elongate with a relatively straight aperture and poorly developed callus on the margins. Convex base. Short teeth. The reticulations on the dorsum are smaller and more even than on *C. arabica*. There is usually a small dark blotch on the spire.

C. histrio Gmelin 1791 Indian Ocean from east Africa to Western Australia, 70mm. Inclined to be rather hump-backed. Strong marginal callus. Moderately strong teeth. Convex base. Base colour white or very pale blue-white; reticulations on dorsum large and clear-cut with the brown lines confluent and indistinct; large purple-brown spots on the callus and one on the spire; base white; teeth red-brown.

C. maculifera Schilder 1932 central Pacific from Fiji and Tonga to Tahiti and northward to Hawaii and Midway, 90mm. The largest of the complex. Strong marginal callus. The brown lines on the dorsum are replaced by solid brown colour, even more than in *C. histrio*; quite large reticulations. Its most distinguishing feature is a dark brown blotch on the columella.

C. grayana Schilder 1930 Red Sea, Gulfs of Aden and Oman and the Persian Gulf, 65mm. Convex base. The dorsum is humped posteriorly and the markings on it are between those of *C. arabica* and *C. histrio*. It seldom has a blotch on the spire. The faint bands which can be seen across the dorsum on most of the *C. arabica* complex are strongest in this species.

C. scurra Gmelin 1791 Indo-Pacific from east Africa to Clipperton Island and Tahiti, 50mm. Cylindrical and elongate. Very round base. Large dorsal reticulations; marginal spots purple to purple-brown; no blotch on spire; teeth red-brown.

C. depressa Gray 1824 Indian Ocean from east Africa to Java, and also the central Pacific from the Marshalls, to Clipperton Island and from New Caledonia to Tahiti, but has not been found in the Borneo, Philippines, New Guinea and Australian area, 55mm. Heavily callous and expanded margins give it a depressed look and a round outline. Very convex base. Teeth relatively coarse and large. The reticulations are small and clear cut, white against a red-brown background; margins with strong, brown spots and purple-brown clouding which tend to run over on to the white base; teeth red-brown.

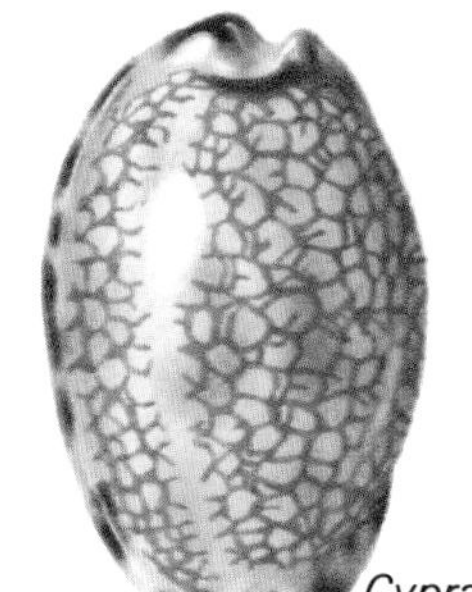

Cypraea histrio

Cypraea eglantina

Cypraea arabica

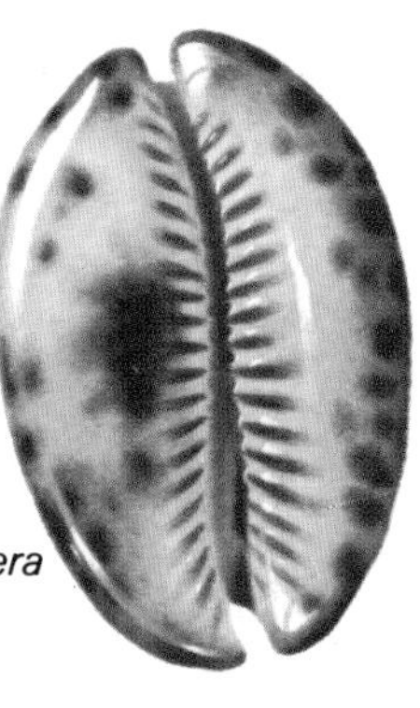
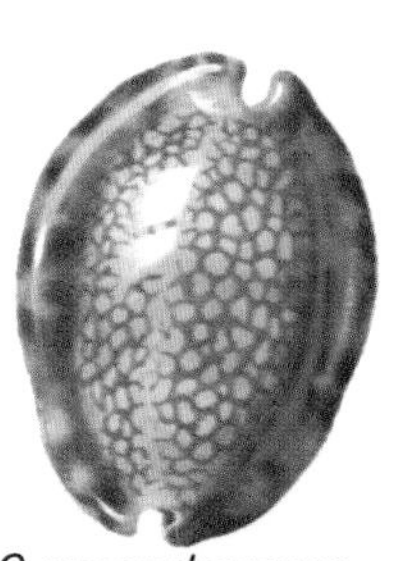

Cypraea depressa

Cypraea scurra

Cypraea maculifera

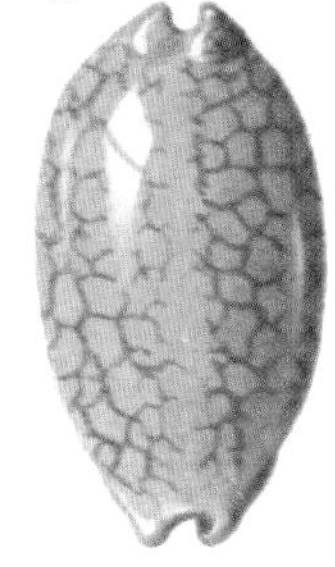

Cypraea grayana

C. tigris L. 1758 Indian, west and central Pacific Oceans, 100mm. Probably the best known of all the cowries and the most variable in size, shape and colour. It is one of the largest cowries, up to 130mm, the largest specimens being found in Hawaii. Its base is white and the dorsum usually cream-white with purple-black spots, sometimes so sparse as to give the shell an almost albino appearance and sometimes so heavy as to make it look black. It is the only cowrie which I have ever found blatantly exposed on a reef during the day. All the others have been fairly well hidden in or under coral and rock or marine growths.

C. pantherina Lightfoot 1786 Red Sea and Gulf of Aden, 65mm. Anterior end of aperture widens out markedly. This shell varies little in shape but considerably in colour. Base white; dorsum can be from white with smallish dark purple-brown spots and faint pale blue clouding, to darkish purple-brown heavily mottled with deeper colour; dorsal line is markedly conspicuous and contrasting.

C. camelopardalis Perry 1811 Red Sea and Gulf of Aden, but not at the northern end of the former, 50mm. Similar in shape to *C. pantherina,* but is much rarer and does not vary in colour to any extent. Base, margins and extremities are a white or pale flesh colour and the dorsum a pink-fawn with discrete white to pale blue spots; columellar sulcus is a rich purple with flesh-coloured teeth.

C. vitellus L. 1758 Indian Ocean, west and central Pacific to Tahiti, 50mm. This cowrie varies considerably in size but is similar in shape and markings to *C. camelopardalis* except that its dorsum is a light brown and the spots white. It has no colouring on the columella sulcus. It is also a much more common shell.

Closely related to *C. vitellus* are two rare cowries, neither illustrated:
C. nivosa Broderip 1827 from the central Indian Ocean north-east to Thailand, 60mm. The spotting on the dorsum is not white but a paler brown than the surrounding colour. Near the margin some of the spots are white and the base is pink-brown. The extremities are pale brown instead of cream or white.
C. broderipii Sowerby 1832 east South Africa, 90mm. It is one of the rarest of cowries, less than ten specimens being known. It is white with an orange-red reticulated dorsum.

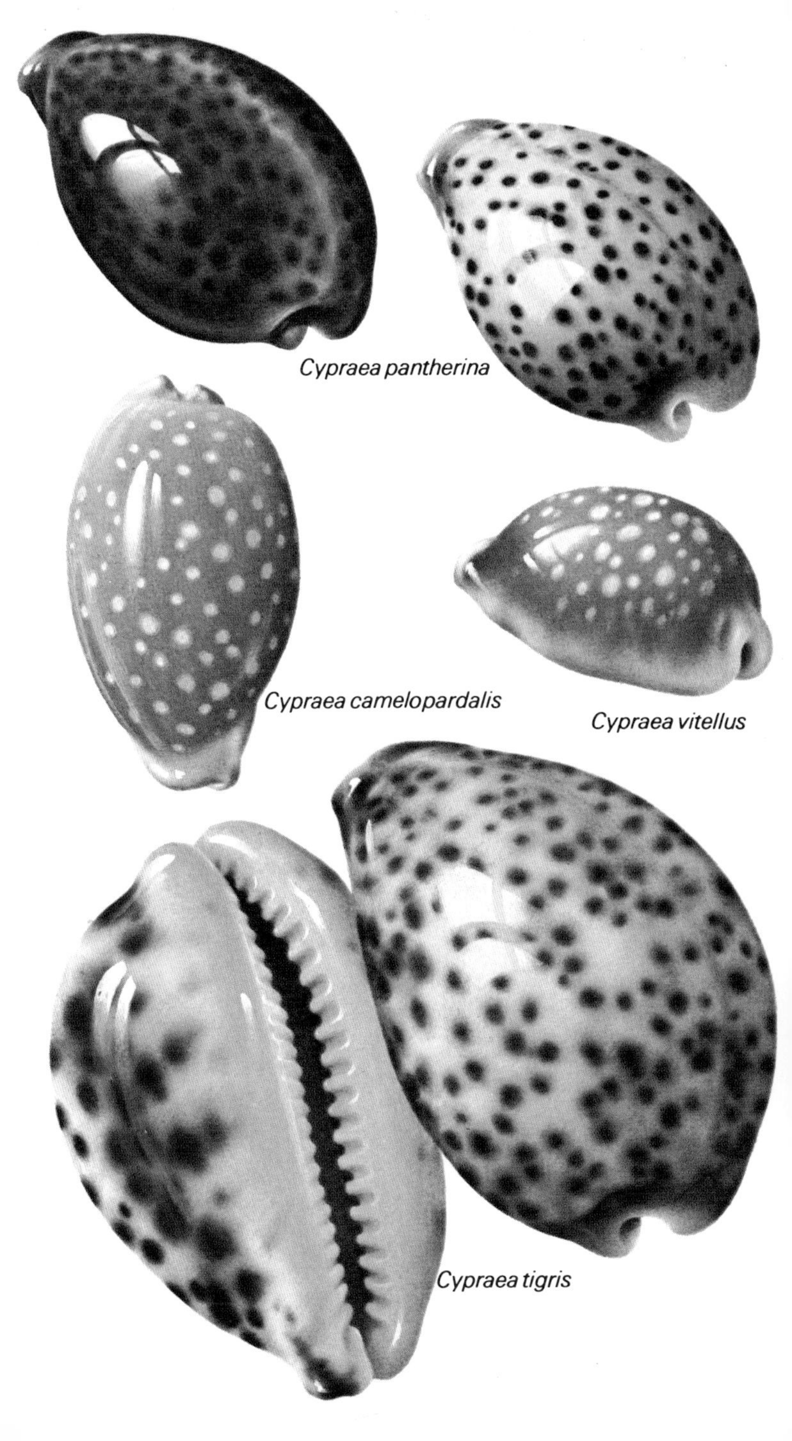
Cypraea pantherina
Cypraea camelopardalis
Cypraea vitellus
Cypraea tigris

The three species of cowries on this page are very similar, and all are from tropical America, two from the Atlantic side and one from the Pacific. One, *C. cervus*, can grow to a greater length than any other cowrie, although *C. tigris* may be of greater volume.

C. cervus L. 1771 Florida, the Bahamas and Mexico, 130mm. It is somewhat inflated with a convex base and only a little callous on the labial margin. Its underlying colour is blue-grey, and the dorsum and sometimes the columellar side of the base is heavily clouded with brown, the grey showing through as profuse, indiscrete spots except on the base itself. Three pale bands, also indiscrete, cross the dorsum. The teeth are a rich dark brown and the interior blue. Some of the spots on the sides may be faintly ocellated.

C. cervinetta Kiener 1843 southern end of the Gulf of California, northwest coast of South America and the Galapagos Islands, 100mm. Despite living on the other side of the continent it is very similar to *C. cervus*. It is, however, smaller and more cylindrical. It is not as inflated as its neighbour, and its extremities are more produced. Its teeth are more sharply defined and the banding on the dorsum is more distinct, as also are the ocellations of the marginal spots. Spots on the dorsum are also frequently ocellated. Its general colour, however, is the same.

C. zebra L. 1758 Caribbean and down the east coast of South America to south Brazil, 90mm. Not as inflated as *C. cervus* though often more so than *C. cervinetta*. Colours are generally rather lighter than in either of the others, and the interior is a paler blue. The spots on the margins are much more strongly ocellated. A variety from northern South America which is smaller and darker has been named *C.z. dissimilis* Schilder 1924.

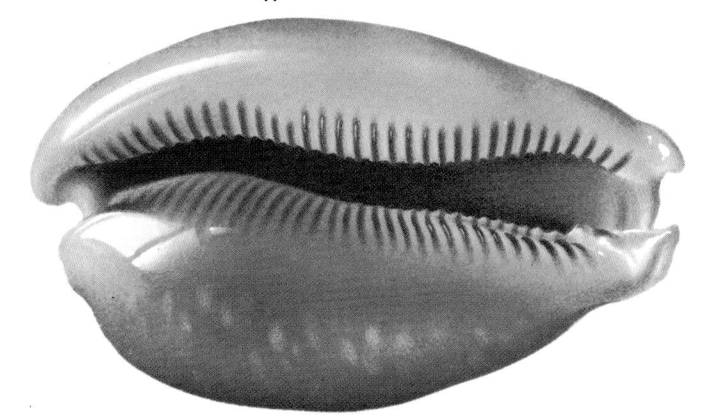

Cypraea cervus

Cypraea zebra

Cypraea cervinetta

C. mappa L. 1758 Indian Ocean, west and central Pacific but excluding north and west Australia, south Indonesia and the Hawaiian Islands, 75mm. It varies considerably in size, shape and colour, but is invariably handsome. It has a most unusual dorsal sulcus from which it gets its name, and can be heavily inflated or sub-cylindrical. The colour of the base can be white or a rich pink; teeth white or orange; dorsum white or more usually brown with a darker reticulated pattern, and unreticulated spots; slightly callous margins are fawn with darker, indiscrete spots.

C. stercoraria L. 1758 west Africa from Cape Verde Islands to Loanda, 85mm. A somewhat gross shell, very variable in size. It is inflated with the lateral calluses extended at both the extremities, which are markedly produced. Its teeth are coarse, crossing the columellar sulcus and the conspicuous columellar fossula, and reaching to the margins of the extensions at the anterior end of the aperture. Teeth white with black interstices; base white or fawn; margins blue-grey and spotted with brown and with dark clouding at the upper edge; dorsum fawn to blue mottled with brown; small spire blotch. Dwarf specimens are found in the islands in the Bight of Biafra.

C. turdus Lamarck 1810 Red Sea, Gulf of Aden and Oman, the east African coast to Mozambique and Madagascar, 45mm. Variable in shape and size but usually depressed, with callous margins. White with small olive brown spots profusely covering the dorsum which give way to larger purple spots on the sides and margins; base white.

C. lynx L. 1758 Indian, west and central Pacific Oceans, 30mm. It varies in size and shape over its vast range, where it is common, but is easily recognizable. It has a ridge on each side of the base running the length of the shell. The dorsal sulcus is conspicuous. Creamy dorsum is covered with small, purple spots and blotches, and a few larger spots of the same colour; base white and interstices of teeth orange.

C. mauritiana L. 1758 Indian, west and central Pacific Oceans but not in north and west Australia nor in most of the China Seas, 65mm. Often found where rough water is common. It is a very heavy, solid shell with strongly callous margins and a humped back. Base and sides almost black and teeth dark brown with white interstices; dorsum dark brown, reticulated with pink-brown and a deep mahogany, dorsal line.

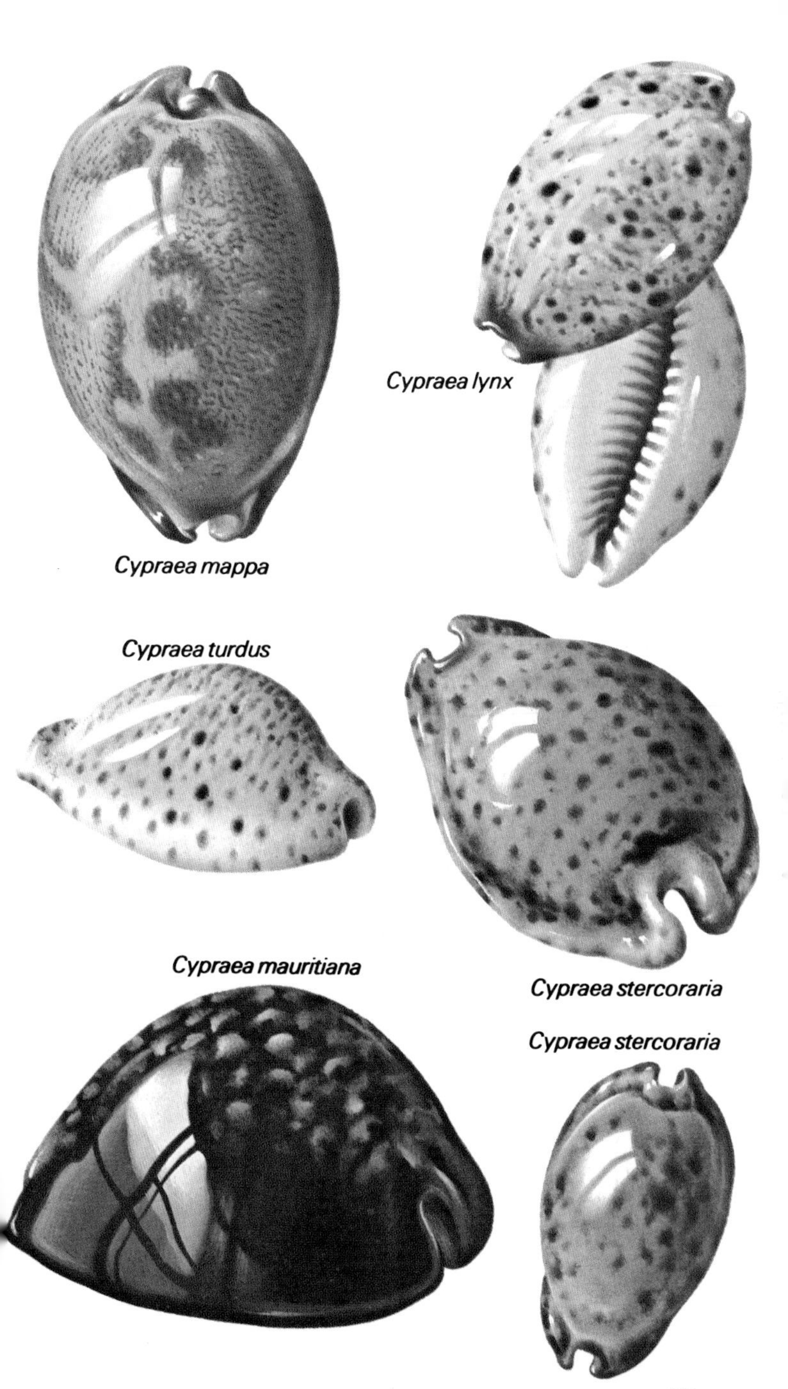

Cypraea mappa

Cypraea lynx

Cypraea turdus

Cypraea mauritiana

Cypraea stercoraria

Cypraea stercoraria

C. testudinaria L.1758 central Indian and Pacific Oceans from Mombasa and Madagascar to Tahiti including Sri Lanka, Malaysia, the north Indonesian Islands, the Philippines and north New Guinea, 105mm. It is the third largest of the cowries. Cylindrical and heavy. Base fawn, light at aperture and teeth; dorsum also fawn with dark brown spots, irregular light and dark brown blotches covering much of the area especially at the extremities. All but the area adjoining the aperture is profusely covered with minute, white spots which often barely show unless examined very closely or with a magnifying glass. They appear to be in the glaze and look as if they are imperfections in the specimen being examined.

C. aurantium Gmelin 1791 central Pacific including the Society, Ellice, Gilbert, Marshall, Caroline, Guam, New Caledonia and Fiji Islands; it has also been found in the Philippines; 80mm. Though a comparatively rare shell it is included here as it is among the cowries much as *Conus gloriamaris* is among the cones – not by any means the rarest, but probably the most sought after. Inflated and heavy. Dorsum rich peach orange; base and extremities white with orange teeth.

C. argus L. 1758 Indian, north, west and central Pacific Oceans, 90mm. Another very beautiful cowrie which varies considerably in size. Cylindrical with only slightly callous margins. Base fawn with two, dark brown blotches on either side of aperture, those on the right (outer lip) side being sometimes absent or nearly so; teeth edged with brown; dorsum lighter fawn with four, darker bands, darkest at top of shell. Whole of dorsum down to callus profusely covered with dark red-brown rings of varying thickness – the 'Argus' Eyes' or 'Pheasant's Eyes' from which it takes its name.

Cypraea testudinaria
Cypraea aurantium
Cypraea argus

C. limacina Lamarck 1810 Indo-west Pacific to Samoa and from north Australia to Japan, 35mm. A fully adult shell has fine nodules on the dorsum, stronger on the sides, unless inhabiting quiet water when, like a juvenile, it is smooth. When fresh it is a rich, deep, red-brown to black which soon fades to a fawn-grey; dorsum is covered with small, white spots again absent in the juvenile; callous labial margin is pitted and white; base white; long, strong teeth and extremities red-brown.

C. staphylaea L. 1758 25mm. Same area as *C. limacina*, to which it is similar, but rounder, more inflated and has tiny nodules generally white and more profuse. It is also dark-coloured when taken alive soon fading through grey to almost white. The main distinguishing feature is the teeth which cross the base to both margins. The extremities are also red-brown.

C. nucleus L. 1758 Indo-Pacific to Tahiti but not north and north-east Australia nor Papua and Solomons area, 30mm. The dorsum is covered with small pustules and fine ridges. The latter seem to be extensions of the teeth which completely cover the base and slightly callous margins; they do not, however, cross the dorsal sulcus. Convex base. Pale brown to off-white, bright and shiny.

C. granulata Pease 1862 Hawaiian Islands, 40mm. The only cowrie not shiny as an adult, though the juvenile is. Rounder in outline than *C. nucleus*, its dorsal pustules are fewer and higher, as are the teeth and their extensions as dorsal ridges. The latter give the margins a strong, toothed effect. Every second or third ridge much bigger than the intervening ones. Brown-grey to white; teeth edged and pustules ringed red-brown.

C. bistrinotata Schilder and Schilder 1937 Indo-Pacific to Tahiti, bar south-east Africa and Hawaii, 20mm. Adults have a pustulate dorsum with a callus posteriorly and three dark brown blotches each divided by the dorsal sulcus. Juveniles smooth. It has four dark brown spots, one at each corner of the base. Teeth are prominent and cross most of the base; anteriorly on the columella they cross the margin on to the edge of the dorsum. One of a complex of five or six species having many similar characteristics. All are small, globose with produced extremities and generally pale brown or tan.

C. cicercula L. 1758 Indian Ocean and the Pacific to Tahiti, including Hawaii, 24mm. Dorsum similarly pustulate but without callus. Instead there is a groove in which is a small dark spot on the spire. No spots on the base. Long teeth. Generally a lighter colour than *C. bistrinotata*. *C. margarita* Dillwyn 1817 (not illustrated) which is smooth on top of the dorsum and has shorter teeth may be a variety of *C. cicercula* or a separate species.

C. globulus L. 1758 Indo-Pacific to Tahiti, bar the Arabian Sea and north-west Australia, 24mm. Smooth dorsum. May have four, dark spots on the base. The anterior teeth on the columella do not cross the margin on to the dorsum. The latter often has small, brown dots, especially on the sides. Like *C. bistrinotata* it has a callus posteriorly on the dorsum.

Also in this complex but not illustrated are:
C. dillwyni Schilder 1922 central Pacific, 15mm. Elongate; light brown; white base; small and very small white spots on the dorsum and a brown blotch at each end.

C. mariae Schilder and Schilder 1927 Philippines and central Pacific including the Solomons, Tonga, Hawaii and the Tuamotu Archipelago, 20mm. Extremities hardly produced. White with small and very small brown rings on the dorsum and sides.

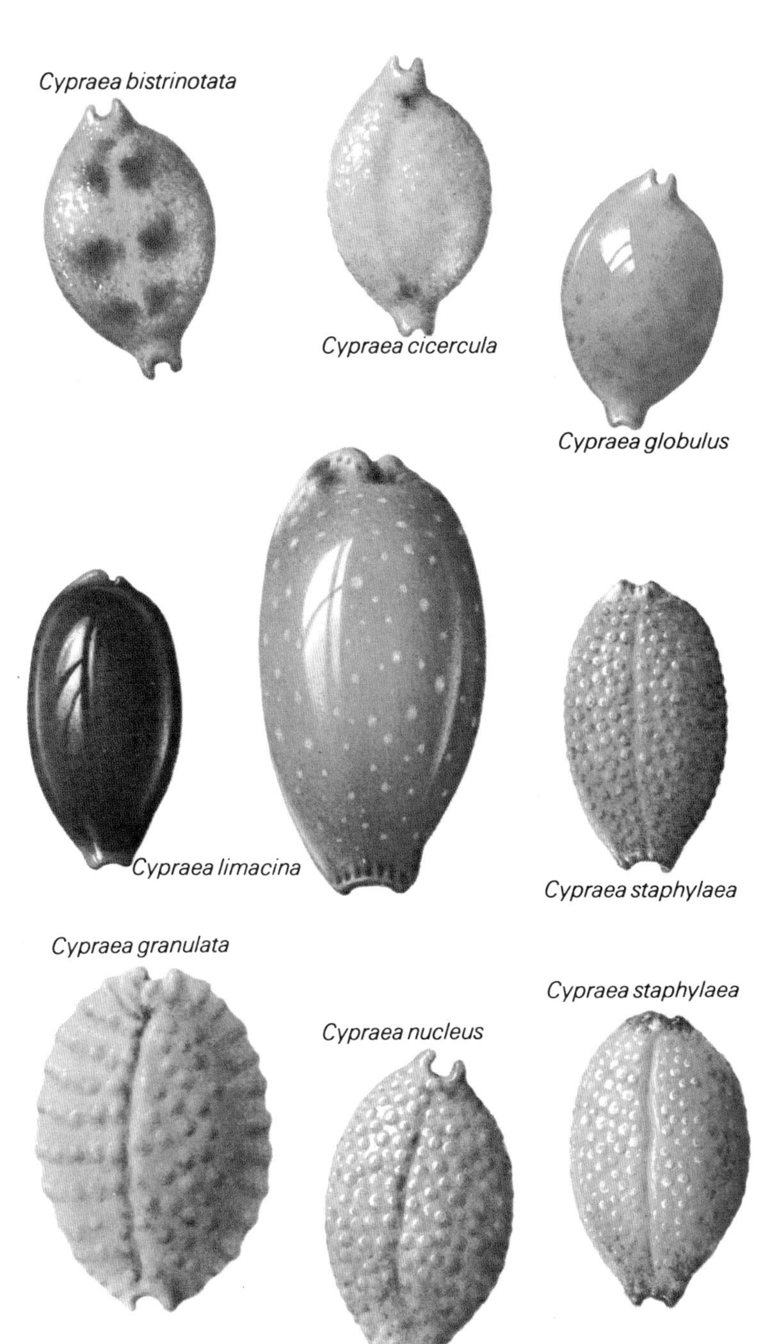
Cypraea bistrinotata
Cypraea cicercula
Cypraea globulus
Cypraea limacina
Cypraea staphylaea
Cypraea granulata
Cypraea staphylaea
Cypraea nucleus

C. kieneri Hidalgo 1906 Indo-west Pacific to Fiji, 22mm. With the two following species, this forms another group of small, very similar cowries. Somewhat inflated, more so posteriorly; a little rostrate anteriorly. Very little marginal callus except on those from Australia. Teeth moderate but very short on the anterior of the columella (chief difference from *C. hirundo* below). Cream or white; dorsum largely clouded with blue and divided in three parts by two, uneven, wavy bands. Degree of marking variable. Margins have more cream than blue area and are speckled red-brown. Two, dark brown spots at each end, others forming a more or less broken band on dorsum which is also minutely speckled with brown; base white.

C. hirundo L. 1758 area as for *C. kieneri* but not south-east Africa or north Pacific east of Japan and the Philippines, 22mm. Similar to *C. kieneri* except that the teeth are fine and cross most of the base.

C. ursellus Gmelin 1791 west Pacific in the triangle south Japan, Tonga and the Great Barrier Reef, 12mm. Like *C. hirundo* but smaller, more inflated; teeth crossing almost the entire base. It is extremely difficult always to distinguish a *C. ursellus* from a small *C. hirundo*.

C. stolida L. 1758 Indo-Pacific to Samoa, 35mm. Very variable in size, shape and colour but always a beautiful shell. Many of the varieties have been given names. May be sub-cylindrical and elongate or almost rectangular. Moderate callus; strongish teeth except in the variety *breventata* Sowerby 1870, where the columellar teeth are short. Base white or cream; dorsum pale blue to pale blue-green; usually a rather square, dark brown blotch on the dorsum which is often more or less joined to other rectangular blotches at each corner; the latter lighter in their centres and on the margin of the shell. Small brown dots more or less profuse over the light-coloured areas and especially on the margins and on each side of the extremities.

C. cylindrica Born 1778 north Australia, South China Sea to Guam, the New Hebrides and New Caledonia, 40mm. Usually elongate and cylindrical, but north-west Australian specimens may be rather inflated. Teeth variable, from very short and small to longer and strong. Blue or pale green-blue with irregular, more or less dark, brown blotches or blotch on the dorsum; basic darker banding is often inconspicuous; profuse, tiny, brown dots; two, larger spots on each side at both extremities; base white.

C. caurica L. 1758 Indo-Pacific to Samoa, and from Japan to north Australia, 55mm. Elongate and usually a little inflated. Labial teeth very strong; more or less developed callus. Cream with three, blue-brown, dorsal bands; the whole dorsum heavily stippled with tiny, brown dots, thickest and confluent on the centre; large and small, brown spots on the yellow-tinged margin; base pale fawn to off-white; teeth lighter.

C. xanthodon Sowerby 1832 east Australia from Sydney northward, 35 mm. Pyriform; moderately callous margins. Dorsum blue-green; profuse, brown speckles; three, indistinct, dark bands. Dark brown spots on margin, one on the spire and one each side anteriorly. Those on the margin are smudged towards, and sometimes on to base. Margin, base cream; teeth darker.

C. subviridis Reeve 1835 northern Australia, New Caledonia and Fiji, 40 mm. Variable; pyriform and inflated from Western Australia, smaller and more cylindrical from the east of its range. Pale blue-cream speckled brown; an irregular blotch of darker brown on dorsum; indistinct purple-brown band; extremities may be tinged pale purple or brown; base white or cream.

Cypraea xanthodon

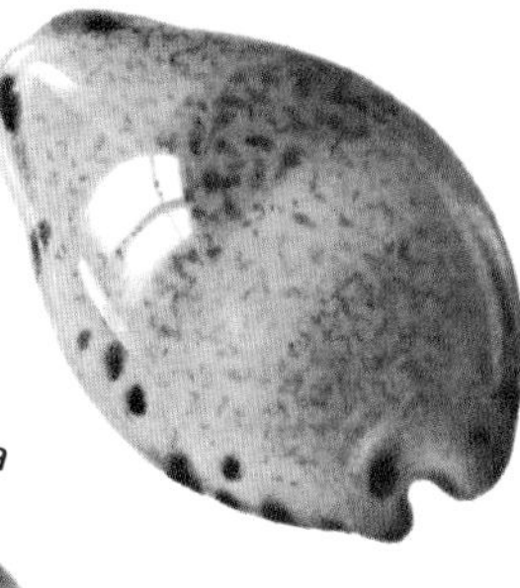

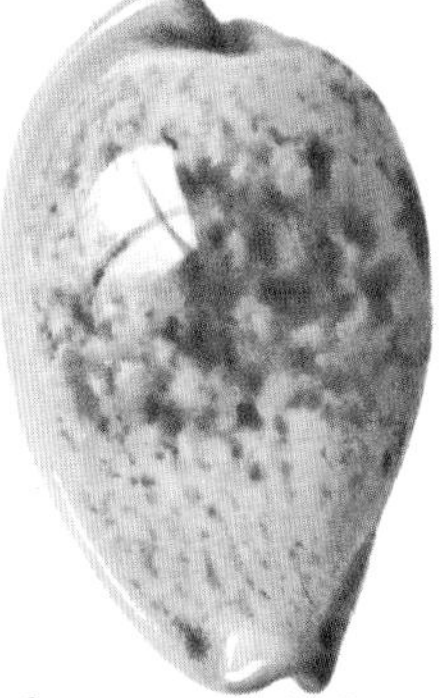

Cypraea subviridis

Cypraea caurica

Cypraea hirundo

Cypraea kieneri

Cypraea cylindrica

Cypraea ursellus

Cypraea stolida

The south Australian Cowrie Complex. There are a number of cowries from south Australia which raise considerable difficulties in classification. Much work has already been done on these, but much more is necessary. I have followed C. M. Burgess who, in his leading work *The Living Cowries* (see Bibliography), has dealt with this complex in detail. The problems arise through many specific and subspecific names having been given because individual characteristics of specimens of the same species vary considerably. It has been noted that some characteristics vary a great deal with the age or state of development of an individual, and also with the depth of water which it inhabits. Characteristics of the animal itself have not yet provided means of specific identification.

C. angustata **Gmelin 1791** south-east Australia and north and east Tasmania, 35mm. Somewhat inflated. Deep water varieties *emblema* and *molleri* both Iredale 1931, are more inflated, thinner and lighter coloured. Dorsum usually not banded; pale pink-brown; margins with discrete, brown spots often extending well up the sides and on to the base.

C. declivis **Sowerby 1870** south Australia from Cape Howe, south of Sydney westwards across the Great Australian Bight and north Tasmania, 28mm. From dark brown to cream or rose and always with pale to dark brown speckling on the dorsum and margins; base white.

C. comptonii **Gray 1847** south Australia, not quite reaching Sydney or Perth. The most variable, especially in colour. Pink-brown, quite dark (var. *trenberthae* Trenberth 1961) through cream (*C. angustata* var. *mayi* Beddome 1897) to pure white (*C. angustata* var. *albata* Beddome 1897). Usually with brown or grey-brown spotting on the margins and sometimes a little brown on the dorsum. Two narrow bands close together across the middle of the dorsum are, however, characteristic except in the pure white variety. There may be a third near the spire and occasionally a fourth anteriorly. In the variety *wilkinsi* Griffiths 1959 the spotting and banding is faint or absent, and in *C. angustata* var. *subcarnea* Beddome is a pale flesh colour.

C. piperata **Gray 1825** same area as *C. comptonii* but including Perth, 25 mm. Cream to white, spotted and/or reticulated with pale brown. Three, broad bands of irregular, broken, darker brown spots; fine light or dark spots on the margin, sides and outer edge of the base; latter white. It is probably as well known by the name *C. bicolor* Gaskoin 1849.

C. pulicaria **Reeve 1846** (not illustrated) south-west Australia only, from the Swan River to Cape Leeuwin; 20mm. Cylindrical. Cream or white with four, narrow, broken, brown bands on the dorsum and brown, spotted margins. The columellar teeth appear to be relatively longer than in the other species in the complex. The main characteristic which distinguishes it is a strongly concave fossula which projects into the aperture.

C. errones **L. 1758** east Indian Ocean and west Pacific from the Strait of Malacca to Samoa, and from Japan to the northern half of Australia, 40mm. Very variable in size, shape and colour. Pyriform to cylindrical, more or less inflated and callous with sunken spire. Aperture wider anteriorly. Generally pale green or blue-green dorsum with three or four, rather faint, bluer bands; heavily speckled with green or green-brown, and often with a dark brown, irregular blotch in the middle of the dorsum and a spot on each side anteriorly.

C. ovum **Gmelin 1791** area as for *C. errones* but not quite as far east, 40mm. Very similar to *C. errones* but with yellow or orange between the teeth, the colour being strong or weak. Generally more pyriform than *C. errones*.

Cypraea angustata var. *mayi*

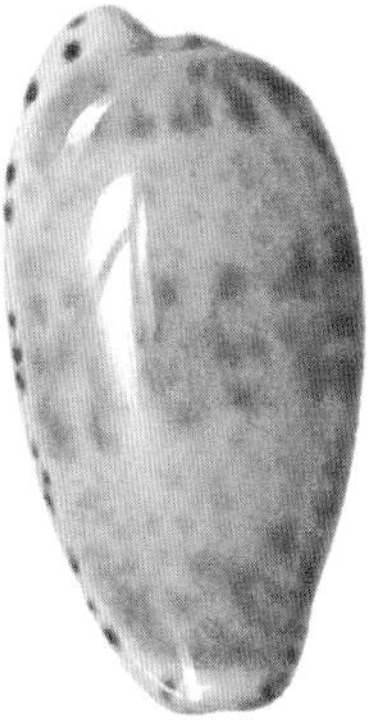

Cypraea piperata

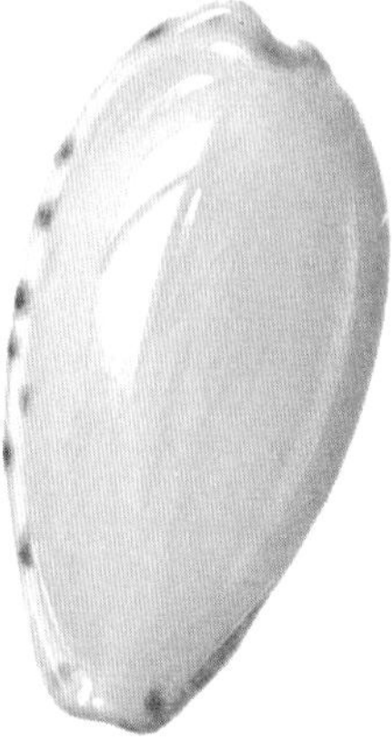

Cypraea angustata var. *molleri*

Cypraea errones

Cypraea ovum

Cypraea ovum

Cypraea comptonii

Cypraea declivis

Cypraea errones

C. moneta L.1758 the Money Cowrie Indo-Pacific from east Africa to the Galapagos Islands, 40mm. Very variable, usually rather flattened and angular; heavily callous sides. Dorsum rich or pale yellow, sometimes a fine and very bright ring like *C. annulus*; sides and margins paler; base white; three, more or less conspicuous, narrow, dorsal bands. Illustrated specimen has the uncommon ring and its sides are unusually white. Teeth less coarse than in *C. annulus* but sometimes cover much of base as low, blunt ridges.

C. annulus L.1758 Indian, west and south Pacific Oceans, 30mm. Variable, generally oval with a somewhat humped dorsum. Smoothly rounded margins, lacking any keel. Dorsum yellow-green or blue-green, faintly banded; it is separated from the callous sides by a narrow ring of bright yellow-orange; flesh-coloured below the ring, becoming paler towards the teeth.

C. obvelata Lamarck 1810 east central Pacific, including the Society Islands, Tahiti, the Marquesas and Jarvis Islands, 25mm. Somewhat similar to *C. annulus* but with a very strong, lateral callus which is so heavy that it forms a groove separating it from the dorsum. Teeth coarse and long. Dorsum pale to very pale blue, narrow banding faint or absent; usually with a thin yellow ring. Callus and base very pale pink, sometimes white.

C. felina Gmelin 1791 Indo-Pacific from east Africa to Samoa, and from Japan to the northern half of Australia, 25mm. Another extremely variable species with named varieties, especially four described below. All have very dark, almost black, discrete spots on their margins. The largest, *C. felina*, is from east Africa, from Zanzibar southwards. Ovate; callous margins. Dorsum cream to pale green with four, blue bands and speckled with tiny, brown dots; margin pale tan with large black spots especially on each side of the two ends; base also pale tan. *C. fabula* Kiener 1843 from the Gulfs of Aden and Oman is much rounder in outline; dorsal markings are darker; its margins have more, bigger and sometimes confluent spots; base lighter or white. *C. listeri* Gray 1825 from Mauritius and the Seychelles, is more cylindrical; dorsum blue-white; blue banding more conspicuous; marginal spotting less heavy; base white with the banding showing on the left side. *C. melvilli* Hidalgo 1906 is from Japan to north Australia; close to *C. listeri* but more cylindrical and generally more green than blue.

C. arabicula Lamarck 1810 west Central America from the southern half of the Gulf of California and Lower California to Peru and the Galapagos Islands, 30mm. High dorsum hump posteriorly; callous, rather sharp margins. Convex base with fine, sharply cut teeth and a deep fossula. Dorsum pale mauve reticulated with dark green-brown; margins broad centrally, pale mauve with smudged dark mauve to black spots which are sometimes profuse and cover the outer third of the base. There is a small, creamy white spot from the spire to the posterior canal.

C. robertsi Hidalgo 1906 same areas as *C. arabicula*, 30mm. Pyriform with callous margins and convex base. Much coarser teeth than *C. arabicula*. Dorsum very pale mauve to white or blue-white with heavy, fine, dark green-brown reticulations. Wide margins are smoky brown with mauve spots which may spill over on to the white base; canals with a dark purple-brown spot on either side and a creamy white spot between.

C. spadicea Swainson 1823 south California from Monterey Bay to San Benito Island, 60mm. Elongate and produced anteriorly. Dorsum is rich brown fairly sharply but unevenly edged with darker brown; margins wide and grey-pink or grey-cream; very faint, smudged spots can sometimes be seen under the callus; base white.

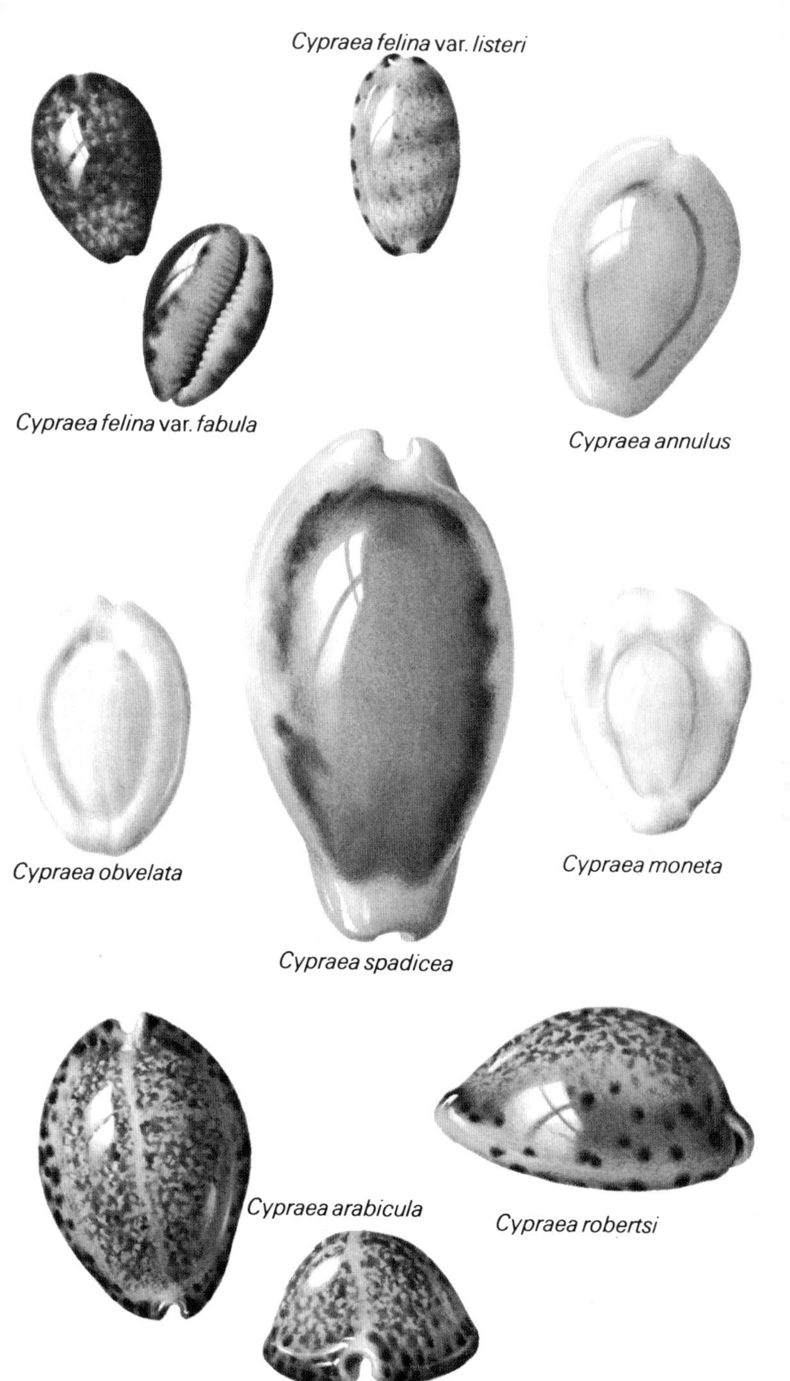
Cypraea felina var. *listeri*
Cypraea felina var. *fabula*
Cypraea annulus
Cypraea obvelata
Cypraea moneta
Cypraea spadicea
Cypraea arabicula
Cypraea robertsi

C. edentula **Gray 1825** South Africa from Cape Town to north of East London, 25mm. One of a group of endemic South African cowries, few if any of which have been taken live. Almost all specimens of this cowrie are toothless. It is hump-backed and its outer lip is callous and extended posteriorly. Its spire is sunken. Base white and dorsum pale brown with darker brown spots.

C. capensis **Gray 1828** South Africa from Cape Town to East London, 25 mm. Another of the group of South African cowries which includes also *C. algoensis* Gray 1825, *C. fuscorubra* Shaw 1909, the rare *C. fultoni* Sowerby 1903, and very rare *C. broderipii* Sowerby 1832 (see page 106). *C. capensis* is a little more cylindrical than *C. edentula* but has the same extended outer lip posteriorly and sunken spire. Fine ridges run around the shell – except across the dorsal sulcus – giving the impression of a thumb print. This impression is heightened by the decidedly flesh colour with a few brown marks and faint banding. There is also a brown spot on the spire.

C. friendii **Gray 1831** south and west Australia from Spencer's Gulf to Shark Bay, 70mm. This species, which is now thought to include *C. thersites* Gaskoin 1849, is very variable in shape and colour. *C. friendii* is found at the north-eastern end of the range, at shallower depths, and is narrower than *C. thersites*. The margins project anteriorly and posteriorly to form a flange, larger on the outer lip, at each end of the shell. Base and margins are mostly dark brown; teeth white; columellar groove white with brown markings. *C. thersites*, found at the eastern end of the range and in deep water, is more inflated. Base white around the aperture – the dark brown of the margins spreading on to the outer edge of the base. Cream background of dorsum is almost completely covered by a very deep brown.

C. decipiens **Smith 1880** north-west Australia, 50mm. Pyriform, and margins only slightly callous. Both the spire and outer lip project posteriorly. Base and teeth brown; dorsum white; almost obscured by a network of light and dark brown markings.

C. hesitata **Iredale 1916** south-east Australia from the Bass Strait to Sydney, in deep water, 95mm. This once rare shell is found in three forms, the normal *C. hesitata*, the dwarf form *beddomei* Schilder 1930 (which may be the female) and the albino variety *howelli* Iredale 1931. The shell is very pyriform with the anterior extremity produced and the posterior sharply turned through 45° towards the left or columellar side. Deeply sunken spire. White, profusely marked on dorsum with light brown blotches and darker spots, the latter becoming more discrete on the margins; extremities are also light brown. *C. beddomei*, 50mm, is similarly coloured.

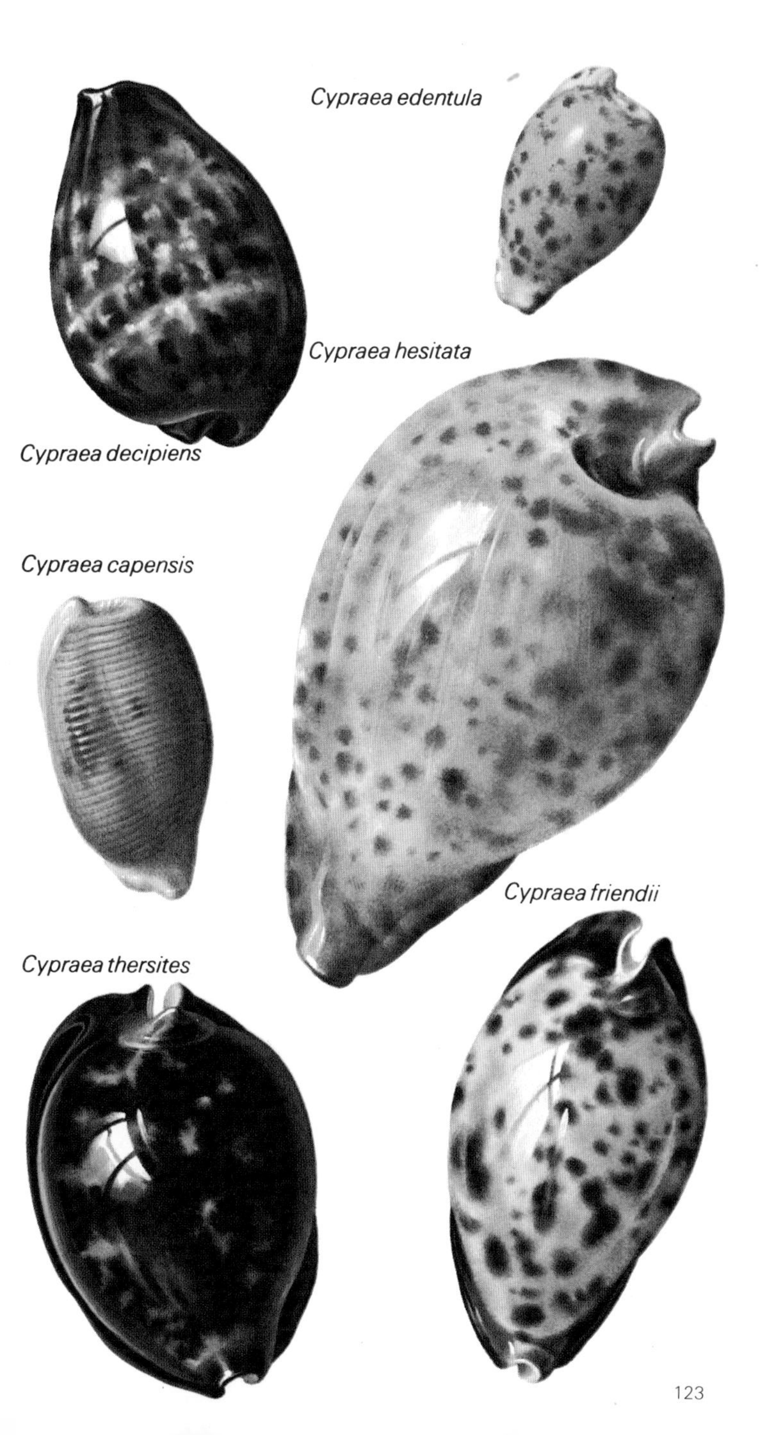
Cypraea edentula
Cypraea hesitata
Cypraea decipiens
Cypraea capensis
Cypraea friendii
Cypraea thersites

The false cowries are related to the true cowries. They include the families Ovulidae or Amphiperatidae, Eratoidae and Lamellariidae.

Family: Ovulidae

Ovula ovum L.1758 Indo-west Pacific, 120mm. Smooth with no spire but extended posterior canal and coarsely ridged lip. Pure shiny white; chocolate or red-brown interior. Found on coral reefs. Used a great deal in the South Seas as jewellery and charms, and for canoe decoration.

O. costellata Lamarck 1810 Indo-Pacific, 50mm. Similar to above, but posterior canal is somewhat notched on body side, and interior is rose pink.

Calpurnus verrucosus L. 1758 Indo-west Pacific, 35mm. Cowrie-like bar two ribs crossing dorsum giving it an angular shape, and tubercle at each end of dorsum. Outline of base rather angular, with very fine, spiral striae. Toothed lip is coarsely ribbed. White; pink-white at extremities.

Volva brevirostris L. 1758 South China Sea, 40mm. Long, thin, spindle shape. Narrow aperture except near siphonal canal where lip expands a little. Lip thickened and smooth like columella. Faint, spiral lirae at extremities. Flesh, red or yellow-pink, depending on colour of gorgonian on which it lives.

V. volva L. 1758 central Indo-Pacific, 130mm. Rounded body whorl with fine, axial grooves; expanded, thickened lip; very long, slightly curved, obliquely ridged canals; smooth columella. Flesh-pink or brown-pink; lip white or paler pink; extremities of canals with brown tinge.

V. sowerbyana Weinkauff 1881 Indian Ocean and west Pacific, 30mm. Slightly inflated body, with fine, axial striae. Short and obliquely striate canals. Thickened lip has furrow behind. Columella with about twenty, oblique, incised striae anteriorly; thick, broad rib posteriorly. Body dark flesh-pink, faintly banded centrally with cream; lip white; extremities may be orange.

Cyphoma gibbosa L. 1758 the Flamingo Tongue, south-east USA and West Indies, 30mm. Chunky, callous shell with heavy, encircling ridge. Obsoletely toothed lip and smooth columella. Apricot-cream; long, rather narrow oblong of white on dorsum; callus paler; base almost white. Lives on gorgonian.

Jenneria pustulata Lightfoot 1786 west Central America, 25mm. Depressed; pustulate dorsum. Base has coarse ribs from aperture to margin, fifteen per side and most with a short, small rib between. Blue-white; pustules bright orange ringed with dark brown; base blue-white overlaid with dark brown; ribs white ending at margin with orange, and each joining a pustule.

Family: Eratoidae

Trivia sanguinea Sowerby 1822 Lower California, 14mm. Inflated; high dorsum with small crease along mid-line. Raised, coarse, radial, white riblets, about twenty-three a side, extending across base into aperture; slightly expanded margins. Purple-brown; darker blotch centrally on dorsum; inside of columella, lip white.

T. monacha da Costa 1778 Mediterranean to British Isles, 15mm. Chubby little shell with fairly high dorsum. Encircled by about twenty-five, coarse riblets. Thickened lip backed by shallow channel. Flesh; purple spots, one at each end and one in middle of dorsum; base white. *T. arctica* Montagu 1803 (not illustrated) from the same area is similar but lacks the purple spots.

Jenneria pustulata

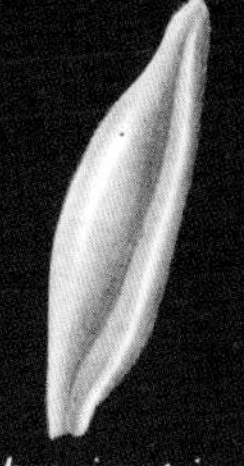

Volva brevirostris

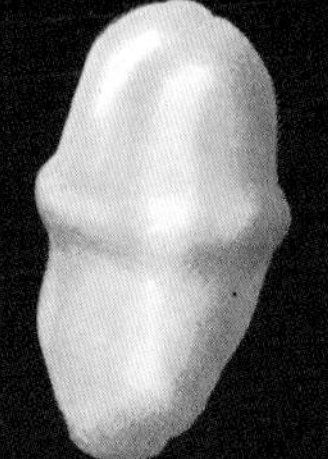

Cyphoma gibbosa

Trivia monacha

Volva volva

Ovula ovum

Trivia sanguinea

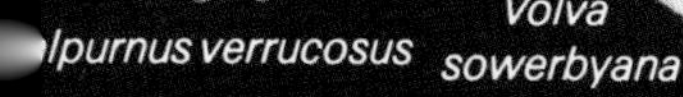

lpurnus verrucosus

Volva sowerbyana

Ovula costellata

Superfamily: Tonnacea

Family: Cassidae

The helmet shells are solid with large body whorls and short spires, usually with blunt knobs, varices, and a thickened, toothed outer lip. There is usually a broad shield beside the plicated columella or over the parietal area. They have thin, horny opercula. The animals live in sandy areas in the tropics and temperate zones, in shallow and deep water. They prey on sea urchins and other echinoids. There are about sixty species.

Cassis cornuta L. 1758 Indo-Pacific, up to 350mm. Very solid and heavy with a short spire of about seven whorls. Pitted all over with mostly rather long, narrow pits between small, spiral ridges. Angular shoulder has five to seven, flat, protruding knobs or blunt spines. The suture generally outlines these knobs on the earlier whorls. Body whorl with two medium and one small smooth band with lumps or very blunt knobs. The posterior ends of earlier lips show as varices on earlier whorls – one about every two-thirds of a whorl. Outer lip thickened, widened, and the edge recurved. Siphonal canal twisted and turned up vertically. Narrowly umbilicate. Large parietal shield almost obscures the body whorl when looked at from below. Outer lip with up to twelve, blunt teeth. Columella has up to fifteen plaits on the lower two-thirds. Sculpture and colour show through the parietal shield at the posterior end of the columella. Anterior end, columella and surrounding part of the shield are indented. White with light brown shading and sparse brown spots; purple-brown marks on the smooth bands; outer lip has about seven, brown 'squares'; teeth and outer lip white; interstices of teeth, the interior, columella and inner area of shield orange-brown; rest of shield pinky-white. The males tend to be smaller than the females with a slightly less expanded shield.

C. flammea L. 1758 Caribbean, 125mm. It has a low spire, about seven whorls and is axially ribbed. A row of blunt knobs on the shoulder and one of smaller knobs a little below it, one or both of which may show on earlier whorls. A row of nodules immediately below suture and another between this and the shoulder. Earlier varices about every two-thirds of a whorl. Thickened, callous lip is dentate – about ten teeth. Columella with about twenty, long, raised plicae. Short siphonal canal is twisted vertically. Colour shows through much of the parietal shield. White with brown mottling and darker brown, axial, zigzag markings; outer lip has about six, dark brown spots, two or three of which show on early varices.

C. tuberosa L. 1758 Caribbean to Brazil, 230mm. Very low spire. Covered with very small, axial riblets. A row of blunt knobs on the shoulder, and two others with much smaller knobs or bumps. Early varices. Thickened, callous outer lip with about eleven teeth. Short, upturned siphonal canal. Coarsely and strongly lirate columella. Parietal callus is triangular in shape. White to blue-grey with axial, dark purple-brown, zigzag lines. About seven, rather obscure, brown spiral lines which end in dark brown, rectangular blotches on the outer lip; the interior and teeth dirty white; brown in the interstices; columella and inner area of shield dark brown; lirae white; part of parietal shield a metallic brown-pink. The illustration is of a rather young shell on which the colour is particularly strong.

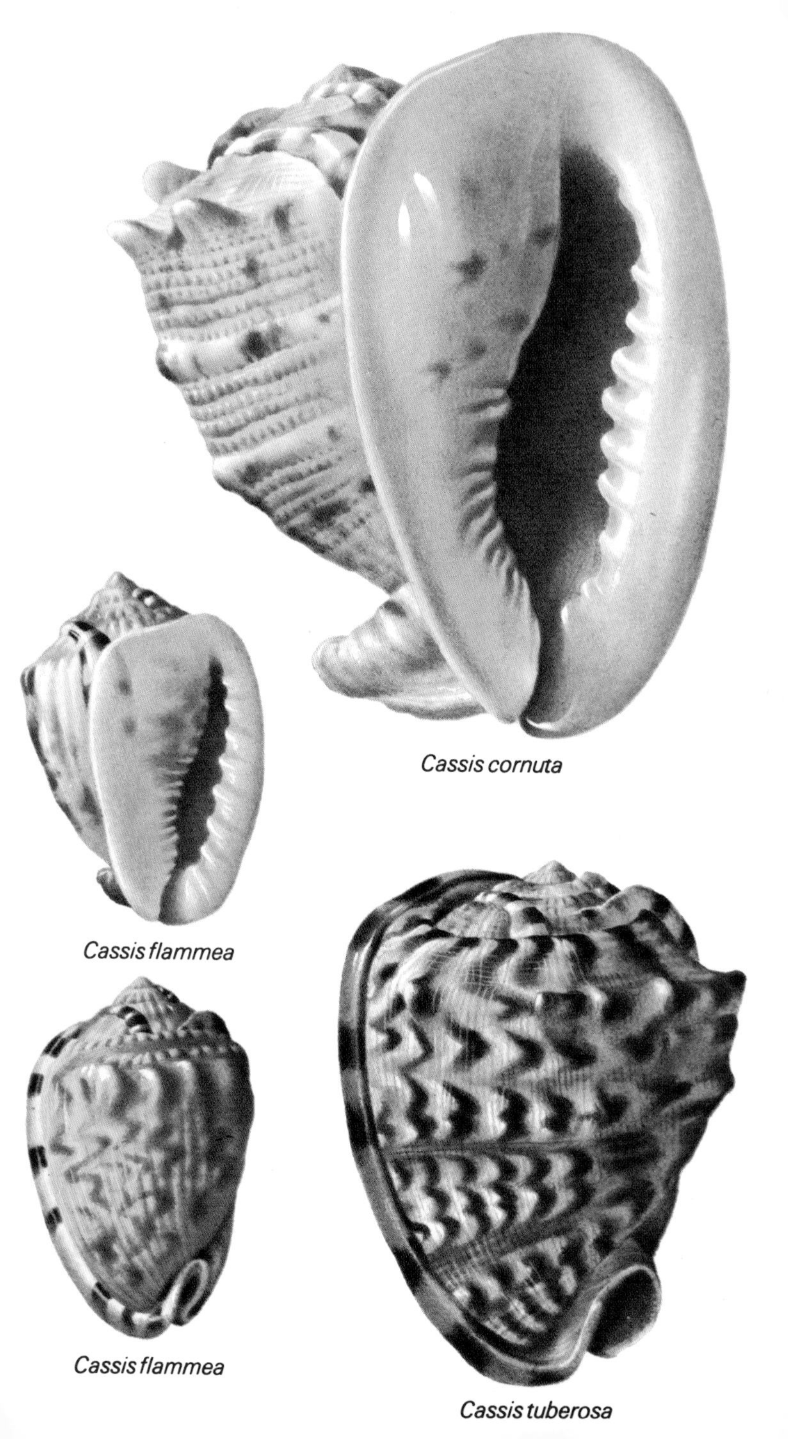

Cassis cornuta

Cassis flammea

Cassis flammea

Cassis tuberosa

C. tessellata Gmelin 1791 west Africa, 260mm. Rather light and fragile for its size. Low spire and about seven whorls. Former varices about every two-thirds of a whorl. Body inflated with three rows of small knobs, the ones on the shoulder row being relatively sharp. It has one other row of very small knobs a little above the first large row, and another of small riblets immediately below the suture. It is otherwise smooth apart from axial growth lines. Inside lip turned in and back on itself; nine teeth, the centre one rather elongate. Columella with eight, long, thin folds, the largest in the middle. Four teeth at the bottom of the parietal shield. The callus on the parietal wall is thin and may be pustulate from covering grains of sand or other foreign matter. Siphonal canal twisted vertically. Narrowly umbilicate. Light brown; knobs with darker brown or white blotches and other spiral rows of brown and white, alternate marks lower on the body whorl; outer lip, interior and columella white; parietal shield white but with much of the underlying colour pattern showing through.

C. nana Tenison-Woods 1879 east Australia, 60mm – the smallest of the Cassidae. Very low spire with a short, pointed apex. About five or six whorls with two, spiral rows of small, sharp knobs on and below the shoulder; sometimes up to three other rows with blunt nodules lower on the body whorl. Early varices show and there are fine growth lines. Outer lip inverted and toothed, the teeth and columella with fine folds. Thin parietal wall, channelled towards the anterior end and lirate below the channel. Anterior end of columella has a shelf within the aperture. Grey-white with light brown, axial flame marks on the spire, and three rows of white and orange-brown spots on the body whorl, coinciding with the rows of nodules. Outer lip white with brown tinge at each end; columella white; parietal shield is brown with markings behind it.

C. madagascariensis Lamarck 1822 Caribbean, Bermuda and south-east USA, up to 350mm. Very short spire of about ten whorls, and inflated body, especially ventrally. Spiral cords and ridges, the largest below the shoulder on the body whorl which carries six, blunt spines. There is a second row of small knobs and may be another ridge with 'bumps'. Suture roughly follows the shoulder on earlier whorls; varices about every two-thirds of a whorl. Outer lip is turned in then back on itself and has about eleven, blunt teeth. Upturned siphonal canal. Columella has many strong ribs and folds extending into the parietal wall. Parietal shield rather pear-shaped; underlying knobs show through as they do also within the aperture. Creamy white with pale brown clouding, especially on the spire; outer lip and parietal shield pink; columella and adjacent area of parietal shield dark brown; lirae white; interstices of teeth and outer lip brown. *C. madagascariensis* and *C. tuberosa* are very similar but can easily be distinguished by the shape of the parietal shield, which in the former is somewhat pear-shaped with a rounded outer edge, while in the latter is triangular with a straight edge.

Cassis tessellata

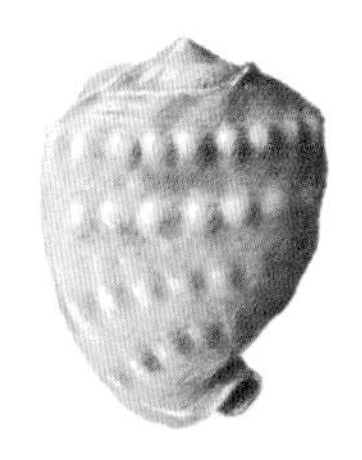

Cassis nana

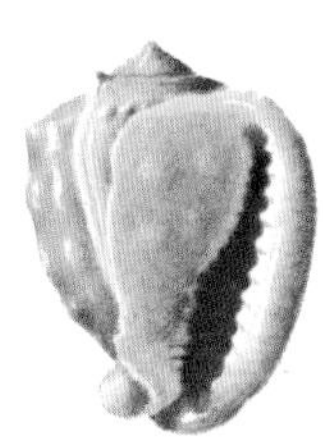

Cassis madagascariensis

Cypraecassis rufa L. 1758 Indo-west Pacific, 180mm. Solid with low spire, sharp apex and angulate shoulders. Body whorl has three or four rows of blunt knobs, posterior row biggest, decreasing anteriorly. Spiral cords and bands, some with small knobs and some, near anterior, with axial ribs; overall fine axial cords. Thickened, broad, recurved lip, with twenty-two elongated teeth. Heavily lirate columella is toothed; large parietal shield. Siphonal canal is turned up vertically. Red-brown; pale orange-brown bands; columella red; teeth white; interstices dark brown; parietal shield pink-brown with vague darker bands.

C. testiculus L. 1758 tropical east and west Atlantic, 80mm. Solid with short spire and deep suture, having small nodules below it. Body whorl has axial riblets and very fine spiral striae at each end, smooth centrally. A row of nine, small, blunt nodules on shoulder and one row a little below it. Lip is thickened, callous, backed by a deep, narrow channel, and has twenty-five teeth, large and small mostly alternating. Columella has twenty-four long ribs; heavily callous parietal shield; siphonal canal turned up vertically. Pale brown with darker brown spots; lip, interior and columella white with metallic brown nacreous overlay; outer edge of lip has seven pairs of dark brown spots.

Phalium glaucum L. 1758 Indo-west Pacific, 120mm. Early whorls roundly shouldered with fine, spiral and axial cords; last three angular and crenulate; usually one or more varices. Expanded, rotund, smooth or malleate body whorl. Narrow ridge below suture; shoulder crenulations becoming weak nodules. Thickened lip, backed by deep channel, with three or four, strong, sharp spikes anteriorly; inside lip are up to twenty-five teeth. Columella has wide shield at anterior end and is weakly spirally wrinkled. Narrow, deep umbilicus; siphonal canal turned up vertically. Grey; lip orange; columella and shield creamy pink to white; interior rich dark purple-brown; umbilical area white.

P. strigatum Gmelin 1791 north China Seas and Japan, 110mm. Solid with moderate, concave spire. Varices every two-thirds of a whorl; spiral cords on earlier whorls; axial riblets becoming obsolete on penultimate whorl. Body whorl with cords and grooves below suture, covered with incised, spiral grooves, deeper nearer anterior end but barely visible on upper body whorl. Lip is thickened, recurved, backed by deep narrow channel, and has twenty teeth. Columella has shield at anterior end; upper part of parietal shield and columella with incised, spiral striae, lower half with coarse, irregular folds. White; brown axial streaks; lip with six, pale brown blotches; interior tan.

P. decussatum L. 1758 Indonesia and south-east Asia to Taiwan, 70mm. Moderate spire; six or seven former varices; finely cancellate including upper part of parietal shield. Thickened, recurved lip backed by deep, narrow channel, has twenty teeth. Lower half of columella and shield with irregular, coarse, spiral ridges; upper half with four small teeth over cancellations. Posterior end of lip has two small ribs showing on later varices. Two colour forms: one with spots like *P. areola*, one (illustrated) with stripes like *P. strigatum*. Blue-grey; stripes or spots dark brown; anterior part of columella and shield white; interior brown; lip, later varices white with brown squares.

P. areola L. 1758 east Africa and central Indo-Pacific, 70mm. Shaped like *P. decussatum* above. Spire and shoulder of penultimate whorl cancellate. Body whorl has fine, incised, spiral lines visible below suture and near base, but barely so centrally. Lip, columella and parietal area as *P. decussatum*. Blue-white with rows of brown, square-shaped spots, their axial sides curved towards the lip; lip and outer edge of shield with a pink-brown tinge.

Phalium decussatum

Phalium strigatum

Cypraecassis testiculus

Phalium glaucum

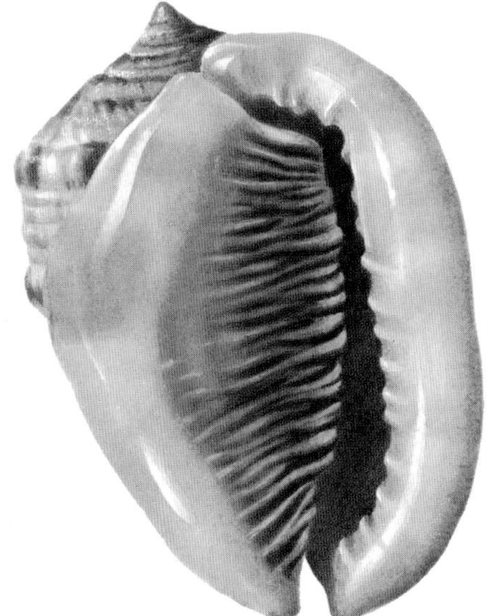

Phalium areola

Cypraecassis rufa

P. pyrum **Lamarck 1822** New Zealand, Tasmania, south Australia and South Africa, 90mm. Variable with many named forms. Generally moderate spire; usually globose but sometimes elongate. Early whorls have spiral cords, sometimes cancellate. Shoulder develops small nodules on penultimate whorl, becoming blunt, low, axial knobs on body whorl, but may be smooth. An incised spiral line below suture develops on last two or three whorls. Three or four weak striae near base of body whorl. Lip recurved, denticulate or almost smooth. Lower columella and shield have four folds; upper parietal wall has smooth white callus. Dull brown-grey; red-brown areas; four bands of arrow-shaped brown marks; lip, end of canal with purple marks; interior brown.

P. labiatum **Perry 1811** 75mm. Three subspecies: from Australia and New Zealand *P.l. labiatum* Perry 1811; *P.l. iredalei;* and *P.l. iheringi.*

P. labiatum iheringi **Carcelles 1953** South America. Seven or eight whorls; rather elongate with incised spiral striae on early whorls disappearing on later whorls which are roundly shouldered. Body whorl has seven, faint, blunt nodules on shoulder. Recurved, bluntly denticulate lip; columella and shield have four small and one large plait at base; thin parietal shield. Grey-cream; brown marks below suture and three rows of brown, arrow-shaped marks, backed by lighter area; aperture white; lip with brown marks.

P. labiatum iredalei **Bayer 1935** Natal to the Cape of Good Hope, 70mm. Variable from smooth-shouldered (illustrated), to knobbed, larger-shouldered form *zeylanica* Lamarck. Rather dark in colour; columella smooth with fold at bottom. Illustrated specimen has round shoulders of *iredalei*, and blue-white arrow-head spots on dorsum usually found on *zeylanica*.

P. thomsoni **Brazier 1875** south-east Australia, north New Zealand, 90mm. Rather light; spire usually shorter than illustrated. Early whorls cancellate; later whorls angulate, up to three spiral cords on shoulders, sometimes nodulose. Spiral incised striae on base of body whorl. Recurved lip is generally smooth, but may have very weak teeth. Posterior end of columella has a few weak folds; outer edge of columellar shield arched; lower columella folded; shield with many weak folds oblique to columellar folds. Dull white; four rows of square brown spots on body whorl, two on earlier whorls; five dark brown marks on lip.

P. glabratum **Dunker 1852** 60mm. Three subspecies: *P.g. glabratum* Dunker 1852, Philippines, Indonesia; *P.g. angasi*; *P.g. bulla* Habe 1961, China, Japan, Hawaii.

P. glabratum angasi **Iredale 1927** east Australia. Thin, elongate with moderate spire; early whorls with finely beaded, spiral cords; round shoulders. Body whorl has faint spiral striae and five, incised spiral lines at base. Lip thickened and dentate, more strongly anteriorly. Columella has faint plaits at top and strong folds on lower half and shield; latter with deep indentation over umbilicus. Tan or white; dorsum side has faint brown tinge.

P. bisulcatum **Schubert and Wagner 1829** Indo-west Pacific, 70mm. Very variable with many named forms. Heavy or light; spire moderate but variable; globose or narrow; spirally striate, smooth or striate below suture and at base. Lip with strong or weak, small teeth; columellar shield somewhat rugose. White or cream, may have blue-grey tinge; may have five rows of more or less prominent, squarish, light brown spots; lip with or without five groups of dark or light brown marks; columella and shield white; interior white or purple-brown. Illustrated are: spirally striate form with pale or absent markings, *diuturnum* Iredale 1927; one described by Schubert and Wagner; and *sophia* Brazier 1872 from east Australia with sharply angled shoulders and generally smooth, the brown squares tending to turn into axial streaks.

Phalium labiatum iheringi

Phalium pyrum

Phalium labiatum iredalei

Phalium bisulcatum form *sophia*

Phalium bisulcatum

Phalium thomsoni

Phalium glabratum angasi

Phalium bisulcatum form *diuturnum*

Phalium bandatum Perry 1811 Japan, Philippines, Indonesia and north Australia, 120mm. High concave spire. Early whorls with beaded spiral cords disappearing on last two whorls leaving faint striae; small knobs develop on antepenultimate whorl. May have nil to three former varices. Thickened, recurved lip has twenty elongate teeth, largest centrally. Columella with five small folds anteriorly. Columellar shield strongly folded with raised lirae. White; five spiral bands of pale brown; axial flame marks; darker brown squares at intersections; lip with six purple-brown spots, three short spikes posteriorly. Subspecies *P.b. exaratum* Reeve 1848, Indian Ocean islands, has spiral grooves, less shouldered whorls and reduced or absent spines on lip.

P. granulatum Born 1778. Three subspecies: *P.g. granulatum* Born 1718, Caribbean; *P.g. undulatum*; *P.g. centriquadratum* Valenciennes 1832, west Central America.

P.g. undulatum Gmelin 1791 Mediterranean and east Atlantic Islands, 110mm. High spire; slightly angular shoulders. Strong spiral cords, beaded on early whorls and posterior of body whorl; axial riblets near lip. Thick, wide lip has seventeen teeth. Columella smooth posteriorly, pustulate or granulate anteriorly and on shield. Callous pad where lip joins body. Creamy brown; brown, axial flame marks; interstices of cords dark brown; outer lip with broad and narrow brown bands; inside lip, columella, shield, pad white. *P.g. granulatum* has shorter spire, more spiral cords, no flame marks or spiral lines. *P.g. centriquadratum* as *P.g. granulatum* but sometimes with knobbed shoulders and rather low spire.

Casmaria erinaceus L. 1758. Synonym *C. vibex* L. 1758. Three subspecies: *C.e. erinaceus*; *C.e. kalosmodix* Melville 1883, east central Pacific; and *C.e. vibexmexicana* Stearns 1894, west Central America.

C.e. erinaceus L. 1758 Africa eastward to west Pacific, 70mm. Two forms, typical having nodulose shoulders, and large smooth-shouldered form *vibex*. Moderate spire, rounded shoulders, faint axial striae, and axial folds near lip. Thickened, recurved lip is smooth bar five or six, small, sharp spines anteriorly. Smooth columella, two folds anteriorly; shield with spiral ridges. Cream-brown; purple tinges; aperture white; purple tip to up-turned siphonal canal; purple-brown blotches on outer lip; body whorl has faint yellow-brown squares and axial flares, yellow-brown marks below suture; growth striae have some minute brown specks on inner side. Occasionally a former varix (as illustrated). This specimen also has pustulate columella and callous parietal shield.

C. ponderosa Gmelin 1791 Indo-Pacific, 50mm. Varieties at extremes of its range have subspecific names. Stubbier than *C. erinaceus*. Moderate spire; sometimes nodulose shoulder; thickened, recurved lip with small, sharp teeth. Cream; two, pale brown, spiral bands; dark brown marks below suture and anteriorly on body whorl (*C. erinaceus* lacks specks); early whorls blue-white; up-turned siphonal canal has purple-brown spot; dark brown oblong marks on outer lip. Form *turgida* Reeve 1848 (illustrated) is rather large, lighter weight and smooth with teeth only on anterior two-thirds of lip, and axial red-brown flames.

Galeodea echinophora L. 1787 Mediterranean, 110mm. Medium spire. Five spiral ridges, two posterior rows bluntly nodulose, fainter on next two; blunt nodulose ridge on shoulder of penultimate whorl; spiral cords between suture and shoulders and anteriorly; smooth between body whorl ridges. Lip a little thickened, flared, obsoletely toothed within. Columella smooth bar ridges, smooth shield and thinly callous parietal wall. Grey-brown; nodules white, brown between; aperture white.

Phalium bandatum

Galeodea echinophora

Casmaria erinaceus erinaceus

Casmaria erinaceus erinaceus form *vibex*

Casmaria ponderosa

Casmaria ponderosa form *turgida*

Phalium granulatum undulatum

Genus: *Morum*

Shells in this genus of the Cassidae are solid, coarsely-sculptured and lack the turned-up siphonal canal of the preceding genera.

Morum oniscus L. 1758 Caribbean, 25mm. Flat spire; spiral cords and axial ribs on body whorl. Three adjacent cords below suture are raised – axial ribs giving a tubercled effect – two cords in middle of whorl, and one anteriorly. Thickened lip has about fifteen teeth. Columella, parietal shield callous, strongly rugose or pustulate. White speckled with brown; apex, interior white.

M. grande A. Adams 1855 south Japan, 70mm. Solid, rather elongate with moderate spire. Coarse, spiral cords becoming lamellated on later whorls; coarse axial ribs; pointed tubercles forming at their intersections. Lip is expanded, thickened, not recurved, has many folds and teeth. Columellar shield and parietal callus rather thin, pustulate and rugose. Short, slightly recurved siphonal canal. Dirty white; four, faint, interrupted, brown bands showing most clearly on lip edge; inside lip, interior, columella, shield, parietal callus white.

M. macandrewi Sowerby 1889 south Japan and China, 50mm. Moderately low spire and inflated at shoulder narrowing sharply anteriorly. Spiral cords, axial ribs and lamellated, axial growth lines. Ribs and cords form sharp points at their intersections. Thickened, wide lip, with about twelve large teeth, more small ones between. Narrow columellar shield, parietal wall with thin callus; both pustulate and rugose as is columella. White; dark grey-brown bands and mottling; lip with ten, short, black streaks; inside lip, teeth, columella and shield white; brown bands showing through parietal callus.

M. ponderosum Hanley 1888 Indo-Pacific, 40mm. Solid, heavy with low spire; fine spiral cords; nodulose shoulders. Body whorl has ten, broad, spiral ridges with small cords between; faint lamellate growth lines. Thickened lip has about fifteen, small teeth becoming obsolete at ends. Columella with fine, raised lirae and smooth, callous parietal wall. White; four, obscure, red-brown bands; sparse, small spots; lip white with small, brown spots; interior, lirae white.

M. tuberculosum Reeve 1842 Lower California to Peru, 17mm. Similar to *M. oniscus* L. 1758 of Atlantic coast of Central America. Flat spire and pointed apex. Cancellate, finely on spire; spiral cords. Five, spiral rows of heavy, blunt knobs – about six per row – absent on parietal wall. Narrow aperture. Lip turned in with some eighteen teeth. Smooth columella, narrow thin callus. Dark brown, almost black; white or yellow dots; lip edge, interior white or yellow.

M. cancellatum Sowerby 1824 south Japan and China, 45mm. Similar in sculpture to *M. macandrewi* but smaller and more delicate. White or cream; about four, more or less indistinct, brown spiral bands; interior white.

Family: Ficidae

A small family (one genus) of tropical shells. They are fig-shaped, thin and low-spired with a large, long body whorl, long, wide aperture and siphonal canal. They live on sand usually in fairly deep water.

Ficus gracilis Sowerby south Japan, 150mm. Small spire with spiral cords and smaller axial cords. Light brown; many, axial, darker brown, wavy lines; interior rich chestnut brown; lip pale blue-grey.

F. ficoides Lamarck 1822 Indo-Pacific, 100mm. Almost flat spire. Siphonal canal relatively shorter than the above species. Cancellate with fine spiral and

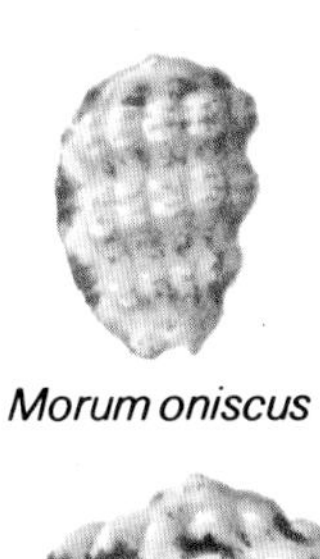
Morum oniscus

Morum grande

Morum macandrewi

Morum ponderosum

Morum tuberculosum

Morum cancellatum

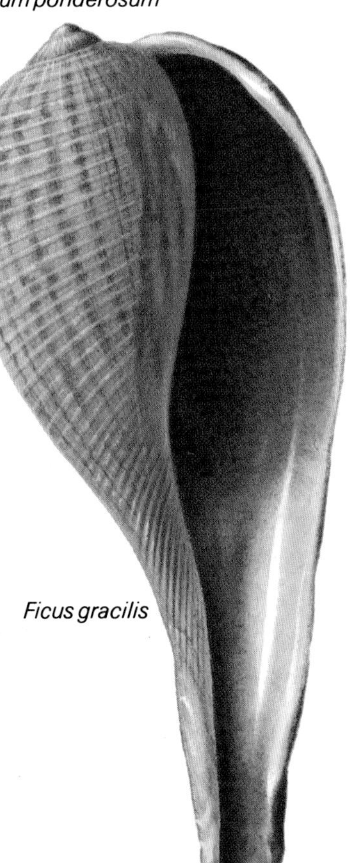
Ficus gracilis

Ficus ficoides

smaller axial cords. Pink-fawn; about five, narrow, pale or white bands; dark brown spots and blotches especially on the pale bands; interior white with a violet tinge becoming brown towards the interior.

Family: Tonnidae

Not a large family, they have thin, rather fragile shells, some large, with short spires and very inflated body whorls. They are sculptured with spiral cords or ridges. They live mostly in the tropics, on sandy areas beyond the reefs, and are carnivorous, living mainly on echinoderms and crustaceans.

Tonna variegata Lamarck 1822 Indo-Pacific, 160mm. Globose, low-spired with seven whorls and deep suture. Body has about sixteen, prominent, broad ribs, five showing on penultimate whorl. Interstices channelled, posterior ones having faint cords. Deeply and narrowly umbilicate. Crenulated lip. Small columellar shield and thinly callous parietal area. Creamy white to oatmeal with pale cream-pink or fulvous red staining and indistinct axial streaks; a few ribs spotted with red-brown especially near the suture – very faint in the illustrated specimen; inside lip white; interior orange.

T. tessellata Lamarck 1816 west Pacific to South China Sea, 80mm. Moderately high spire; about seven whorls and deep sutures. Body whorl has about fourteen, rather narrow, rounded ribs, four on penultimate whorl. Wide, smooth interstices, which sometimes have a single thread. Lip fluted with about ten pairs of teeth on the inner edge. A small parietal shield. Body whorl an opaque white; ribs white with brown spots; penultimate whorl with pink tinge; earlier whorls with dark pink spiral band to the apex.

T. cepa Röding 1798 Indo-Pacific, 130mm. Ovate with moderately high spire; about seven whorls and deeply channelled sutures. Spiral cords on early whorls, becoming obsolete on the penultimate whorl. About sixteen, broad, flat ribs on body whorl divided by narrow grooves. Rounded and malleate shoulders. Narrowly umbilicate. Unevenly marked with light browns, purple-brown, and creamy white, spiral smears, especially on the shoulders; inside lip, small columellar shield and thin parietal callus white; interior brown.

T. luteostoma Küster 1857 west Pacific, Japan to New Zealand, 200mm. Globose and low-spired; about seven whorls and deep suture. Broad, rather flattened, rounded ribs, about seventeen on the body whorl and five showing on the penultimate whorl; rather narrow, excavated channels between some posterior ones with a single cord. Crenulate lip; small columellar shield over, but not closing, the deep umbilicus; thin parietal callus. White or light tan with alternating white and red-brown axial stripes across the ribs with colour gradations between them; some ribs have long streaks of browns without white; very shiny; lip and columella white; parietal callus clear and faintly white; interior brown. This shell is rather variable in its colouring.

T. dolium L. 1758 (not illustrated), Indian Ocean, west Pacific, 150mm. Very variable. It is globose and has from ten to twenty, broad, flat, spiral ribs which may or may not have intermediate threads between them on the posterior half of the whorls; two to four of the ribs show on the penultimate whorl. Deep sutures, fluted outer lip and umbilicus is open and deep; no parietal shield. White to fawn, with rectangular orange-brown spots on the ribs; apex brown; interior brown.

Tonna variegata
Tonna tessellata
Tonna cepa
Tonna luteostoma

T. allium Dillwyn 1817 Indo-Pacific, up to 90mm. Rather high spire for a *Tonna* giving an ovate, rather than globose, outline. Rather widely-spaced ribs, about fourteen on the body whorl and three showing on the penultimate. Ribs widest apart at the posterior and closest at the anterior end. Interstices concave and smooth except for fine growth striae. Deep suture. Crenulated lip may have up to twelve pairs of teeth. Small, thin columella and parietal shield. Narrowly umbilicate. Opaque white and may have a brown tinge on the ribs; apex and lower half of early whorls purple.

T. sulcosa Born 1778 Indo-west Pacific, 120mm. Moderate spire; seven whorls. Flat ribs, about twenty on the body whorl and four on the penultimate; ribs are of varying width as are the shallow channels between them. Moderately deep suture. Lip thickened and a little expanded, with about twenty teeth, some paired. Small, thin columellar shield over a very narrow umbilicus; thin parietal callus. White with three or four brown bands covering about two ribs each; apex dark purple.

T. perdix L. 1758 the Partridge Tun, Indo-Pacific, 200mm. Narrow, high-spired and rather thin with rounded shoulders and not very deep sutures. There are narrow, shallow grooves between flat ribs, about twenty on body whorl and eight on penultimate. Slightly thickened lip; thin, small parietal callus; deeply umbilicate; siphonal canal twisted towards aperture. Brown with dark brown and white smears on ribs, especially near suture, base and lip; the white marks are to the right of the brown ones. In some specimens, as illustrated, the brown and white marks cover almost the whole shell. Columella, callus white; interior brown.

T. maculosa Dillwyn 1817 (not illustrated) the Atlantic Partridge Tun, east tropical America, 130mm. It is very similar to the above, but the spire is lower and has more spiral ribs. The rectangles of colour are more blurred.

T. galea L. 1758 tropical Indo-Pacific, west Atlantic and Mediterranean, 200mm. Globose and low-spired with about seven whorls. Body whorl has some fifteen to twenty, broad, flat ribs; on the upper half up to three smaller ribs between, and up to five between the top two ribs. Deep suture; two or three cords at the top of the whorl within the sutural channel; about three main ribs show on the penultimate whorl. Fluted outer lip is slightly expanded and obsoletely toothed. Narrow umbilicus almost closed by small columellar shield; thin parietal callus. Light milk-chocolate brown, axially streaked with lighter brown, also light at the suture; apex dark purple; interior and columellar shield white; brown marks on inside of lip and edge of siphonal canal.

T. tetracotula Hedley 1919 (not illustrated) east Australia and New Zealand, 200mm. Rather similar to *T. sulcosa*, of which it may be a subspecies from temperate water. Globose, it has about twenty, flattened ribs which become smaller towards the anterior end of the whorl. Between the ribs are threads varying from four at the posterior end to one or none at the anterior end. Deep suture; lip finely denticulate within, denticles in pairs; deep and narrow umbilicus. White, it may or may not have brown spiral bands and these may not completely encircle the body whorl; part of the body whorl may be pale cream-brown; area round siphonal notch may be grey-brown; inside of lip white; interior tan; columella and thin parietal callus white.

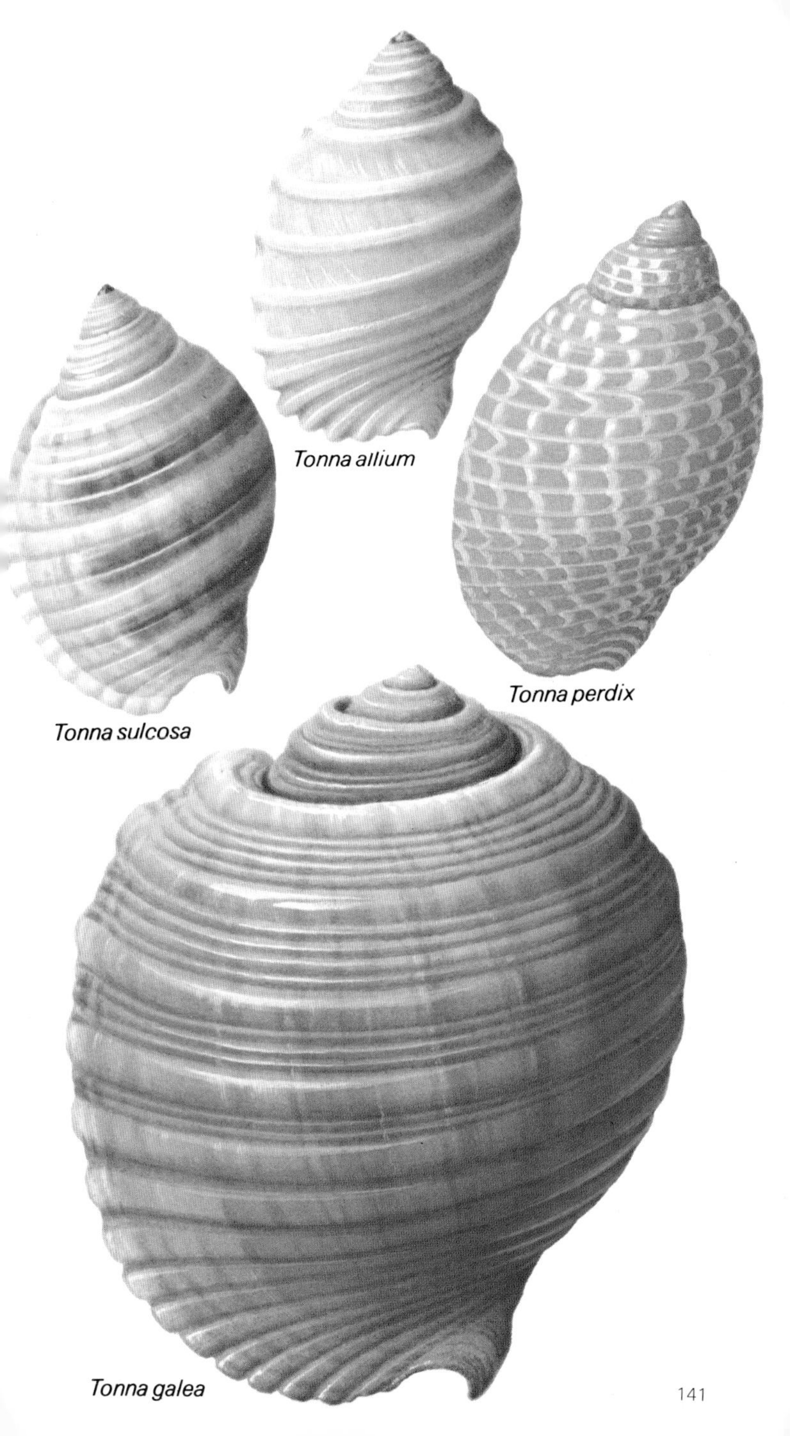
Tonna allium
Tonna perdix
Tonna sulcosa
Tonna galea

T. olearium L. 1758 south Japan and South China Sea to the Philippines, Indonesia and Malaysia, 200mm. Very similar to *T. galea* but with a higher spire. About seventeen main ribs on the body whorl which are narrower and slightly more rounded and may have threads or small cords between all the main ribs. The shoulders are more rounded; the siphonal canal is longer. It is a darker brown; the lip is edged with dark brown; the early whorls are lighter.

T. cerevisina Hedley 1919 (not illustrated) south-west Pacific, up to 240mm. Shape similar to *T. galea*. About twenty rather flat ribs on body whorl with shallow channels between some of the anterior channels which have small intermediate threads, as also have the five or six ribs showing on the earlier whorls. Coarse growth lines near the aperture. Crenulate lip. Columella with small parietal shield partly covering deep umbilicus; parietal wall glazed. Colour is variable from dark brown with light spots and streaks, to off-white with about five, brown, spiral bands and brown spots on the white areas between; sometimes off-white with rather sparse brown spotting.

Malea ringens Swainson 1822 the Grinning Tun, west tropical America, can grow to 240mm but usually about 100mm. Solid and heavy with low spire. About seven whorls with eighteen flat ribs on the body whorl, three showing on the penultimate. Shallow sutures and rounded shoulders. Outer lip strongly constricted before flaring out and flattening; outer edge fluted; inner edge with about seventeen long teeth, opposite the outside constriction. Columella with a triple-ridged boss in the middle and five, high, narrow, folds, the first two and last two joined. This sculpturing gives the shell its apt common name. Columellar shield is rather uneven and faintly rugose, almost sealing the umbilicus. Broad parietal callus and reverted siphonal canal. Putty coloured; paler on inside of outer lip, columella, callus and early whorls; interior pale brown.

M. pomum L. 1758 the Pacific Grinning Tun, Indo-Pacific, 75mm. Solid, with a low spire, about seven whorls and sutures a little indented. Body whorl with about twelve, smoothly rounded ribs and shallow interstices; three ribs show on the penultimate whorl. Restricted before the outer lip which is fluted, thickened and somewhat flattened, and shows coarse growth lines. Inside lip has ten or eleven, strong, tooth-like, folds. Columella and parietal area callous, underlying channels showing through at the top of the columella. Lower end with about four folds and one large rib above the siphonal notch. The umbilicus is covered by the parietal callus. Cream-fawn with white squares and oblongs on the ribs and a few darker fawn markings; outer area of lip white becoming orange-brown within the aperture; columella and callus white.

Malea pomum

Malea ringens

Malea ringens

Tonna olearium

Family: Cymatiidae

Rather solid shells, angulate with spiral ribs and cords, knobs, tubercles and varices. They are operculate and the siphonal canal is often long. They live in sandy and rocky areas in deep and shallow water in the tropics worldwide. The periostracum is sometimes very thick and hairy and the operculum is horny. They are carnivorous, living on echinoderms and molluscs. They have a long veliger or free-swimming stage, after they are hatched and before developing into the final form. This is likely to account for some species being found in the Atlantic and Mediterranean as well as in the Indo-Pacific oceans —e.g., *Ranella olearium* L.

Genus: *Charonia*

There are about a dozen species of tritons. They have high spires and large apertures. Some grow to a large size, all are tropical or sub-tropical.

Charonia tritonis L. 1758 Triton's Trumpet, Indo-Pacific, 400mm. Sometimes used, particularly in the Pacific, as a horn. High, pointed spire; coarse, spiral cords and axial ribs on early whorls. Spiral ribs below suture – two on earlier whorls, three on the body whorl being beaded – are broad and flat with a small, narrow rib between. Outer lip flares to form large aperture and expands to form a low ridge before the lip which recurves. These axial ridges and lips show as varices on earlier whorls, one every two-thirds of a whorl, and therefore line up axially on every alternate whorl. Scalloped lip has about fifteen, spiral ribs running into the interior, posterior ones forming pairs of denticles on the lip. Concave columella is strongly and coarsely lirate. Narrowly umbilicate. Short siphonal canal. Creamy white; purple and brown, rounded, scale-like markings on spiral ribs; lip pinky-white; interior of aperture and inner ribs orange; denticles white; columella orange-pink, purple-brown between the lirae.

C. variegata Lamarck 1816 the Atlantic Triton, Florida to Brazil, 380mm. Similar to the above but squatter. The whorls are sometimes unevenly swollen and occasionally bulge over the suture which descends in an uneven spiral thereafter. Outer edge of lip is scalloped but less projected, and toothed with rib-like teeth which are mostly in pairs, about ten pairs in all. Coloration similar to *C. tritonis* except that the teeth on the lip are white and each pair is in a rich brown blotch; inside of aperture orange-pink; interior white.

C. lampas L. 1758 Mediterranean, up to 300mm. Spirally corded; two rows of blunt nodules per whorl, lower row sometimes hidden in places beneath suture. Varices about every two-thirds of a whorl. Body whorl with two, broad, flat ridges on which the cords show and earlier nodules almost disappear, and with other lesser bands. Expanded, thickened, scalloped lip with about twelve, blunt, rib-like denticles. Columella with one large and one smaller fold posteriorly; three or four, oblique, large ones anteriorly; many, fine, obsolete ones between. Umbilicus sealed by parietal shield. Siphonal canal relatively larger than in preceding species. White; broken, brown bands axially and spirally on body whorl; teeth on lip brown; brown blotch at base of columella.

Charonia variegata
Charonia tritonis
Charonia lampas

Cymatium gutturnium **Röding 1798** Indo-Pacific, 90mm. About six whorls with spiral cords and axial riblets or folds, and larger ribs, about eight on the body whorl, carrying pointed nodules where they are crossed by the spiral cords. Slightly constricted suture. Former varices every 240°. Outer lip thickened and callous with seven strong teeth. Columella and parietal wall heavily callous, three or four folds at the top; one strong rib near the lower end; about three smaller ones at the top of the long, angular siphonal canal. White with brown axial marks along the large axial ribs; inside lip, columella, callus, and interior of siphonal canal, orange-red, or may be white or yellow.

C. (Ranularia) pyrum **L. 1758** Indo-Pacific, 80mm. Solid and heavy. Sculptured with beaded spiral cords. Ribs have pointed nodules where they intersect on the angular shoulders and below; about eight on the body whorl each with some four to six nodes – two or three showing on earlier whorls. Outer lip with thick, heavy varix; strongly toothed internally – about seven teeth – with much smaller teeth – about eight pairs – just within lip. Heavily plicate columella and parietal callus; long, contorted siphonal canal. Red-brown; varices with darker and lighter markings; inside lip, interior white with pink areas round the teeth; columella orange-red with white plicae. Coarse, heavy periostracum.

C. lotorium **L. 1758** Indo-Pacific, 100mm. Solid, heavy and rugged. Smooth suture, high spire and body whorl just over half the length of the shell. Heavy varix every 240°. Spiral cords; fine threads. Shoulders angulate with heavy knobs, each carrying about three cords, three or four between each varix. Outer lip has about seven teeth. Columella with heavy, callous rib at posterior end, a small one at top of long siphonal canal, and lirate between. Narrow columellar shield, lirate at edge where it crosses the fasciole. Yellow-brown; varices dark brown and white; inside lip and interior white; edge of columellar callus stained with dark brown and columella with pale brown marks.

C. (Monoplex) parthenopeum **Von Salis 1793** worldwide in warm waters, 150mm. Solid, fairly high spire. Irregular, weak former varices. Whorls angulate with fine spiral threads, nodulose ribs, and fine axial growth striae and folds. Outer lip with heavy varix; interior callous with six plicae of long, thin teeth. Columella with narrow parietal shield and lirate with a tooth posteriorly. Short, slightly recurved siphonal canal. Dark brown with lighter areas; varices white and very dark brown; inside lip pink with white denticles on a dark brown base; columella dark brown with white lirae and tooth; interior white.

C. (Septa) pileare **L.. 1758** Indo-Pacific, 100mm. High-spired and elongate with varices about every 240°. Spirally beaded ridges of varing size and intermediate threads. Two ridges on the shoulder bear small blunt nodes. Outer lip with heavy varix and about eight pairs of long teeth. Columella heavily lirate. Shades of light and dark tan; varices darker brown and white; inside of lip, interior and columella deep rich red or may be pale orange. Very hairy periostracum.

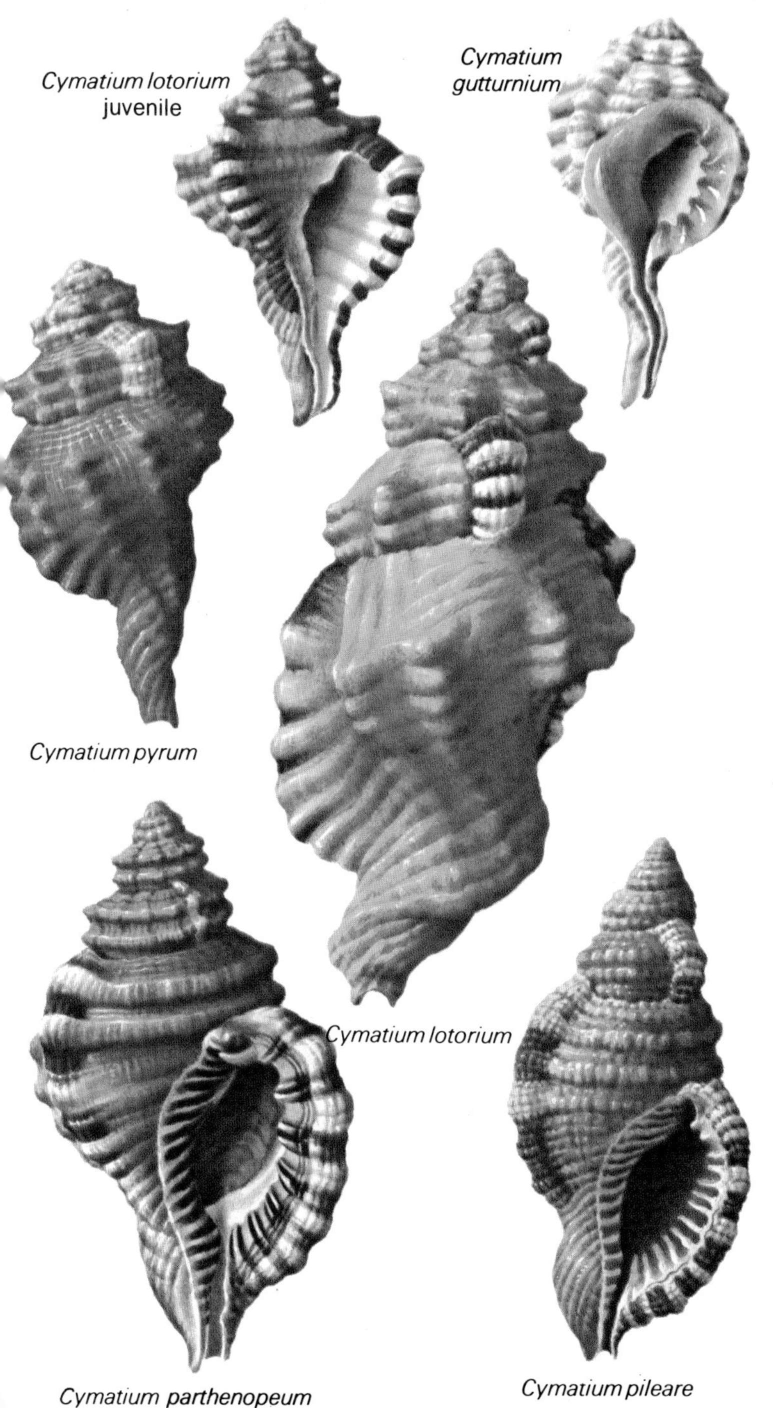
Cymatium lotorium juvenile
Cymatium gutturnium
Cymatium pyrum
Cymatium lotorium
Cymatium parthenopeum
Cymatium pileare

C. (Septa) hepaticum Röding 1798 Pacific and Indonesia, 50mm. Broad ribs, wide interstices and axial riblets. 'Lump' on shoulder in middle of dorsum. Lip has thick varix and nine teeth; lirate columella; spirally channelled interior. Cords red or orange-brown; interstices mostly black; varices white, black and red; lip edge pink; interior white; columella red, lirae white.

C. (Turritriton) gibbosum Broderip 1833 west Central America, 40 mm. Constricted at suture and expanded at shoulder with expanded varices. Fine, beaded spiral cords and axial riblets. Shoulder with one large and three small knobs, lower cords having two or three small nodules. Lip with large varix has four weak teeth. Columella obsoletely plicate with small tooth, and callous over lower end of penultimate varix and much of dorsum. Red-brown: five, indistinct, purple-brown, spiral bands; early whorls grey-purple; varices with white area on aperture side, edged with brown; nuclear whorls obvious, smooth, white with dark brown line below suture.

C. (Cymatriton) nicobaricum Röding 1798 Indo-Pacific, 50mm. Spiral nodulose cords and fine threads. Three to five large knobs on shoulder between each pair of varices; latter every 240°. Lip with heavy varix and seven to fourteen long teeth; columella lirate. Grey, red stains, sometimes indistinct cream band; varix dark grey and white; lip, teeth, lirae white; interior, columella yellow or orange-yellow.

C. (Gelagna) clandestinum Lamarck 1816 Indo-Pacific, 45mm. Light, no varices and six inflated whorls with twenty spiral cords, seven on penultimate whorl, and smooth wide channels between. Faint axial ribs which disappear on body whorl. Interior channelled; lip expanded, not thickened; columella smooth. Pale brown; cords red, inside of lip, interior, lower columella white.

C. (Mayena) australasia Perry 1811 south Australia and New Zealand, 90mm. Very fine spiral threads; blunt nodules on shoulders, twelve on the body whorl; oblique varices every 200°. Lip heavily dentate; columella with one fold; bottom of earlier varix protrudes into top of aperture behind columella. Dark brown: varices brown, white; inside lip, columella, interior white.

C. labiosum Wood 1828 Indo-Pacific and Caribbean, 30mm. Fine, beaded spiral threads and six spiral cords; four axial ribs between varices, forming blunt nodules where cords cross. Lip with well-developed varix; aperture with six blunt teeth; columella with blunt tooth posteriorly and anteriorly; columellar shield edge faintly plicate. Red-brown; aperture, inside lip, columella white.

C. (Argobuccinum) argus Gmelin 1791 South Africa in deep water, 60mm. Spiral threads; five rows of small nodules on body, two on earlier whorls. Varices every 200°. Lip thickened to weak varix with eight pairs of teeth. Columella with three plicae posteriorly; outer edge of parietal callus plicate anteriorly. Light brown; dark brown stripes across the light brown nodules, one above and two below without nodules and with dark brown lines between stripes; inside lip, columella, interior white. Spire and nodules often worn to white, as illustrated.

C. (Cabestana) dolarium L. 1758 South Africa, 40mm. Angulate with spiral ridges carrying fine incised line centrally; a small cord between ribs; axial ribs, ten on body, becoming blunt nodules where ridges cross; two ridges show on penultimate whorl. Flared, crenulate outer lip, with seven teeth, some bifurcated. Smooth columella. Brown or off-white; inside lip, aperture, columella white.

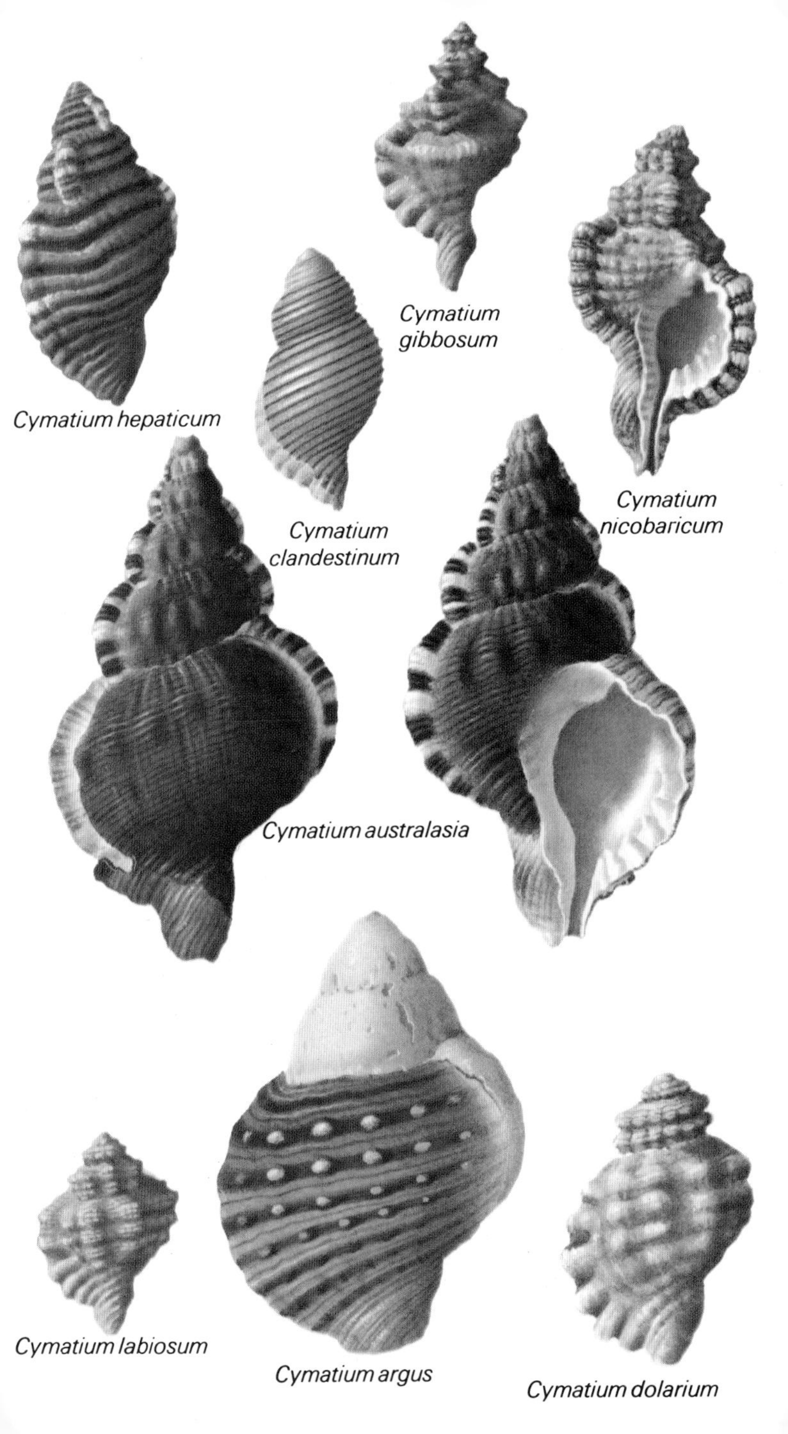
Cymatium hepaticum
Cymatium gibbosum
Cymatium clandestinum
Cymatium nicobaricum
Cymatium australasia
Cymatium labiosum
Cymatium argus
Cymatium dolarium

C. (Cabestana) spengleri Perry 1811 east Australia, New Zealand, 120mm. Spiral ribs and cords of different widths, all closely beaded. Close axial riblets; spiral, raised lines; fine threads. Angular shoulders have six, small knobs. Varices every 240°. Lip with thick varix, denticulate, seven pairs of teeth. Weakly furrowed columella has bifurcated tooth posteriorly. Narrow parietal shield, pustulate anteriorly; open canal. Creamy tan; black in some interstices; inside of lip, interior, columella white.

C. muricinum Röding 1798 Indo-Pacific and Caribbean, 80mm. Solid; long siphonal canal; coarse spiral and axial cords and threads, jumbled below suture. Varices every 240°. Seven, main cords on body whorl, three on penultimate. Axial ribs nodulose where cords cross. Lip with strong varix, heavily callous inside, as is parietal shield. Canal angled at about 45°. White; grey and brown mottling, basically in spiral bands; inside of lip, columella, parietal callus cream; interior purple-brown.

C. (Fusitriton) laudandus Finlay 1927 south New Zealand, 110mm. Deep water. Round shoulders; fine, spiral cords, axial ribs, and nodules at intersections. Former varices indistinct; lip with small varix. Columella has a large, callous tooth posteriorly; extended siphonal canal. Cream; cords chestnut; interior, columella white, sometimes pink or purple tinge; grey-brown periostracum (illustrated).

C. (Ranella) olearium L. 1758 Mediterranean to South Africa, 210mm. Varices every 200°. Fine spiral threads and low ridges, six showing on upper whorls. Axial, fine, close striae. Twenty ribs on upper whorls, ten indistinct on last three. Upper whorls have small, sharp nodules where ridges and ribs intersect; on last three whorls, especially centrally, pointed nodules on some ridges become larger and coarser. Lip with edge flattened, slightly expanded behind varix, seventeen teeth. One, strong, columellar tooth posteriorly; small parietal callus. Siphonal canal moderate, a little curved. White; light brown stains; interior, columella white. Was believed to be confined to Mediterranean and Lusitanian areas, but the illustrated specimen was trawled off south Zululand, South Africa.

C. africanum A. Adams 1855 South Africa, 60mm. Elongate or stubby. Spiral ridges on upper whorl, obsolete and noduled on shoulder of last few. Six nodules on body; varices every 200°. Lip with heavy varix, seven teeth. Columellar tooth posteriorly; umbilicate. Red-brown; lip, interior, columella white.

Gyrineum gyrinum L. 1758 Indo-Pacific, 50mm. Compressed. Spiral threads; larger cords, seven on body whorl, three on earlier; axial threads; ribs, twelve on body whorl; cords and ribs form small nodules at intersections. Varices every 200°. Lip has seven denticles, some bifurcated. Columella has tooth posteriorly, weak plicae. Dark brown; white band on dorsum; nodules with red tinge; varices orange-brown and white; lip, columella white; interior dark brown, white spiral band; nuclear whorls dark, conspicuous.

G. natator Röding 1798 Indo-west Pacific, 40mm. Dorso-ventrally compressed; varices almost aligned each side. Ten, spiral cords on body, four on early whorls. Fine raised threads; axial ribs, fifteen per whorl; rounded shiny nodules at intersections. Lip with strong varix, seven teeth. Columella with posterior tooth, obsolete plicae. Grey-tan; dark brown bands and nodules; varices brown and white; inside lip, columella and interior white.

Biplex bitubercularis Lamarck 1816 Indo-west Pacific, 40mm. Slightly inflated. Varices every 180°. Fine spiral and axial threads. Spiral ridges, ten on

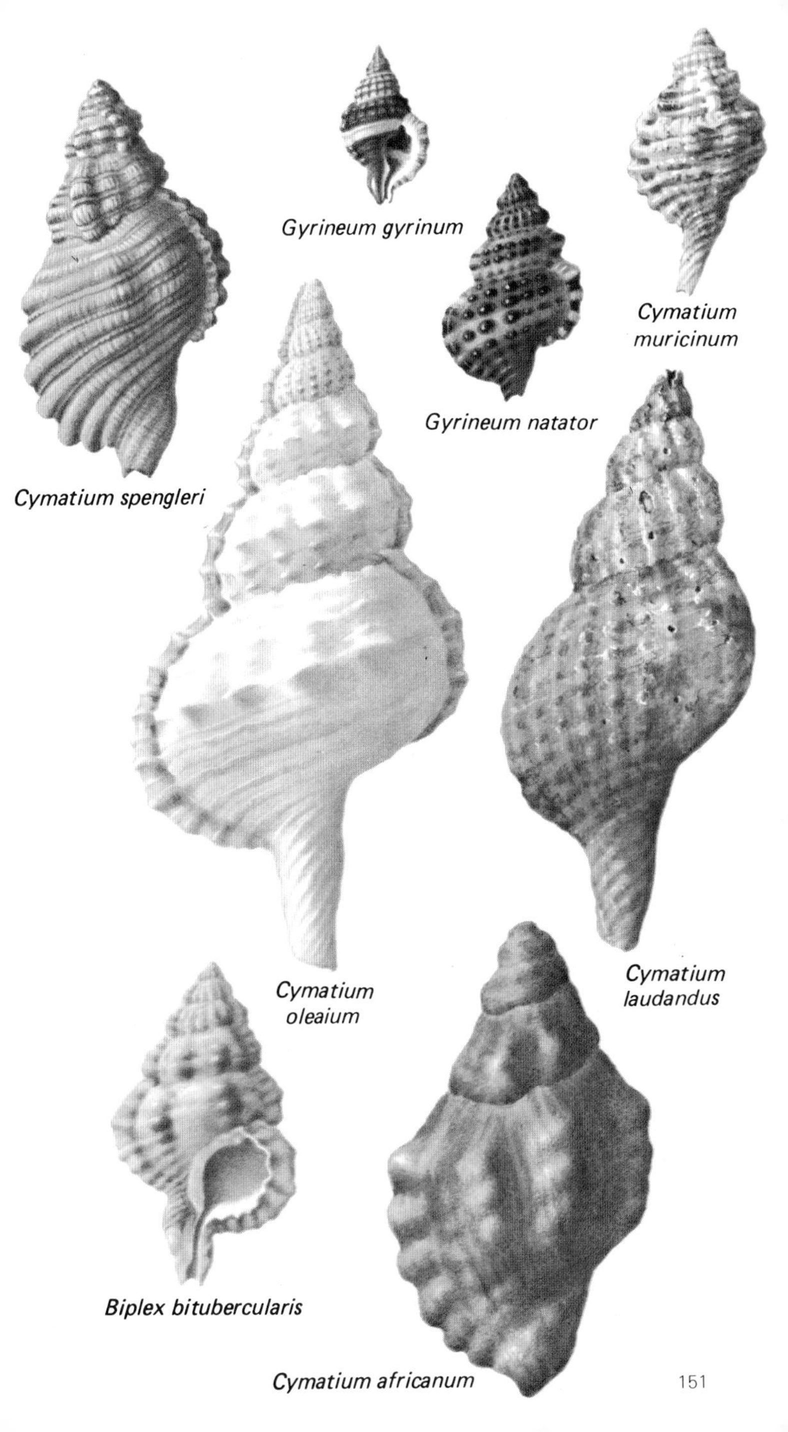

Gyrineum gyrinum

Cymatium muricinum

Gyrineum natator

Cymatium spengleri

Cymatium laudandus

Cymatium oleaium

Biplex bitubercularis

Cymatium africanum

body whorl; axial ridges, five between each varix on upper whorls, two or three on last two whorls. Nodules where ridges and ribs cross, larger on two or three ridges in middle of whorl. Lip with nine teeth; columella plicate at top and bottom, rugose centrally; twisted siphonal canal. Pale fawn; varices cream; larger nodules red; aperture white.

B. jucundum A. Adams 1854 north Australia, Indonesia, Malaya, 25mm. Better known as *B. pulchella* Forbes 1852. Very flattened dorso-ventrally. Varices every 180°, thin, expanded, running together. Spiral, alternately wide and narrow cords, five on body whorl, with deep narrow interstices; broader cords end in points on varices. Whorls have axial riblets, twenty-five on body whorl, forming nodules at intersections with spiral cords. Lip, columella smooth; slender, moderately long siphonal canal. Cream or grey-brown.

B. perca Perry 1811 central Indo-west Pacific, 80mm. Flattened dorso-ventrally. Expanded, flattened varices every 180°. Spiral threads and ridges, seven on body whorl, ending in blunt points on varices; axial ribs, twenty on body whorl. Blunt nodules where ridges and ribs cross. Rugose or obsoletely toothed lip; columella with double tooth posteriorly, obsoletely plicate; extended, slightly curved siphonal canal. Pale grey-brown, stained brown on the larger whorls; nodules cream; aperture white.

Distorsio anus L. 1758 Indo-Pacific on reefs, 80mm. Pointed spire. Irregularly inflated whorls giving distorted shape. Nodulose, spiral ridges; axial riblets. Crenulate lip with row of seven teeth near edge pointing outward, a second row of bigger teeth pointing inward; a channel between rows. Heavily rugose parietal shield fluted round edge. Very large columella has deep wide notch centrally. Siphonal canal ridged along both sides, sharply recurved. White; brown spiral bands and blotches; aperture, shield shiny white with pink and brown areas.

D. reticulata Röding 1798 Indo-Pacific, deep water, 55mm. Similar to above but less extreme. Spiral cords; axial ribs; sharp nodules at intersections. Flattened lip; ten, very small, outer teeth; inner teeth at end of ridges crossing lip. Thinly callous pareital area. Columella with long tooth at top; large, rectangular notch centrally; lower area strongly plicate. Cream; teeth, interior white.

D. clathrata Lamarck 1816 Caribbean and west Africa, deep water, 60mm. Similar to *D. reticulata* but sculpturing much coarser, and nodules become blunt spines on middle of whorls. Narrower lip, ten outer and inner teeth, that opposite columellar notch being much bigger than the others. Columellar sculpturing very similar, but with ridges instead of rows of small nodules across parietal area. Cream with brown clouding; interior and columella white; parietal callus shiny brown.

Family: Colubrariidae

False tritons. Long, narrow, with varices, short siphonal canal and horny operculum. Not well studied and may be more correctly placed among the Buccinidae.

Colubraria obscura Reeve 1844 Indian and tropical Atlantic, 60mm. Narrow, high-spired, solid and rather flat. Varices every 240°. Spiral rows of granules, twenty rows on body whorl; axial growth lines. Short siphonal canal. Lip with heavy varix and about thirteen teeth. One columellar tooth posteriorly. Cream; pale brown marks on varices and some nodules.

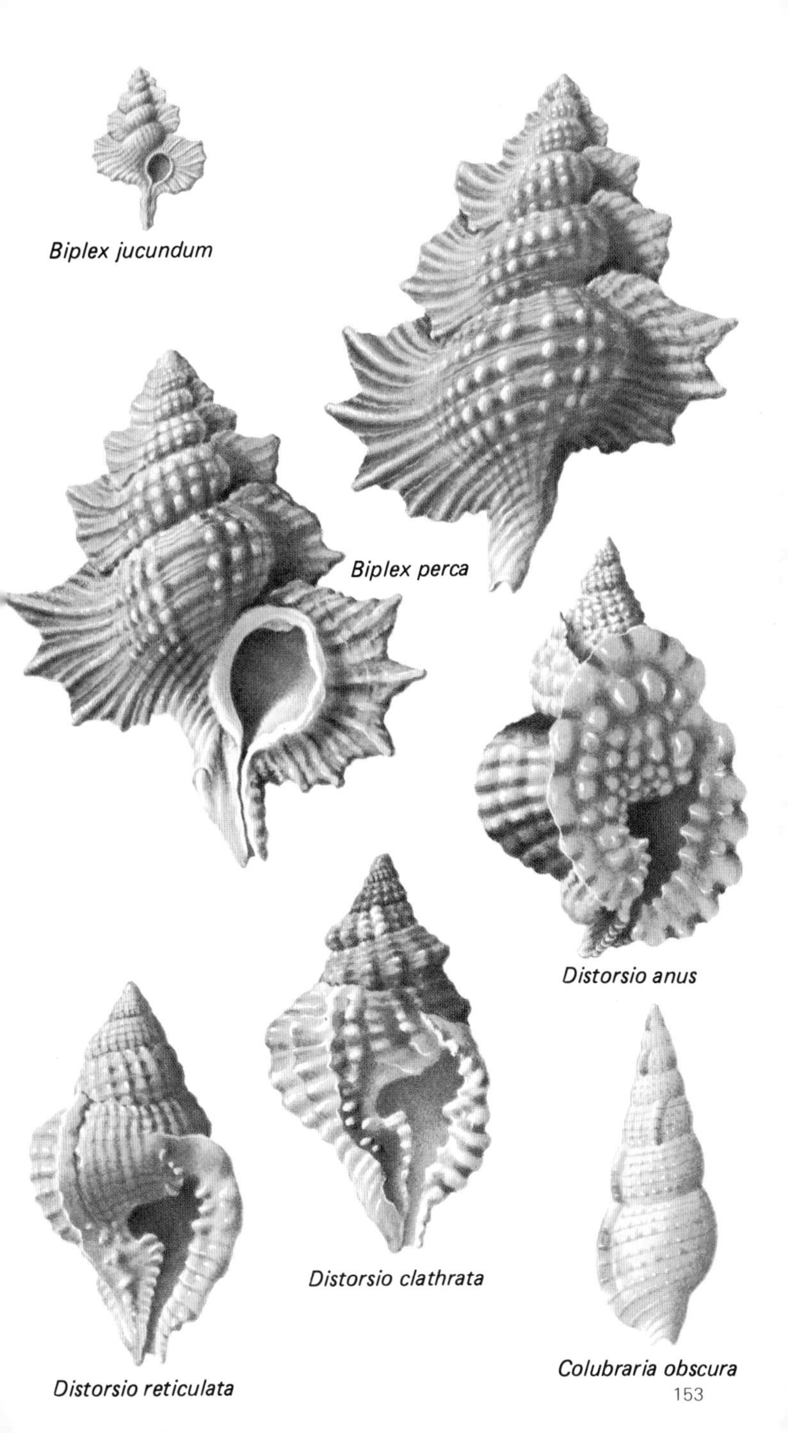
Biplex jucundum
Biplex perca
Distorsio anus
Distorsio clathrata
Distorsio reticulata
Colubraria obscura

Family: Bursidae

The frog shells are close relatives of the Cymatiidae. They live mainly in tropical areas among rock and coral in shallow water. Most are nodulose, have coarse heavy varices and strong anterior and posterior canals.

Bursa caelata **Broderip 1833** California to Peru, 50mm. Rectangular and flattened; moderate straight-sided spire; and five whorls. Fine, spiral, faintly beaded ridges; six larger rows of nodules, that on broadest part of the body whorl largest, and, like the next anteriorly, a double row. Lip fluted and strongly toothed. Columella and parietal area lirate and pustulate. Short, deep siphonal canal; deep, open, curved anterior canal; weak fasciole. Dark red-brown; nodules shiny; interior, columella white, pale chestnut clouding.

B. rosa **Perry 1811** Indo-west Pacific, 40mm. Squat, solid, moderate spire, two varices per whorl and fine spiral cords. Three, uneven spiral ridges, two posterior ones coalescing to form two or three blunt knobs between varices. Narrow, nodulose lip with nine, small, strong teeth. Columella with three, strong teeth anteriorly, lirate behind. Long, extended, almost tubular posterior canal; deep, curved anterior canal. Creamy white; red-brown broken lines on ridges; lip, narrow parietal area creamy yellow; teeth white; interior mauve.

B. bufonia **Gmelin 1791** Indo-west Pacific, 80mm. Solid, heavy, fairly high spire, two varices per whorl. Coarse, granular spiral ridges with three, coarse, blunt knobs on shoulders between varices. Expanded, fluted lip, with nine teeth on inner and outer edge. Coarsely lirate columella. Deep, extended, tubular siphonal canal; deep, sharply curved, almost closed anterior canal. Coarse, strong fasciole. Creamy white; brown markings; lip, parietal area cream; interior white.

B. granularis **Röding 1798** Indo-west Pacific, 60mm. Slightly flattened; high spire; suture a little constricted. Spiral rows of corded nodules, fine striae between. Lip has fourteen strong teeth. Lirate columella. Short, deep posterior canal; deep, curved, open anterior canal; weak fasciole. Shades of red-brown; white beading on varices; lip, columella creamy white; interior white.

B. bubo **L. 1758** Indo-Pacific, 260mm. Solid, heavy, coarsely sculptured with fairly high spire. Large, rough spiral cords and knobs at shoulder, five between varices. Slightly constricted suture and circular aperture. Flaring, toothed lip with internal denticle bordering posterior canal. Concave, plicate columella with tooth posteriorly. Short, deep, open posterior canal; short, deep, nearly closed, twisted anterior canal; wide, callous parietal shield. Creamy white; light brown flecks and splashes; aperture cream to straw.

B. rubeta **L. 1758** Indo-Pacific, 250mm. Similar to *B. bubo* except spire is slightly higher. Outer edge of lip denticulate, fourteen teeth on inner edge, coarsely lirate within. Columella profusely plicate, plicae stronger anteriorly and covering most of the expanded parietal shield. Creamy white with brown markings; aperture red or orange-red; teeth, plicae white or pale yellow.

B. foliata **Broderip 1825** Indian Ocean, 90mm. Solid, moderate spire, two flattened varices per whorl forming a flange on each side. Fine, nodulose spiral cords and three rows of axially flattened spines, posterior row largest; four spines between varices; rows end with short spines on varices. Oval aperture; expanded, flattened, fluted, denticulate lip; columella profusely plicate; short, deep, open posterior canal; short, recurved anterior canal. Cream or pale tan with darker brown shading; spines with dark brown and white rectangular spots; aperture white with orange-red on lip and columellar shield; interior grey-brown.

Bursa caelata
Bursa rosa
Bursa bufonia
Bursa granularis
Bursa bubo
Bursa rubeta
Bursa foliata

Order: Neogastropoda

Superfamily: Muricacea

Family: Muricidae

The murex live among rock and coral, mostly in shallow water, in the tropics. They are carnivorous, preying on other molluscs by boring a hole through their shells and eating the contents. They are usually highly sculptured with spines or fronds on early varices. A number of genera and many species.

Murex troscheli **Lischke 1873** north Australia through Philippines to Japan, 140mm. Moderate spire; deep suture; three varices per whorl. Narrow spiral cords with small threads between; about twelve threads on body whorl; small ridges becoming obsolete towards aperture. Lip is backed by solid varix on which the cords end in long, sharp spines, longer ones more or less at 90° to shell surface, smaller ones pointing more toward aperture. Very long, almost sealed siphonal canal bearing similar spines. Lip has blunt tooth anteriorly; smooth columella has small shield. Cream white; cords light brown; aperture white.

M. pecten **Solander 1786** the Venus Comb Murex, Indo-west Pacific, 125mm. Synonym *M. triremis* Perry 1811. Moderate spire; deep sutures; spiral cords, large and small alternately. Axial ribs and small lamellae, nodulose at intersections. Three varices per whorl with about sixteen long spines and a number of short ones at an angle to the others; spines on shoulder of body whorl and on spire point posteriorly. Long, straight, almost sealed siphonal canal; finely toothed lip with large tooth anteriorly, and backed by small varix; smooth columella. Creamy white, biggest spines a little darker; lip edge has tiny brown marks which show on earlier varices; aperture white.

M. mindanoensis **Sowerby 1814** Philippines, 63mm. High spire and strong spiral ridges, some with a weak cord between. Three varices per whorl and three strong axial ribs between varices: latter with short, sharp, hollow spines. Finely dentate lip; columella has three pleats anteriorly; long, recurved siphonal canal. Light tan-yellow to brown; canal darker; aperture white.

M. nigrispinosus **Reeve 1845** Indo-west Pacific, 90mm. Moderate spire; fine spiral striae and microscopic lamellae. About six, large spiral ridges on body whorl; four or five axial ribs between varices, nodulose at intersections; three varices per whorl having strong, rather straight spines, those on shoulder largest. Lip has strong varix, with three large and about six intervening small teeth; dentate inside lip, sometimes with larger internal tooth. Siphonal canal long. White; ends of spines dark grey; aperture white.

M. trapa **Röding 1798** central Indo-Pacific, 110mm. Moderate spire; strong spiral ridges and axial ribs on early whorls, ribs becoming obsolete on body whorl; uneven axial striae. Three varices per whorl with larger and shorter spines, some curving posteriorly. Dentate lip with not very strong varix and large, prominent, flattened tooth posteriorly. Long, straight siphonal canal. Blue-grey, tinged brown on early whorls; aperture white; dark grey spots at base of interior of large spines; interior brown with blue-grey spiral streaks.

M. tribulus **L. 1758** Indo-Pacific, 120mm. Moderately high spire; spiral cords and axial ribs, nodulose at intersections. Three varices per whorl; varices spinous, those on shoulders longest. Lip with weak varix; dentate inside lip has large tooth anteriorly. Long, almost straight siphonal canal. Light tan; aperture white; lip edge with tiny brown marks which show less on earlier whorls.

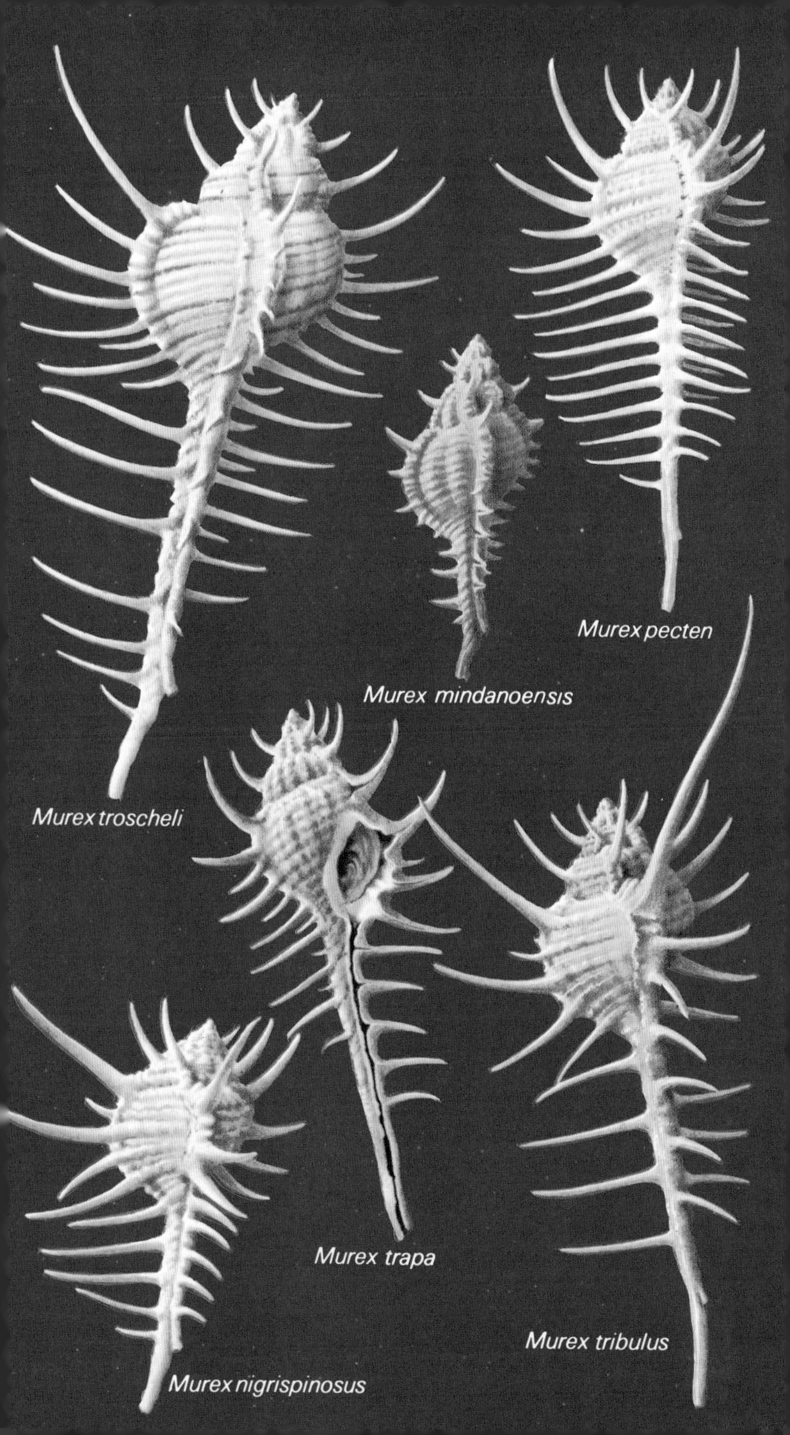
Murex troscheli
Murex mindanoensis
Murex pecten
Murex trapa
Murex tribulus
Murex nigrispinosus

Haustellum haustellum L. 1758 Indo-Pacific, 110mm. Solid body; low spire; long, slender and almost closed siphonal canal. Axially ribbed; three varices per whorl, three axial ribs between varices, or four on body whorl. Blunt spines where ribs cross sharply angled shoulders. Spiral, raised threads which also cover most of the canal, and three or four, spiral rows of nodules on ribs on the anterior half of the body whorl. Wide aperture; finely dentate lip, lirate within; concave columella, rugose anteriorly. Cream with dark brown blotches; threads red-brown; aperture pink, interior white.

H. tweedianum Macpherson 1962 east Australia, 65mm. Similar to the above but smaller with a relatively shorter siphonal canal which has short spines. The spiral threads are wider apart and finely beaded. White with brown clouding and touches of mustard yellow on the varices and threads.

Bolinus cornutus L. 1758 west Africa, 150mm. Solid, strong body with long, almost closed siphonal canal; short spire; angular shoulders; slightly constricted suture; fine, spiral ridges and axial striae. About seven varices per whorl which are not very raised except between shoulder and suture; the posterior ends of earlier lips show as low, wavy ridges on the spire. Each varix carries a strong, more or less hollow spine on the shoulder, a second on the centre of the body whorl, and two or three short ones on the siphonal canal. Wide aperture; coarsely dentate lip, ridged internally; concave, smooth columella. Strongly developed parietal shield. Cream to white with brown, generally spiral clouding; parietal area, outer edge of shield and between teeth on lip tan.

B. brandaris L. 1758 Mediterranean, Portugal and west Africa, 90mm. Similar to, though smaller than, the above. Spines much shorter and less pointed; base sculpturing relatively coarser. Tan; aperture slightly darker tan. It was from this animal that the Romans produced the Royal Purple dye.

Murex bellegladensis Vokes 1963 Florida and the Caribbean, 70mm. Moderate spire and extended siphonal canal; spirally ridged. Three varices per whorl each with three or four, long, sharp spines; one to three, axial ribs between varices. Slender and delicate siphonal canal with three rows of spines. Off-white, sometimes with a pink tinge between varices.

M. kiiensis Kira 1962 Japan, 50mm Moderately high spire; constricted suture; straight, slender siphonal canal. Three varices per whorl bearing short, strong spines; two or three, axial ribs between varices. Fine, sharply raised, spiral threads overall. Very finely dentate lip; almost closed siphonal canal with a very few, small spines posteriorly. White with a very pale, rather indistinct, brown band on the body whorl and on the canal; raised threads are mostly red-brown on the last three whorls; aperture white.

Siratus pliciferoides Kuroda 1942 Japan, 110mm. Moderately high spire; constricted suture. Three small varices per whorl bearing short, strong spines, the largest on the shoulder. Finely spirally striate; two axially elongate knobs on the shoulder between each pair of varices. Dentate lip; smooth columella; almost closed, spinous and somewhat recurved siphonal canal is extended, but not comparably with the other species on this page. White with some light brown shading; aperture white.

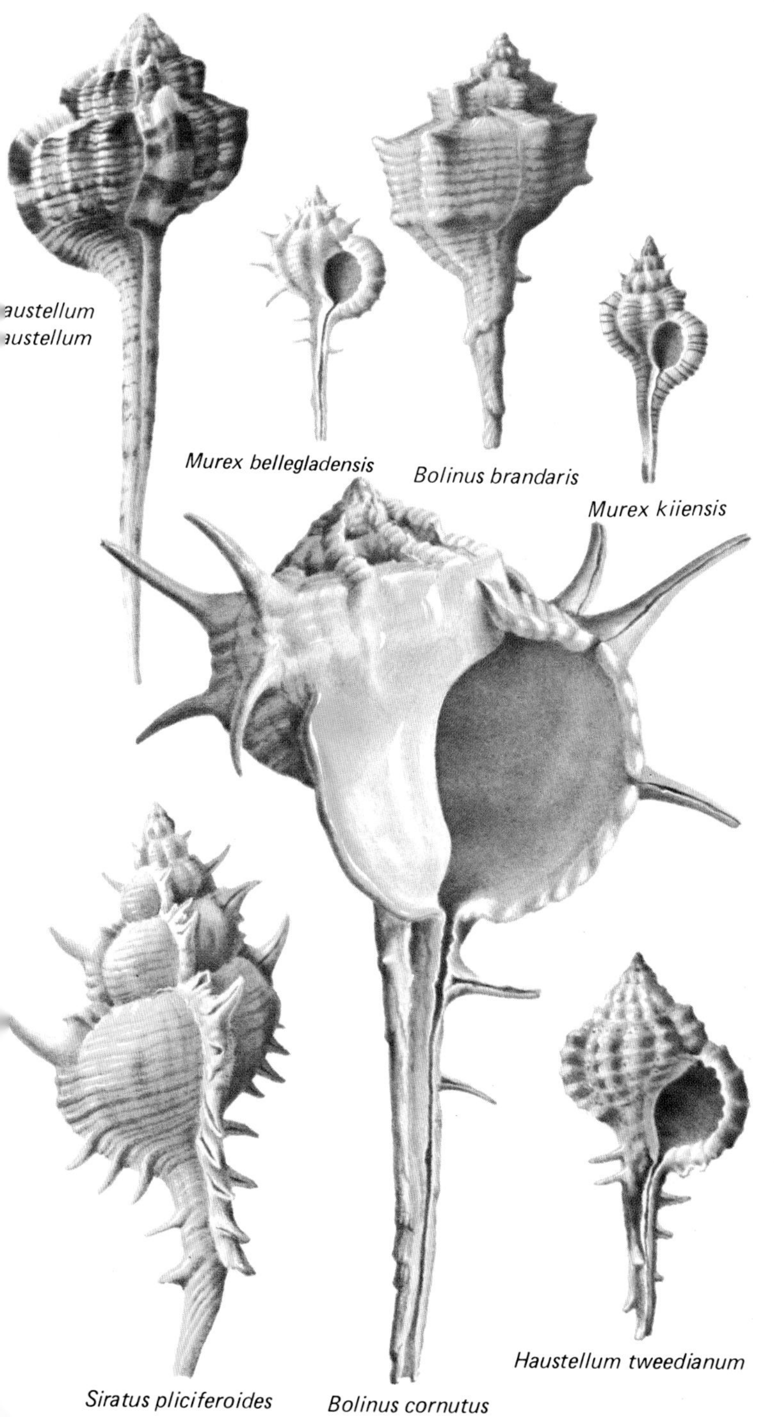

austellum
austellum
Murex bellegladensis
Bolinus brandaris
Murex kiiensis
Haustellum tweedianum
Siratus pliciferoides
Bolinus cornutus

Chicoreus brevifrons **Lamarck 1822** Caribbean and south Florida, 150mm. Moderate spire; three varices per whorl; fine spiral cords; two nodulose axial ribs between varices. Solid varices covered with fairly long, foliate spines from largest on shoulder to end of moderate, recurved siphonal canal. Dentate lip; columella has spiral ridge posteriorly. Shades of light brown.

C. ramosus **L. 1758** Indo-Pacific, 300mm. Largest Indo-Pacific murex. Rather short spire; angular and inflated whorls with three varices per whorl; fine spiral striae. One or two small knobbed axial ribs between varices, latter with recurved, frond-like spines, ten from shoulder to end of recurved siphonal canal, which is quite long for the genus. Dentate lip with large tooth anteriorly. White; aperture tinged pink.

C. cornucervi **Röding 1798** north-west Australia, 110mm. Synonym *M. monodon* Sowerby 1825. Reasonably high spire; roundly angled shoulders; deep sutures; fine, spiral ribs; and finer, uneven, axial riblets. Three varices per whorl; varices very weak with frond-like recurving spines, three of those on lip narrower and longer; one or two obsolete ribs between varices. Lip has spines growing directly out of edge and one large tooth anteriorly. Smooth columella has small, blunt, posterior tooth; fairly long, somewhat recurved siphonal canal. Usually brown; dark brown spines; uncommonly white as illustrated; aperture white; columellar edge pink.

C. palmarosae **Lamarck 1822** Indo-Pacific, 100mm. Rather elongate; high spire. Three varices per whorl and strong, spiral, beaded cords. Strong varices have frond-like spines down to base of quite long siphonal canal, those on shoulders much bigger and more fronded; two axial ribs between varices. Lip has ten, small denticles, backed by channel and row of blunt, rounded teeth. Smooth columella has short, oblique plicae on outer edge. Deep, almost closed, recurved siphonal canal. Pale brown; darker cords; ends of fronds white, sometimes pink; aperture and columellar plicae white.

C. axicornis **Lamarck 1822** Indo-Pacific, 70mm. Moderate spire; very large solid varices, three per whorl. Fine, spiral, minutely beaded threads and obsolete axial riblets. Two coarse ribs between varices; latter with one, long, hollow, blunt spine, very slightly foliate, on shoulder, one shorter lower down, remainder smaller to long, slightly recurved siphonal canal. Finely dentate lip with large tooth posteriorly, inside edge lirate; smooth columella. Cream, clouded or faintly banded with light brown; aperture white.

C. saulii **Sowerby 1841** Indo-Pacific, 125mm. Solid and rather elongate with high spire and fine, minutely beaded spiral cords. Three varices per whorl, having eight, moderately long, foliated spines, smaller ones between each pointing slightly forward; a large and small axial ridge between varices. Dentate lip; smooth columella, spiral ridge posteriorly; deep, long, recurved siphonal canal. Pale yellow-brown; cords usually stained dark brown; interior white.

C. clausii **Dunker 1879** west Africa, Brazil, 55mm. Squat, angulate; fine, beaded, spiral cords and ridges. Three varices per whorl; one, large, coarse, slightly lop-sided, axial ridge between varices. Lip and varices have hollow, non-foliate spines, those on shoulder long and recurved, two others of moderate length and two very small. Dentate lip; columella has two spiral ridges posteriorly; deep siphonal canal of moderate length and rather open. Cream; brown clouding; dark brown apex; white band near base, taking in second largest spines.

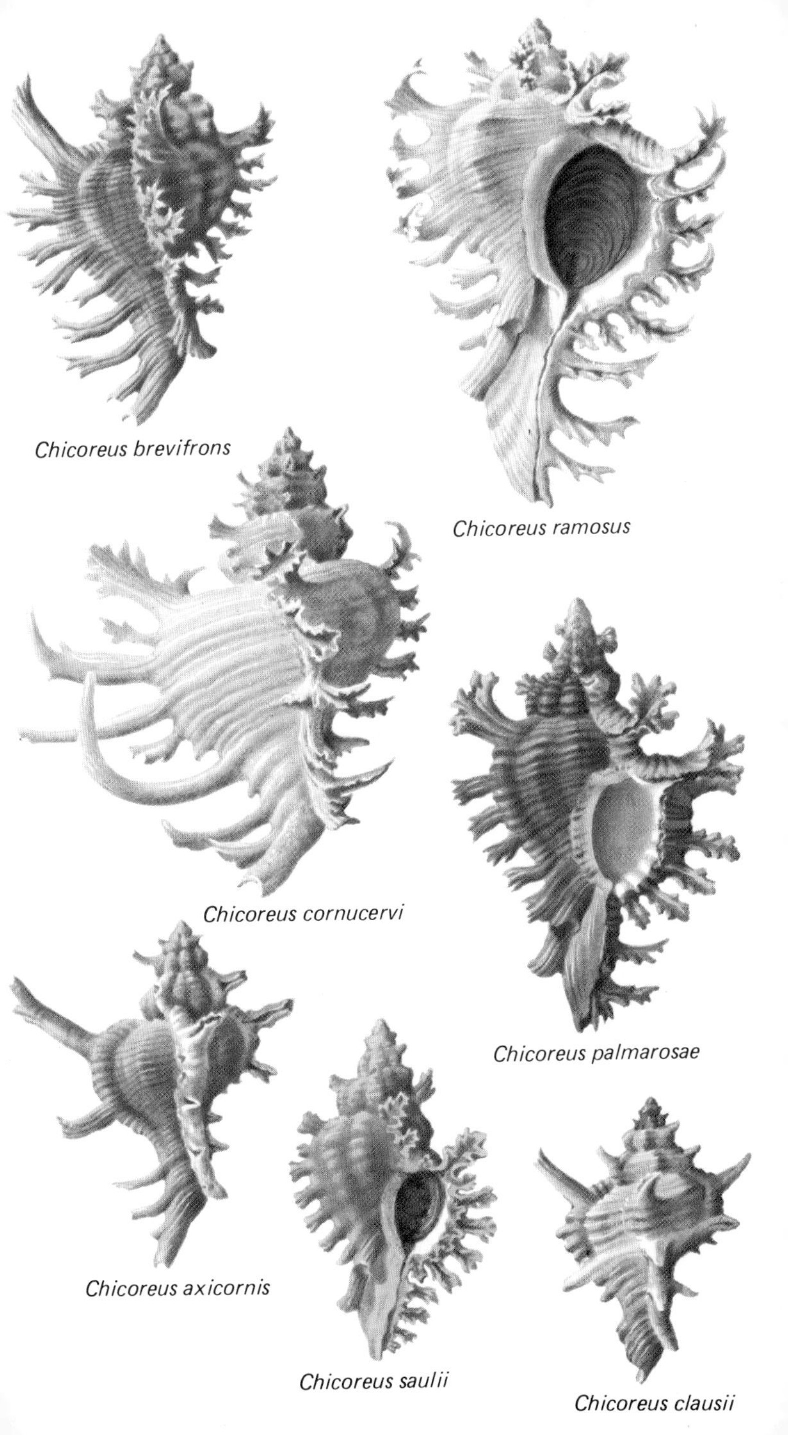
Chicoreus brevifrons
Chicoreus ramosus
Chicoreus cornucervi
Chicoreus palmarosae
Chicoreus axicornis
Chicoreus saulii
Chicoreus clausii

C. rubiginosus Reeve 1845 Australia, 90mm. High spire. Fine, spiral, granulose cords; three varices per whorl; fairly long, straight, frondose spines with smaller ones between. Edge of lip dentate. Columella has one spiral ridge posteriorly and faint plications; deep, recurved, open siphonal canal. Rusty brown; darker spiral ribs and fronds; sometimes cream, or orange-red; aperture white; edge of columella pink. This may be a variety of *C. torrefactus* Sowerby 1841 (see below), but the latter usually has shorter, less frondose spines and the shell is more elongate.

C. territus Reeve 1845 Queensland, Australia, 70mm. Sculptured with finely beaded, spiral ridges, each with a beaded thread on either side, and pustulate channel between; ridges end in spines on varices, three to each whorl; one nodulose rib between varices and sometimes a much smaller one as well. Spines are open and joined together to form a frilly curtain from suture to top of quite long, almost straight siphonal canal. Dentate lip; columella with posterior tooth. Normally white, cream, grey or brown, with white aperture. However, I have a specimen the colour of cream toffee, inside and out, illustrated here. A more typical specimen is illustrated on page 175.

C. torrefactus Sowerby 1841 Indo-Pacific, 100mm. High spire; many, close-set, granulose, spiral cords and small ridges ending in fronds on varices; three varices per whorl. Very small, axial ridges one large, one smaller, and sometimes a third and/or a fourth. Solid varices; very short, frondose spines. Dentate lip. Columella with one posterior ridge and a notch anteriorly. Deep, recurved and almost closed siphonal canal. Brown-grey; spiral ribs, spines darker; interior white; columella pink or orange, darker at the edge.

C. damicornis Hedley 1903 south-east Australia, 60mm. Thin with high spire; constricted at suture. Rather shouldered with fine, spiral, granulose threads and cords. Three varices per whorl; one to three axial ridges between each; strong varices with long, hollow spine bifurcated at tip, on the shoulder, and a number of shorter webbed spines below, some very small. Finely dentate lip; smooth columella; short, well-developed anal canal; short, recurved posterior canal. Cream, occasionally stained with brown.

C. capucinus Röding 1798 south-west Pacific, Samoa to Malay Peninsula, 65mm. High spire and six whorls with three varices per whorl. Strong, spiral ridges and may have two, weak, flat, broad, axial ribs between varices; varices have no spines. Bluntly dentate lip; columella with one small posterior denticle; moderate siphonal canal. Dark brown; aperture grey-brown; interior white.

C. brunneus Link 1807 Indo-Pacific, 75mm. Solid and heavy with many, close-set, finely beaded, spiral cords. Three varices per whorl. Very thickly foliate, foliations wide but not long. One, large, blunt knob on shoulder between varices. Finely dentate lip; columella with blunt tooth posteriorly; short, closed siphonal canal. Dark grey-brown or brown-black; aperture white, edged with orange. *C. brunneus* is, however, variable and a form often found in Singapore is much less heavy and coarse. The foliations are fewer, less frondose and usually longer, and the knob between the varices is smaller. The aperture and columella are dark purple-blue. Illustrated bottom left.

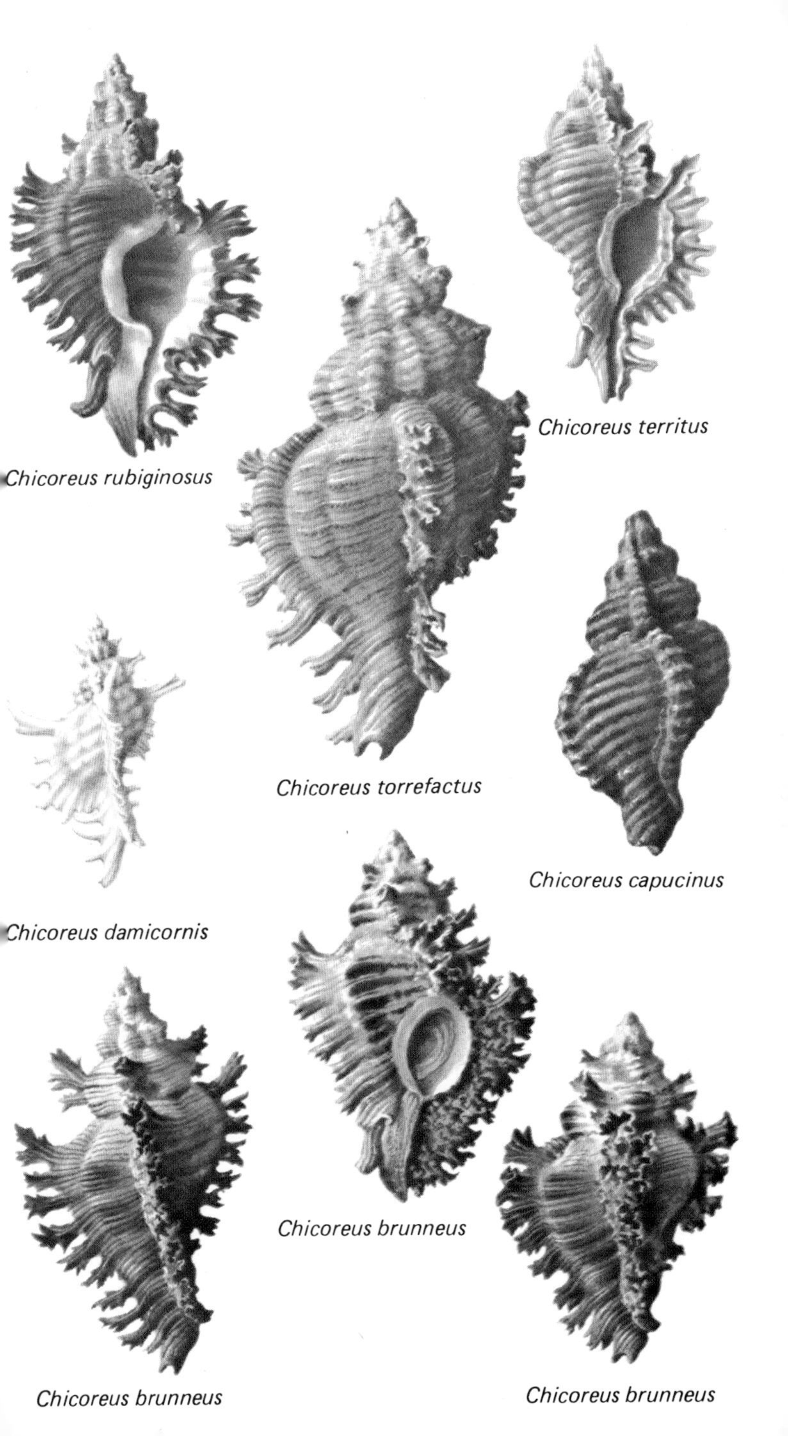

Chicoreus rubiginosus

Chicoreus territus

Chicoreus torrefactus

Chicoreus capucinus

Chicoreus damicornis

Chicoreus brunneus

Chicoreus brunneus

Chicoreus brunneus

Chicoreus laciniatus Sowerby 1841 Philippines, 60mm. Rather elongate, though with moderately short spire; constricted suture. Three varices per whorl only slightly frondose; two, axial ribs between varices; spiral, lamellate ridges and threads. Finely crenulate lip; smooth columella; broad, open siphonal canal, recurved at tip. White; mauve tinge and very pale brown clouding; darker brown on the short fronds; columella, inside of canal mauve; rest of aperture white.

Hexaplex stainforthi Reeve 1843 north-west Australia, 65mm. Solid, chubby, with moderate spire and slightly constricted suture. Eight varices per whorl with short, frondose spines; spiral ridges and fine threads. Dentate lip; smooth columella; broad, short siphonal canal, almost completely closed and recurved at the tip. White with a yellow, pink or orange flush between dark brown varices; aperture may also be yellow, pink or orange.

H. cichoreum Gmelin 1791 Indo-Pacific, 75mm. Moderate spire; slightly constricted suture. Six, weak varices per whorl with separate, open, more or less foliate spines. Spiral ridges on which spines develop; fine, intermediate threads. Crenulate lip; smooth columella; broad, narrowly open siphonal canal. Narrow, deep posterior canal near end of lip. Deeply umbilicate. White with dark brown banding so that spines alternate brown and white; those on shoulder, the largest, brown as those on canal; aperture white. An all white form occurs.

H. rosarium Röding 1798 west Africa, 200mm. Moderately high spire; suture becomes increasingly constricted anteriorly. Six to seven varices per whorl, bearing short, strong, slightly foliate spines. Spiral ridges and threads. Dentate lip; smooth columella; broad, open, recurved siphonal canal. Deep umbilicus. Light brown with darker brown on most spines; lip white with some rose pink marks; columella rose pink with some white; interior white.

H. regius Swainson 1821 tropical west America, 150mm. Globose with rather short spire and slightly constricted suture. Eight varices per whorl with strong, open spines, those on the shoulder the largest and all but those in one other row are double, one behind the other. Surface rough with fine, spiral threads. Crenulate lip; smooth columella; broad, strong, open and recurved siphonal canal; well-developed anterior canal. May be deeply umbilical. White, some pale brown areas and a band of brown marks, mostly on the varices, on the body whorl and another just posterior to the siphonal canal; aperture pink with a broad, dark brown swathe on the outer edge of the parietal area which shows above the suture on earlier whorls; lip with a few brown marks; aperture side of siphonal canal is dark brown with a pale blue-white overlay.

Phyllonotus pomum Gmelin 1791 south-east USA and Caribbean, 110mm. Solid with moderate spire and slightly constricted suture. Three varices per whorl with blunt knobs where crossed by obsoletely beaded, low, spiral ridges and fine threads; one, short, axial rib between varices, highest at periphery. Crenulate lip; broad, rough-surfaced, open, slightly recurved siphonal canal. Dark or light brown; some off-white and darker brown marks on the varices; shiny aperture from white to orange or yellow; dark brown marks on lip, and one at end of parietal area where it is joined by lip; may be pale brown marks on columella and anterior end of parietal area; interior paler.

Hexaplex stainforthi

Hexaplex cichoreum

Chicoreus laciniatus

Hexaplex regius

Hexaplex rosarium

Phyllonotus pomum

Hexaplex cichoreum

Hexaplex brassica Lamarck 1822 west Central America, Mexico to Peru, 200mm. Low spire and seven angular whorls with about six varices per whorl. Fine, uneven spiral threads. Each varix has a strong, blunt spine on the shoulder – showing on early whorls just above the suture – two or three small spines, and then, on the base of the body whorl and slightly recurved siphonal canal, about six, hollow, sharp spines of varying lengths. Edge of lip and edges of varices with some fifteen small points. Columella with posterior tooth, otherwise smooth. Siphonal canal and end of each varix builds up like a fan. White with three brown bands and other brown shaded areas; aperture white with salmon pink only on the outer lip and parietal wall.

H. kusterianus Tapparone-Canefri 1875 Persian Gulf, 70mm. Squat and heavy with about five whorls and six varices per whorl. Spirally sculptured with fine threads over unevenly-sized, rounded ridges. Coarse, hollow spines on the shoulder of the varices and nodules below, and one row of larger spines near the anterior end. Indented lip; smooth columella has one posterior tooth. Putty-coloured with pink on the columellar edge.

H. erythrostoma Swainson 1831 Gulf of California to Peru, 100mm. Moderate spire and about eight whorls. Fine spiral threads and finely lamellated growth lines. About six varices per whorl with some eight, hollow spines on ridge of earlier, frilled lip; a short ridge with three spines (not hollow) in the middle of the body whorl between each varix. Columella with posterior tooth and expanded, plicate parietal shield. Long, deep, recurved siphonal canal is slightly open. White, sometimes with pink tinges; aperture bright shiny pink.

Muricanthus nigritus Philippi 1845 Gulf of California, 150mm. Moderate spire; about seven whorls; about eight varices per whorl; spirally finely threaded. Varices with hollow spines, largest on the angular shoulder, one smaller and two very small above it, and about twelve of varying size – three quite big – below. Indented lip; siphonal canal is open and built up fan-like with each new growth period; smooth columella. White; spines and end of siphonal canal very dark brown, almost black; spiral brown streaks behind the spines; aperture white.

M. radix Gmelin 1791 Panama to south Ecuador, 100mm. Solid and pyriform, like a small edition of its northern neighbour *M. nigritus*, but relatively more solid and more spinous and the spines more gemmate. Columella has posterior tooth. White; spines, areas behind them and siphonal canal very dark brown to black; aperture white.

M. callidinus Berry 1958 east Central America from Guatemala to Costa Rica, up to 100mm. About nine varices per whorl, all with long thin frond-like spines, longest on the shoulder. Columella has posterior tooth. White; dark brown spines and siphonal canal; brown spiral stripes of varying width; aperture white. Some authors consider this to be a variant of *M. radix* Gmelin 1791.

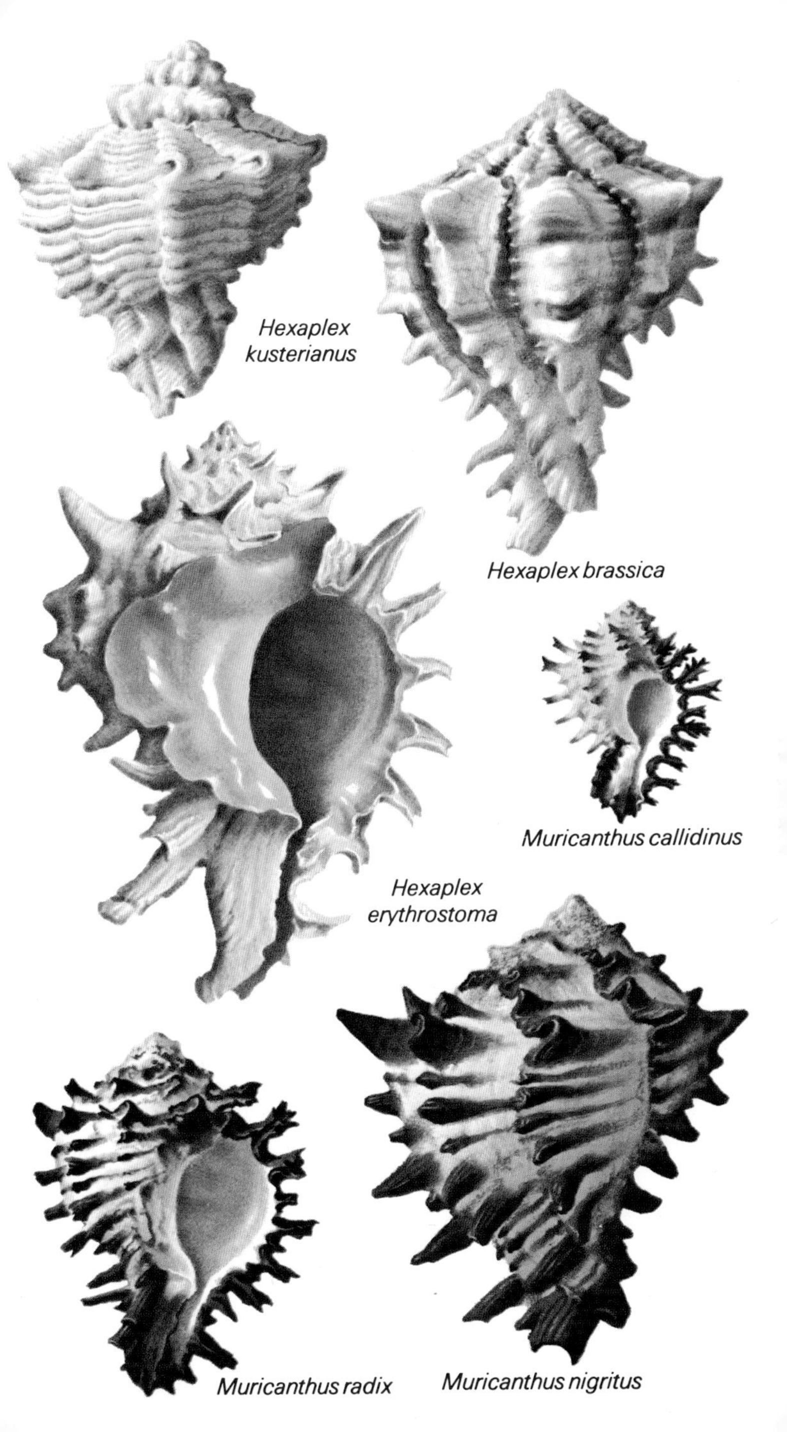
Hexaplex kusterianus
Hexaplex brassica
Muricanthus callidinus
Hexaplex erythrostoma
Muricanthus radix
Muricanthus nigritus

Ceratostoma nuttalli Conrad 1857 California, 55mm. Solid with moderate to short spire. Three varices per whorl and a nodulose axial rib between; spiral ridges and cords; varices solid but with sharp edges. Edge of lip finely denticulate with one, long, pointed tooth near the anterior end; inside lip has about five, blunt teeth. Columella smooth with one blunt tooth posteriorly. Deep, sealed and short siphonal canal. Cream, brown, or banded cream and brown; aperture white.

C. foliatus Gmelin 1791 Alaska to California, 80mm. High spire and about seven whorls constricted at the suture. Strong, well-separated spiral ridges, two large ones on the shoulder. One larger and some lesser axial ribs between varices which are broadly frilled, the frills lamellated on the aperture side. Outer lip with rough edge and one, large, pointed tooth near the posterior end. Smooth columella and short siphonal canal, sealed and sharply turned to the right at the tip. White with light or dark brown banding; the two larger ridges on the shoulder and the area between white; aperture white.

Pteropurpura trialatus Sowerby 1834 California, 80mm. High spire and seven whorls, three varices per whorl. Fine spiral cords and about six, low, blunt ridges on the body whorl and a blunt knob on the shoulder between varices; the latter frilled and foliate, and the frond on the shoulder pointed up, out and slightly backward. Dentate lip and smooth columella. Siphonal canal is long, deep, sealed and slightly curved. Flesh-coloured with dark brown clouding especially between the ridges; aperture white.

P. erinaceoides Valenciennes 1832 south and Lower California and north-west Mexico, 50mm. Moderate spire; about six whorls, three varices per whorl and a large, knobbed, axial ridge between each whorl. About six spiral ridges on body whorl, ending in sharply recurved spines on the varices. Fine spiral cords and very fine lamellae overall. Finely dentate outer lip edge, smooth columella and moderate siphonal canal, sealed and slightly recurved. Shades of red-brown, darker on the knobs and spines; aperture white.

Pterynotus vespertilio Kira 1955 south Japan, 45mm. Delicate and elongate with about six whorls, three varices per whorl. Spirally corded and varices unevenly frilled. Light brown with darker brown spots.

P. bednalli Brazier 1878 north-west Australia, 85mm. One of the most beautiful of the murex, it is delicate and elongate. High spire; about seven whorls; somewhat constricted suture; three varices per whorl and spiral, well-separated ridges. Varices have large, thin, rather smooth-edged and slightly fluted projections, flaring at the shoulders but running together with earlier varices. Outer lip is faintly ridged; smooth columella; siphonal canal moderate, deep, slightly open and recurved at the tip. Glossy cream with pink tinge; pale brown spots and lines on the varices; aperture white.

P. tripterus Born 1778 Indo-west Pacific, 60mm. Moderate spire with about seven whorls, three varices per whorl. Granulose with spiral cords. Varices have fine frilly projection, lamellate on the aperture side; one knob on the shoulder between varices. Outer lip denticulate and about seven teeth within. Columella has seven to ten teeth. Siphonal canal is deep, narrow and recurved at the tip. White or cream; aperture white with a touch of pale yellow-green on the outer edge of the columella.

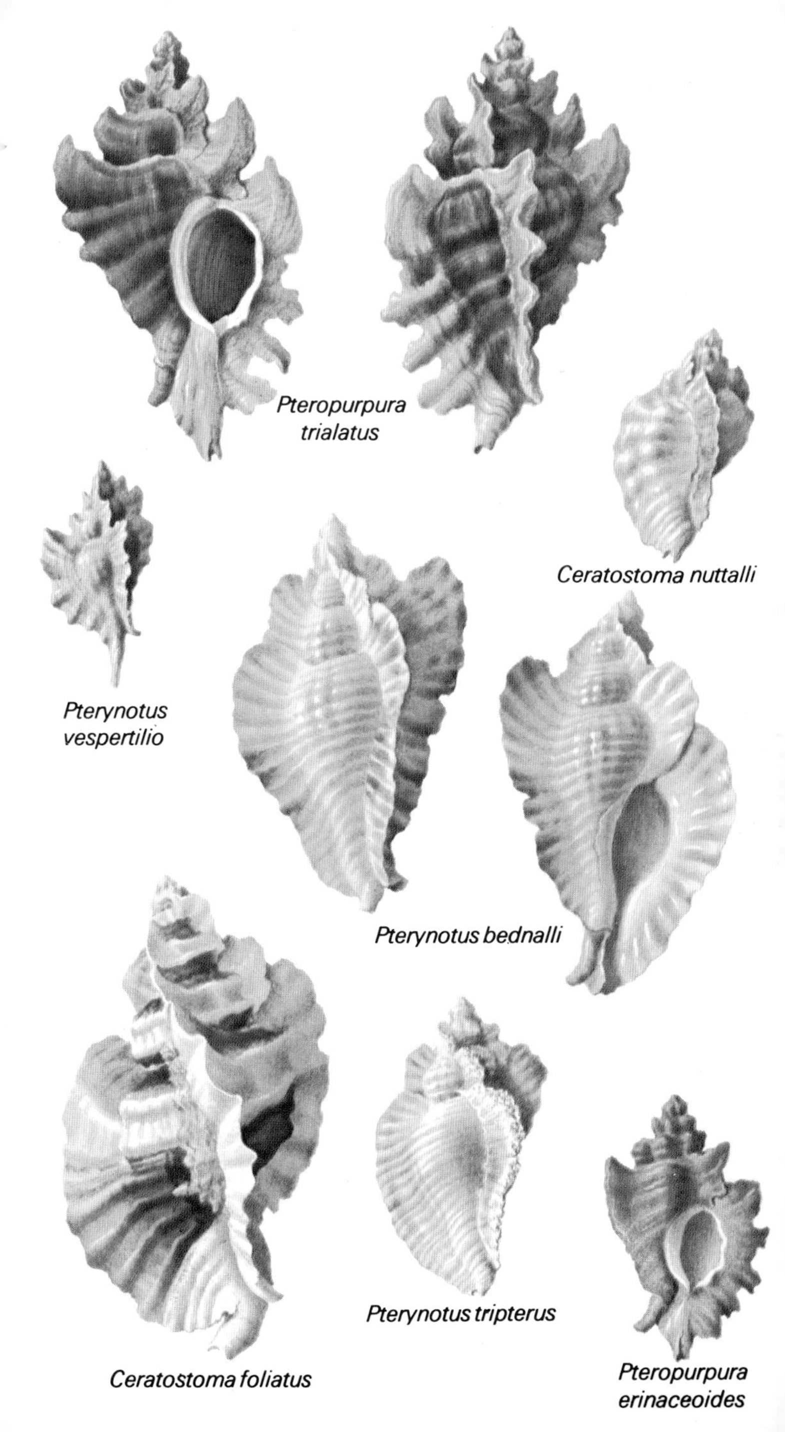
Pteropurpura trialatus
Ceratostoma nuttalli
Pterynotus vespertilio
Pterynotus bednalli
Ceratostoma foliatus
Pterynotus tripterus
Pteropurpura erinaceoides

Siratus motacilla Gmelin 1791 Lesser Antilles, but Reeve records it as being Senegal in west Africa, 60mm. Solid except for the delicate canal. Moderate spire; seven whorls with three varices per whorl; spirally ridged and roughly axially striate; two nodulose ribs between varices have a short, sharp spine on the shoulder of each and another where they join the siphonal canal. Bluntly dentate lip is plicate within and has about three small teeth posteriorly. Smooth columella; plicate edge to narrow parietal shield. Long siphonal canal is straight but angled upward from its base and nearly closed. Cream-white with pale pink-brown bands and blotches.

S. perelegans Vokes 1965 Caribbean and Florida, 60mm. Very similar to its neighbour *S. motacilla*. Solid with long, straight, upturned siphonal canal, but the spines on the shoulder are lacking or minute, and the ridges and ribs are finer; spire slightly lower.

Pterynotus pinnatus Swainson 1822 east Asia, 70mm. Elongate with high spire. About eight whorls and three varices per whorl. Fine, spiral ridges; one rib between varices which have a thin, delicate, continuous, fan-like projection from suture to end of long, curved, rather open siphonal canal. Lip finely dentate on the edge and toothed within; smooth columella. White; rather translucent.

P. elongatus Lightfoot 1786 Indo-Pacific, 100mm. Synonym *M. clavus* Kiener 1842. Graceful and a collectors' item. Elongate with about seven whorls; fine, spiral, minutely beaded cords. Three varices per whorl, with a blunt low ridge between each and with frilly, fan-like projections, slightly pointed at the shoulder and running down to the end of the long, almost sealed siphonal canal; the latter is recurved at the tip. Finely dentate lip. White or creamy white; parietal wall with faint pink tinge.

P. bipinnatus Reeve 1845 Indo-Pacific, 45mm. Elongate and narrow with rather tower-like spire. About seven whorls; spirally and axially finely ridged; the upper whorls have blunt knobs, about seven per whorl. Body whorl has three varices, those two nearest the aperture carrying delicate wavy fins. Lip edge with tiny, sharp teeth, and plicated within, as is the lower half of the columella. Siphonal canal almost as long as the spire and slightly open. White; aperture pale rose.

Ceratostoma fournieri Crosse 1861 Japan, 50mm. Rather low spire. About six whorls with three varices each. Malleate surface with low, blunt, spiral ridges. One large knob between varices which carry a wavy frilly fin. Uneven lip with a projecting tooth near the anterior end. Moderately long siphonal canal is deep, curved and sealed. White, profusely clouded with dark tan.

Pteropurpura plorator A. Adams and Reeve 1849 south Japan and Korea, 40mm. Moderate spire. About six whorls with three varices per whorl, a blunt knob on the shoulder between each. Faint spiral ridges and axial striae. Varices with wavy, winged projections, pointing up and outward at the shoulders. Bluntly dentate lip; smooth columella; sealed siphonal canal. White with spiral tan lines and rows of wavy axial flames; aperture white.

Eupleura muriciformis Broderip 1833 Gulf of California to Ecuador, 40mm. About six angular whorls with a varix about every three-quarters of a whorl. Spirally ridged, with about four knobs on the shoulder between varices, which are more or less strongly developed. Denticulate lip; smooth columella; extended, recurved, open siphonal canal. White to dark grey-brown; varices with large blotch of dark brown; interior purple-brown.

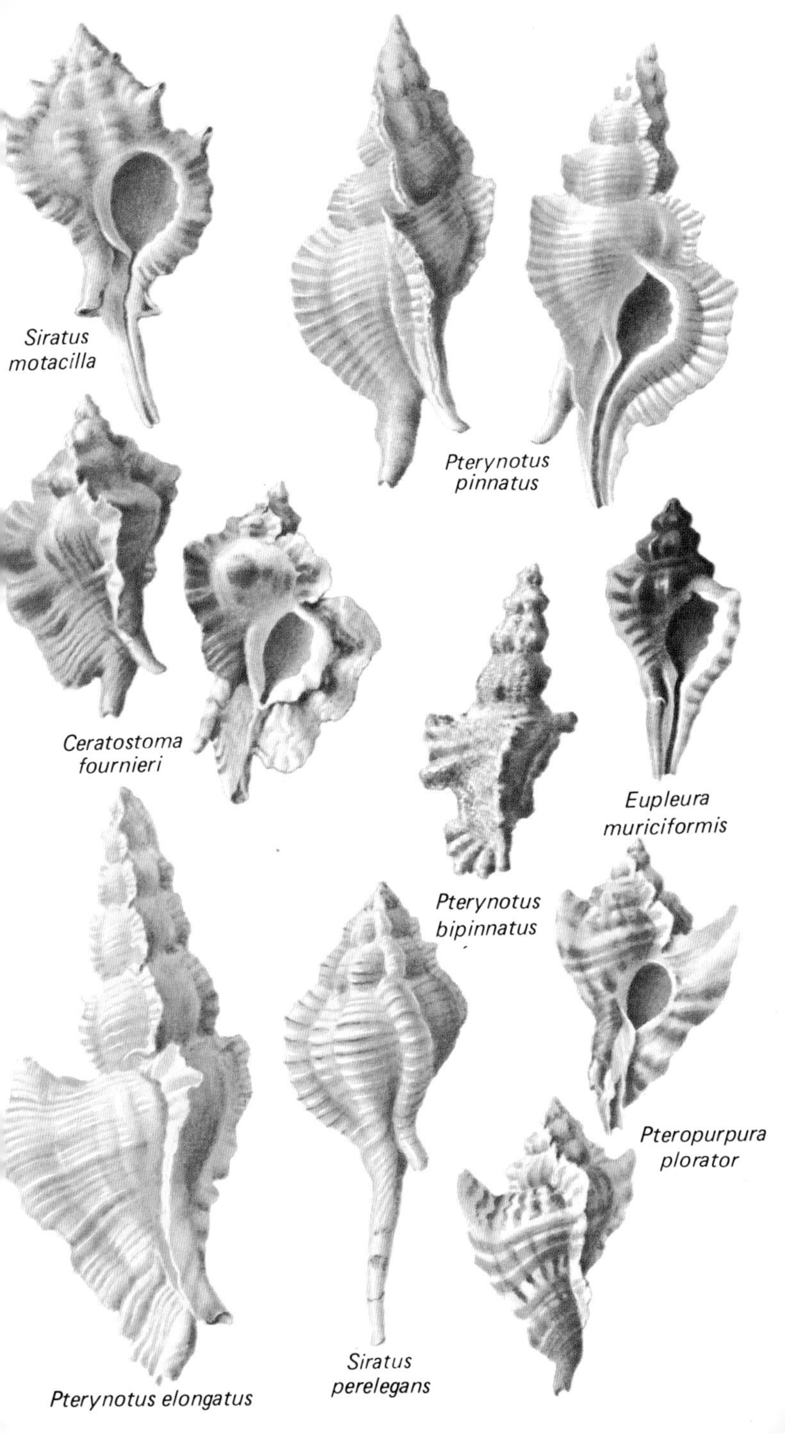
Siratus motacilla
Pterynotus pinnatus
Ceratostoma fournieri
Eupleura muriciformis
Pterynotus bipinnatus
Pteropurpura plorator
Siratus perelegans
Pterynotus elongatus

Homalocantha scorpio L. 1758 Philippines and east Indonesia, 60mm. Low spire and about four whorls with about seven varices per whorl. Very deep, wide sutures crossed by the varices; spirally corded; angular shoulders. Early varices with short, hollow spines which increase in length up to the body whorl; the last two varices have long, rather triangular, hollow projections, the apex being at the varix and the flat 'base' being furthest from the body whorl; they are joined at the lip edge by a small, continuous frill. Finely dentate lip; smooth columella; long, straight siphonal canal. Dark brown to black; light grey or white in places; aperture with purple tinge.

H. zamboi Burch 1960 Philippines, 55mm. Short spire and about five whorls. Deep, wide sutures crossed by varices of which there are about five per whorl. Malleate surface. Last four varices have four, hollow projections, widening towards the tip where they flatten out and may become slightly palmate; at their base they have short blunt 'fingers' pointing towards the finely dentate lip. Smooth inside lip and columella. Long, almost sealed siphonal canal is recurved at the tip. White; aperture pink to pale brown.

Trunculariopsis trunculus L. 1758 Mediterranean and adjacent west Atlantic, 100mm. Moderate to high spire and about seven, angular whorls. Very close, finely lamellate, spiral cords. About five, low, spiral ridges, the one on the shoulder with six to twelve, more or less pointed spines. About six varices per whorl on which are the longest of the spines. Between varices from none to four, low, axial ribs on which there are low, blunt knobs where they cross the ridges. Varices little more than the edge of earlier lips, which are faintly denticulate. Smooth columella with tooth at posterior end. Deep, heavy, open and curved siphonal canal. White with three broad bands of brown or purple-brown; columella white or stained with purple; banding showing through the outer lip into the interior. A very variable shell in sculpture and colouring, but always easily distinguished. Two varieties are illustrated.

Naquetia triqueter Born 1778 Philippines, 55mm. Solid but elongate and six whorls with three varices per whorl. Spirally ridged; three, axial ribs between varices; latter small, those on body whorl with frilly, fan-like projection, narrow posteriorly, getting wider anteriorly and on the straight, almost completely sealed siphonal canal. Lip dentate within; smooth columella. Cream; banded and spotted with brown; aperture white, but with the brown bands showing through.

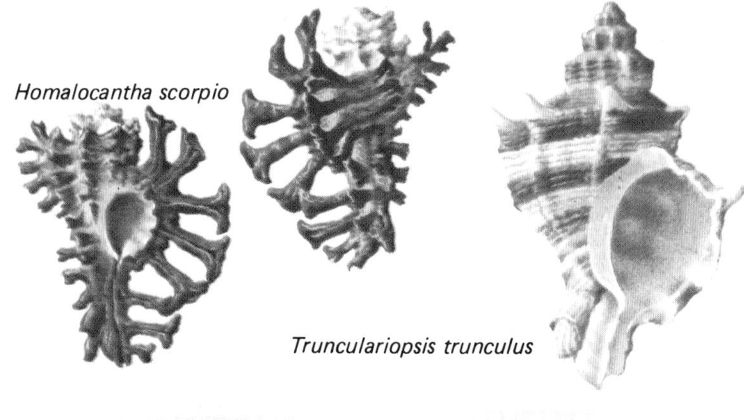
Homalocantha scorpio

Truncularíopsis trunculus

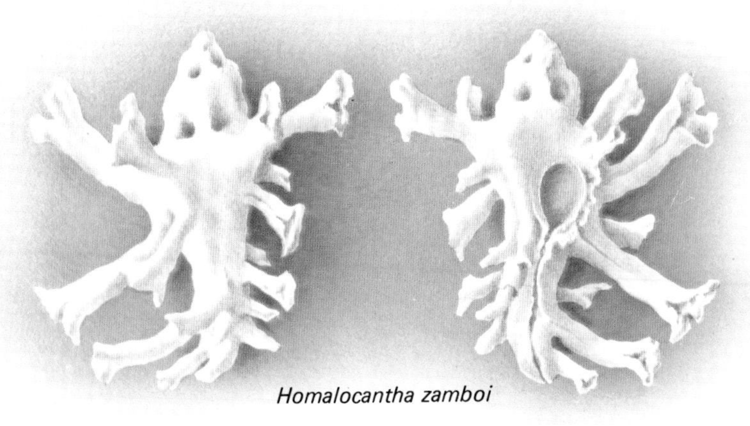
Homalocantha zamboi

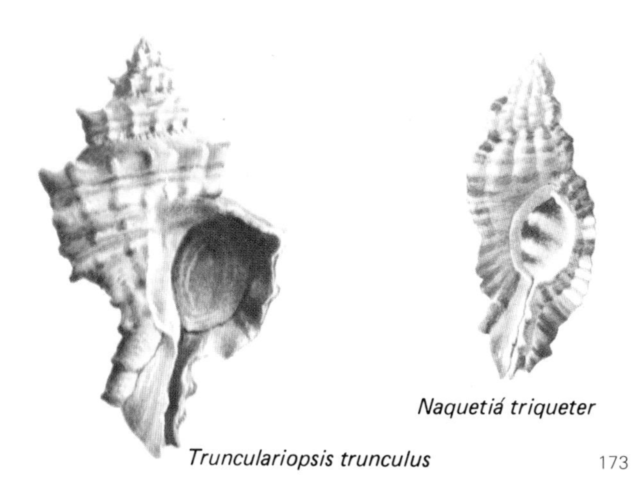
Naquetiá triqueter

Truncularíopsis trunculus

M. uncinarius Lamarck 1822 east South Africa, 20mm. About five whorls; three varices per whorl. A large knob on the shoulder between varices; the latter with short, blunt spines; the largest on the shoulder being hook-shaped. Smooth lip and columella; sealed siphonal canal. Pale brown; aperture white.

M. nodulifera Sowerby 1841 Philippines, 23mm. Moderate spire; sculptured with spiral ridges and axial ribs, nodulose where they intersect. Lip dentate within; columella with two, blunt teeth at the base. Cream, spotted with dark brown; aperture pale yellow.

Pteropurpura festiva Hinds 1844 Gulf of California, 50mm. High spire; about six whorls; three varices per whorl; rough, axial growth lines. A rib with a large, blunt knob on the shoulder between varices. The latter with a fan-like projection folded backward and lamellate on the aperture side; lip finely dentate at the edge and within; smooth columella; deep, sealed, rather short, and recurved siphonal canal. Light brown with fine, spiral, dark brown lines; aperture blue-white.

Ocenebra erinaceus L. 1758 Mediterranean and west Europe, 60mm. About six angular whorls. Sculptured with large, rounded, spiral, lamellated ribs, smaller ones between, and channelled between them. Irregular varices, somewhat ridged between. Dentate lip; smooth columella; deep and sealed siphonal canal. Grey-brown; aperture white.

Vitularia miliaris Gmelin 1791 east Indian Ocean to Pacific, 50mm. Moderate spire; about six whorls; deeply indented suture; oblique, axial, uneven ribs between irregular, very small varices. The shoulder of the body whorl has a shallow channel, which is almost overhung on the upper side. The whole surface granulose. Lip edge rough, slightly expanded, especially anteriorly, four teeth within; smooth columella; short, rather straight, and sealed siphonal canal. Yellow-brown with a row of brown spots on varices; aperture white.

Maxwellia gemma Sowerby 1879 California, 30mm. Stubby; short spire. About five whorls; sutures deep but crossed by oblique varices, about six per whorl; low, spiral ridges; finely dentate lip; smooth columella; short and sealed siphonal canal. White with dark brown on the ridges.

Chicoreus territus Reeve 1845 see page 162.

Murex recurvirostris rubidus F. C. Baker 1897 Florida and the Bahamas, 50mm. About six whorls; moderate spire; three varices per whorl, which may have a spine on the shoulder; two, large, and one, small, axial rib between varices; spirally ridged with a cord between; lip faintly dentate and ridged within; plicate columella; noticeably long, open siphonal canal. Colour variable from cream to red.

Favartia tetragona Broderip 1833 Australia to Fiji, 35mm. Squat and solid; about four whorls; four, solid, oblique varices per whorl; coarse, spiral ridges; finely dentate lip; smooth columella; short, sealed and sharply recurved siphonal canal. White; aperture lavender.

Murex spec. see page 5.

Murex uncinarius

Ocenebra erinaceus

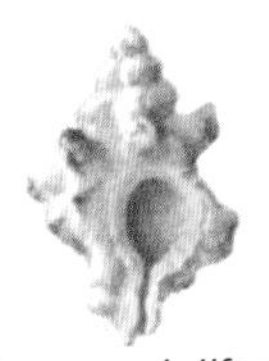

Murex nodulifera

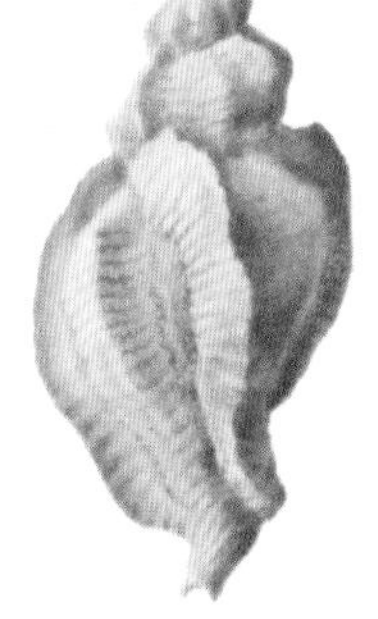

Pteropurpura festiva

Murex spec.

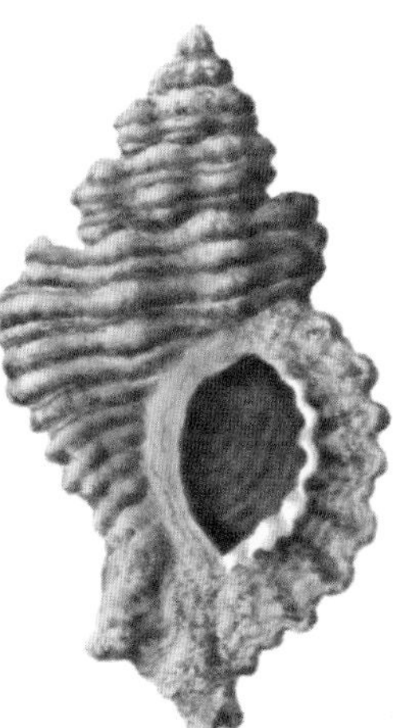

Ocenebra erinaceus

Vitularia miliaris

Maxwellia gemma

Chicoreus territus

Murex recurvirostris rubidus

Favartia tetragona

Family: Thaididae

Rather solid, medium-sized shells, having wide apertures, low spires and no varices. They live in shallow water and are carnivorous, living on other molluscs, especially mussels.

Neorapana muricata Broderip 1832 west Central America, 100mm. Sharply angled shoulder, flat from shoulder to suture, and axially lamellate overall. Smooth spiral cords, five large, bluntly knobbed ridges, plus a small wavy one at suture; largest one at shoulder, three on body whorl. Lip spirally ridged within, a tooth posteriorly. Smooth columella, blunt denticle posteriorly. Narrow columellar shield; strong fasciole; short, open siphonal canal. Cream-grey, possibly sometimes pink; aperture white, faint tinge of pink.

Thais haemastoma L. 1758 Mediterranean, north-west Africa, 80mm. Medium, conical spire; fine, spiral, granulose cords; and shoulder more or less angled. Four spiral rows of knobs; largest on shoulder, next largest below, two small ones below that; row on shoulder has about ten knobs which may have heavy, blunt or obsolescent spines. Lip ridged within; smooth columella with blunt tooth posteriorly; narrow columellar shield; well-developed fasciole; short siphonal canal. Light tan-grey to dark red-brown; lip edge usually dark brown between ridges; aperture and columella pink, orange or red, paler within; parietal wall with blotch of pale yellow and dark brown spiral lines.

Haustrum haustorium Gmelin 1791 New Zealand, 65mm. Low spire and very inflated body whorl; constricted at suture. Coarse, flat spiral ridges; irregular, axial growth lines; very wide, long aperture. Lip runs to bottom of extended siphonal canal, ridged within for a short way. Smooth columella; large parietal shield, lower end level with fasciole. Brown-grey; edge of lip straw with dark purple-brown marking; interior blue-white; columella, shield white, latter with straw and purple-brown clouding posteriorly.

Mancinella bufo Lamarck 1822 Pacific, 60mm. Solid, heavy, with low spire and slightly shouldered. Flat spiral cords of varying widths and four spiral ridges, more or less knobbed, lower two almost obsolete. Bevelled, dentate lip; smooth columella; well-developed but short anal and siphonal canals; moderate fasciole. Callous knob above columella forms one side of anal canal. Brown; interstices of cords cream; cream on ridges between knobs; columella apricot; aperture cream, paler within and dark brown between teeth.

Purpura persica L. 1758 Indo-west Pacific, 100mm. Low spire; close-set, flat spiral cords; seven, larger, more or less knobbed, narrow spiral ridges. Slightly constricted below suture. Wide aperture with lip to end of extended siphonal canal. Dentate lip; smooth columella with small callous lump posteriorly; callous parietal wall; and narrow fasciole. Grey-brown; ridges darker; some cords with dark brown and white dashes; inside of lip with broad, dark purple-brown band; interior pale blue stripes on darker background; columella orange-pink; parietal wall dark brown, yellow and cream.

P. columellaris Lamarck 1822 west Central and South America, 60mm. Solid, heavy, with short spire. Broad, elevated, obsoletely knobbed, spiral ridges, about ten on body whorl; smaller ridges and cords between. Crenulate lip has eight, strong, blunt teeth running as ridges into aperture. Columella with low tooth centrally, callous lump posteriorly. Anal and siphonal canals not well-developed. Parietal wall with two or three faint obsolete folds. Grey-brown; lip orange-tan; interior cream; teeth and columella white; parietal wall red near columella, otherwise cream-tan with some purple marking.

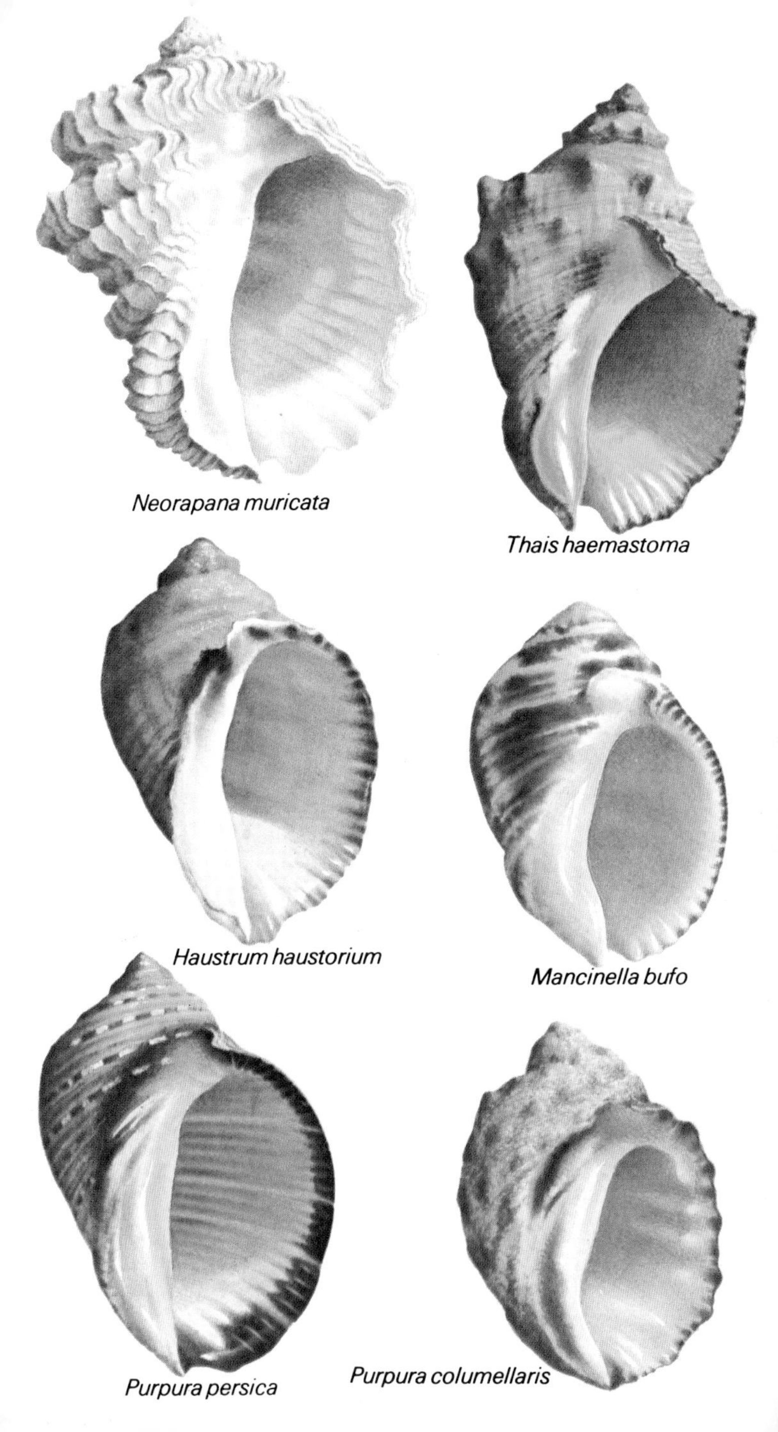
Neorapana muricata
Thais haemastoma
Haustrum haustorium
Mancinella bufo
Purpura persica
Purpura columellaris

Purpura patula L. 1762 Caribbean and south Florida, 100mm. Uneven, flat spiral ribs; axial threads; striations and growth lines. Six spiral rows of blunt knobs, becoming obsolete anteriorly. Wide, dentate lip, smooth columella, shallow recurved siphonal canal and weak fasciole. Dull brown-grey; inside lip edged dark purple-black; interior pale blue-grey; columella orange-pink; parietal wall straw, purple-black, and green-brown.

Thais carinifera Lamarck 1816 east Africa, 50mm. Inflated; very deep suture; close-set, finely beaded spiral cords. Widest part of whorls bear short, wide, flat, blunt projections, sometimes running together, forming uneven keel on last whorl; latter develops deeper suture and begins to drop away from penultimate whorl; it also develops spiral ridge on shoulder and a smaller one below. Dentate lip; columellar shield partly covers narrow, deep umbilicus; strong fasciole. Dirty grey-brown; aperture cream to orange; purple stains deep inside.

Purpura coronata Lamarck 1816 west Africa, 45mm. Solid; globose; low spire; spiral, uneven, beaded cords. Four rows of blunt, rounded knobs, those on two posterior rows larger. Below suture, very coarse lamellations develop from penultimate whorl, which become large and 'wart'-like near aperture. Lip dentate within; smooth columella; short, deep siphonal and anal canals; strongly developed fasciole. Crystalline white, usually with spiral brown lines; aperture cream.

Thais melones Duclos 1832 tropical west America and Galapagos, 50mm. Solid, smooth, globose, with short spire and fine, incised, spiral striae. Finely dentate lip is plicate within and slightly concave posteriorly. Smooth columella, heavy callosity posteriorly; short siphonal canal; small fasciole. Brown; white flashes on shoulder and anteriorly; inside of lip and siphonal canal pale primrose; columella white, bounded by broad purple-brown streak; callosity orange.

Neothais orbita Gmelin 1791 east Australia and New Zealand, 80mm. Solid, not heavy, with moderate spire. Broad, high spiral ridges with deep, wide channels between; fine spiral threads between and on ridges and fine axial lamellae in channels. Coarse sculpture; dentate lip; smooth interior and columella; shallow, short siphonal canal; no umbilicus. Dirty white or cream; aperture edge straw-coloured; interior pink-white; columella white.

Thais kiosquiformis Duclos 1832 tropical west America, 45mm. Very similar to *T. carinifera.* Inflated whorls; deep suture; flat points on shoulders, ten on body whorl. Spiral cords, lamellate below suture; three cords bigger than others, showing obsolete points. Dentate lip; straight, smooth columella; short siphonal canal; developed fasciole; narrowly umbilicate. Brown-grey; white spiral line on shoulder; aperture and interior white; lip edge brown-grey.

T. armigera Link 1807 Indo-west Pacific, 80mm. High spire; three rows of blunt spines on body whorl, shoulder row large and solid, one showing on earlier whorl; spirally corded. Finely dentate lip, plicate within; columella with three pleats anteriorly, obsolete tooth posteriorly: moderate siphonal canal; fasciole. White; brown spiral bands between rows of spines; inside lip edge light brown, pale yellow within this; interior with pink tinge; columella white; parietal area pale yellow, brown edge.

Mancinella mancinella L. 1758 Indo-Pacific, 50mm. Solid, globose and spirally corded. Six rows of small strong spines on body whorl. Finely dentate lip; spirally corded interior; smooth columella, blunt posterior tooth;

Purpura patula

Thais carinifera

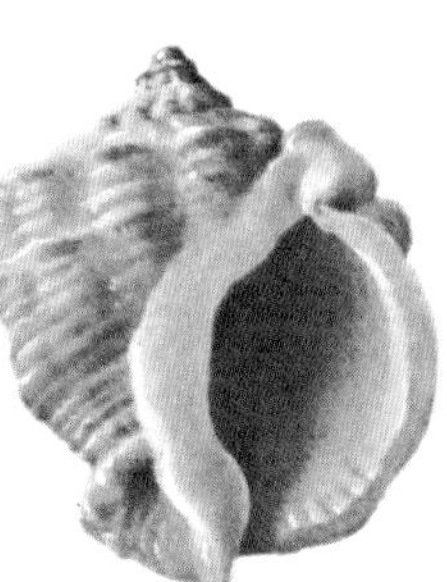
Purpura coronata

Thais melones

Neothais orbita

Thais kiosquiformis

Thais armigera

Mancinella mancinella

obsolete anal canal; short siphonal canal; fasciole. White or light grey, brown banding; aperture orange-yellow; interior dull orange.

Morulina fusca Küster Japan to Singapore, 25mm. Four rows of blunt knobs with spiral cords between rows. Dentate lip, four teeth within. Smooth columella; obsolete plicae on parietal wall; short siphonal canal. White; posterior row of knobs red, then rows alternating black and red; lip straw-coloured with purple blotches; teeth white; columella and interior blue-white; parietal wall purple at top and bottom.

Morula squamosa Pease 1867 Indo-Pacific, 30mm. Low spire. Small spines below suture, two larger rows on shoulder and below, then two smaller rows; two incised lines between rows. Finely dentate lip; smooth columella; open siphonal canal. Grey-brown; coarse oblique white stripes; lip with broad purple band; interior blue-white; columella pinky white; parietal wall blue-white with purple-brown marks on edge and around siphonal canal.

M. margariticola Broderip 1832 Pacific and East Indies, 40mm. Angular shoulders and spirally beaded ridges, two larger. Broad axial ribs, rather pointed on the shoulder. Dentate lip, six teeth within; columella with two or three plicae; short siphonal canal. Dark brown, lighter between ridges; aperture blue-white or mauve-purple; sometimes dark blotch on parietal wall.

M. spinosa H. and A. Adams 1853 Indo-Pacific, 35mm. High spire. Finely spirally ridged; three rows of spines; often long and sharp. Dentate lip, five teeth within; columella with four obsolete teeth; narrow aperture; long siphonal canal. White or grey; spines dark brown; aperture rich violet.

Thais tuberosa Röding 1798 Indo-Pacific, 50mm. Short spire. Spirally ridged; two rows of strong blunt spines on shoulder, and one thick ridge below. Dentate lip, spiral lirae within; columella with three short plicae; blunt tooth bordering obsolete anal canal; short siphonal canal. White; two broad uneven dark brown bands; lip with four brown marks; aperture cream; lirae pale orange; columella cream, large chestnut area posteriorly and small one below.

T. bituberculαris Lamarck 1822 Malaya, Indonesia and Philippines, 50mm. Spirally ridged; two rows of solid, quite sharp, spines on shoulder; two, uneven, thick, nodulose ridges below. Dentate lip, two small and two obsolete teeth within. Columella has a faint plica; narrow shield. Creamy white; dark brown-grey axial stripes; lip cream with dark brown markings; interior and inner side of columella white; parietal wall cream.

T. lamellosa Gmelin 1790 west coast of North America, 125mm. Very variable; high or low spire, angular or rounded shoulders. Bevelled, dentate lip, three to six teeth within. Smooth columella; a blunt tooth low on parietal wall. Cream-grey to dark brown, sometimes banded; aperture white with clouding on lip.

T. hippocastanum L. 1758 Indo-Pacific, 60mm. Spirally ridged; four rows of blunt spines; oblique axial ribs. Dentate lip, four teeth within. Columella with central ridge and one short plica. Cream and purple-brown; white streaks between ribs; aperture blue-white; dark purple-brown band or splashes inside lip, and spiral streaks; columella brown; ridge blue-white; sometimes pale primrose on parietal wall or anteriorly on lip.

Drupella ochrostoma Blainville 1832 Indo-Pacific, 35mm. Four rows of rounded nodules set on twelve axial ribs; two cords between rows. Lip with five or six teeth; columella with two to four. Creamy white with pale yellow areas; nodules dark red-brown; aperture orange or pink.

Morulina fusca

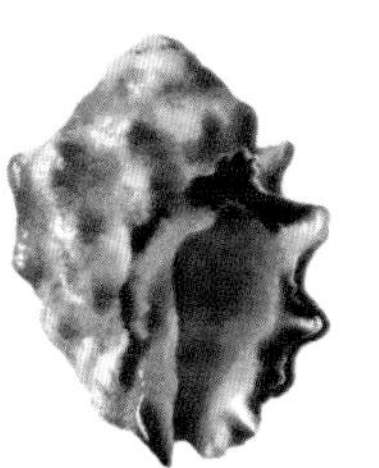

Morula squamosa

Thais tuberosa

Morula margariticola

Thais bitubercularis

Thais lamellosa

Thais hippocastanum

Morula spinosa

Drupella ochrostoma

Concholepas peruvianus Lamarck Peru and Chile, 130mm wide. Like the Haliotidae, this animal lives attached to rock by its strong foot. It has a very expanded aperture and looks a little like one valve of a bivalve. The shell expands in increasing concentric growth from the apex, narrowly on one side and widely on the other, so that the apex overhangs the narrow side. The growth lines cross small, uneven ridges radiating from the apex. Dirty white-grey; interior cream with pale brown clouding and a few blue spots and flecks.

Drupa rubusidaea Röding 1798 Indo-Pacific, 55mm. Solid and rather globose with almost flat spire. Spiral, fine, scaly ridges and five rows of strong spines, longest nearest the aperture and also posteriorly, except the row immediately below the suture where they are smaller; the spines are on low axial ribs, about eight on the body whorl. Lip with about ten teeth; columella with two plicae anteriorly. Callous parietal wall has narrow shield anteriorly. Fasciole has strong spines; no umbilicus. White-grey; points of spines black in illustrated juvenile specimen; outer edge of aperture pale lemon yellow; area of teeth and columella rich pink; interior and band in middle of columella white. Adult and juvenile illustrated.

D. ricina L. 1758 Indo-Pacific, east Africa to Clifton Island and the Galapagos, 30mm. Low spire. Slightly scabrous, spiral ridges. Five rows of spines, longest on the shoulder and nearest the aperture, and more or less joined spirally by a low rib. Lip finely dentate between spines; two double teeth within posteriorly and two single ones anteriorly. Columella has three plaits anteriorly – middle one may be bifid – and one plait posteriorly. Dirty white; spines purple-black; spiral ribs white; aperture white and may have a broken ring of pale yellow round the outside. A subspecies *D.r. hadari* Emerson and Cernohorsky 1973, found in the Red Sea, is larger, up to 38mm, heavier and with a more developed parietal wall.

D. lobata Blainville 1832 Red Sea and Indian Ocean to South China Sea and Western Australia, 32mm. Rather flattened ventro-dorsally. Low spire and apex at an angle of about 30° to the columella. Sculptured with large and small, lamellate spiral ridges, the former ending in four frond-like projections on the lip. There is also a longer, spatulate projection on the lip at the shoulder along which runs the extension of the deep anal canal. Lip with about eight teeth. Coarse fold at bottom of parietal wall. Short, deep, open siphonal canal. Dirty white-brown; interior yellow; edge of lip, outer columella and parietal wall rich chocolate.

D. morum Röding 1798 Indo-Pacific to Clifton Island and Easter Island, 50mm. Solid, slightly flattened, with very low spire. Four rows of blunt nodules on body whorl, striate between. Narrow aperture; lip unevenly thickened internally and with about eight denticles in groups, about four anteriorly, then two or three, then one or two. Columella with three or four plaits anteriorly, from the parietal wall to the interior. Deep anal and siphonal canals. Dirty white-grey; nodules dark brown; aperture rich purple; cream on parietal wall and edge of lip.

D. grossularia Röding 1798 Cocos and Keeling Islands eastward to the Pacific, 30mm. Similar in shape and sculpture to *D. lobata* above, but with a golden yellow instead of brown aperture. The two species overlap only in the Cocos-Keeling, Western Australia areas.

Drupa rubusidaea

Drupa rubusidaea juvenile

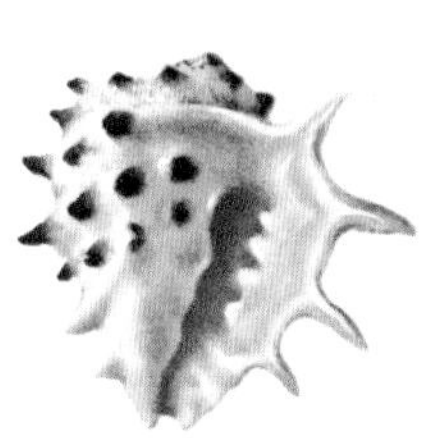

Drupa ricina

Concholepas peruvianus

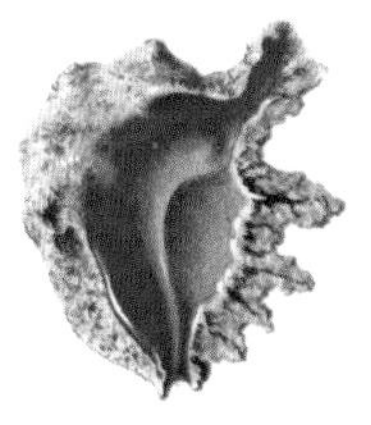

Drupa lobata

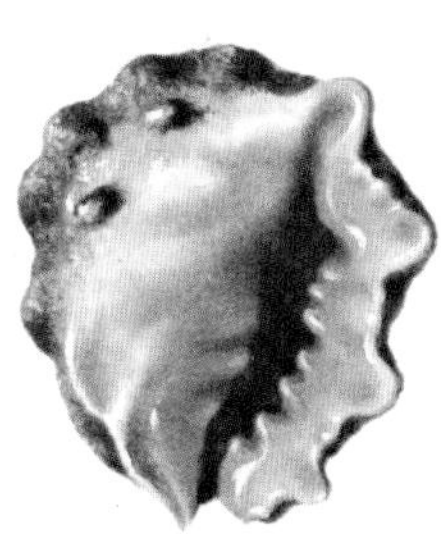

Drupa morum

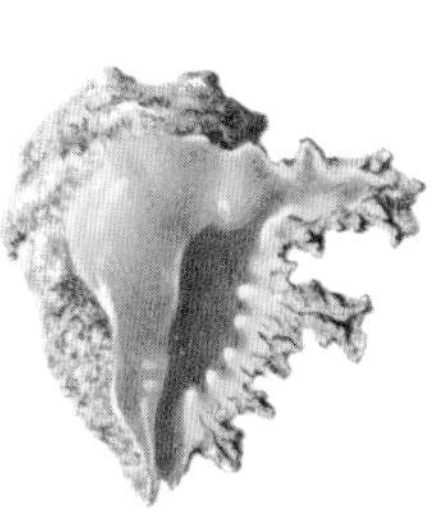

Drupa grossularia

Nassa francolina Bruguière 1789 Indian Ocean, 70mm. Moderate spire; very fine, close, spiral striae. Slightly shouldered and inflated body whorl. Growth lines. Smooth lip has denticle posteriorly opposite a spiral rib on the otherwise smooth columella, which together form anal canal. Short, deep, open siphonal canal; small fasciole. Light red-brown; row of irregular light blotches on shoulder, partly showing at suture; may be some small, light patches further forward on body whorl; lip edge brown; interior cream; columella white backed with brown streak and then pale yellow.

N. serta Bruguière 1789 east Australia and west Pacific, 70mm. Pacific counterpart of the above. Similar except more solid, much coarser, spiral sculpturing, dentate inside lip, and light blotches bigger and running together.

Acanthina imbricatum Lamarck west South America, 55mm. Rather low spire; spiral ridges; channels strongly imbricately scaled. Finely denticulate lip; large, thorn-like tooth – typical of genus – anteriorly. Smooth columella; short siphonal canal; small fasciole. Brick red to brown; lip edge brown; interior brown or white; columella white bordered brown.

Nucella lapillus L. 1758 Atlantic, 65mm. Moderate spire and spirally corded. Dentate lip is ridged within; smooth columella; short, deep siphonal and anal canals. White, yellow or brown; sometimes banded.

Rapana bezoar L. 1758 Japan, 60mm. Short spire. Rounded, spiral, nodulose ridges; strong growth lines cause imbricated effect; shoulder has blunt, open spines. Three ridges and nodulose, axial folds between constricted suture and shoulder. Dentate lip, ridged within; smooth columella; deep, short, recurved canal; deep umbilicus; fasciole. Cream-fawn; interior white.

R. rapiformis Born 1778 west Pacific, New Caledonia to Japan, 100mm. Lighter than *R. bezoar*. Very low spire; expanded body whorl. Fine, incised, spiral lines; axial striae; three, low, slightly nodulose, spiral ridges. Shoulder with fifteen, open, more or less blunt spines, showing on earlier whorls above deep, channelled suture. Wide aperture; denticulate lip, ridged within; wide, shallow channel level with shoulder spines. Smooth columella; extended, open canal; wide, deep umbilicus; imbricately scaled fasciole. Pale brown to off-white; aperture very light brown to cream.

Family: Columbariidae

Columbarium pagoda Lesson 1831 Japan, 60mm. Large nuclear apex and deep, constricted suture. Whorls sharply angled with about ten, flat, slightly upturned, triangular spines. Faint axial striae. Body whorl has strong cord below shoulder, hidden beneath suture on earlier whorls; cord has short, sharp spines near aperture. Very long, open, delicate siphonal canal. Shiny-fawn; siphonal canal darker; apex red-brown.

Family: Magilidae

Rapa rapa L. 1758 Indo-Pacific, 85mm. Fragile with flat or slightly sunken spire and very expanded body whorl. Fine, spiral lamellae overlap suture. Rounded, spiral ridges increase in size and distance apart towards anterior; fine threads between some posterior ridges replaced by fine, axial lamellae anteriorly. Thin lip, crenulated by the ridge ends; callous columella; shield partly covers deep, wide, open umbilicus; strong fasciole. Opaque white or cream.

Coralliophila violacea Kiener 1836 Indo-Pacific, 40mm. Solid; short spire. Uneven, fine, spiral, incised lines. Bevelled lip is finely dentate, finely ridged within. Callous columella has lump posteriorly; strong parietal shield; fasciole; short, recurved, open canal. Off-white; aperture rich violet.

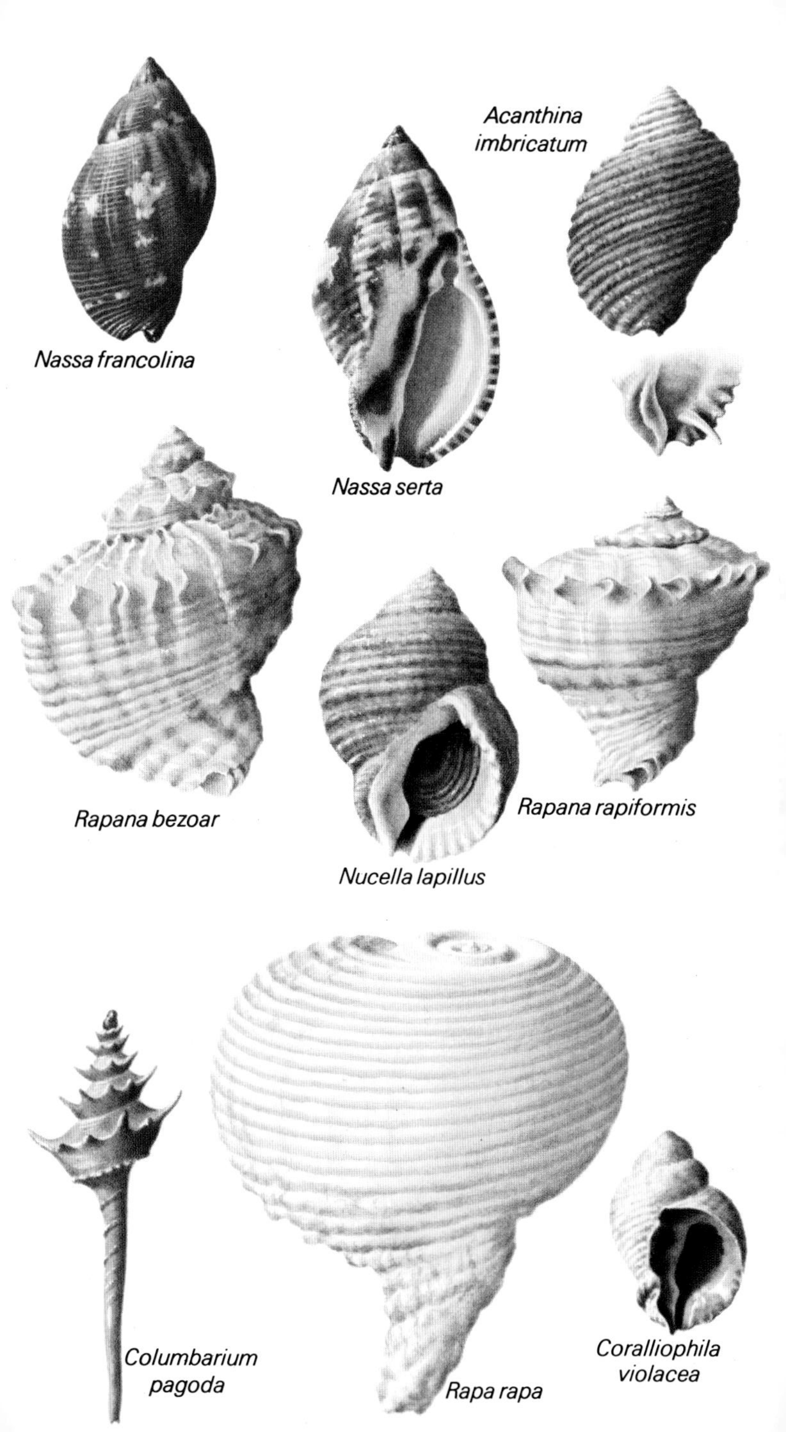
Nassa francolina
Nassa serta
Acanthina
imbricatum
Rapana bezoar
Nucella lapillus
Rapana rapiformis
Columbarium
pagoda
Rapa rapa
Coralliophila
violacea

Latiaxis japonicus Dunker 1882 Japan, 35mm. Moderate spire; whorls with sharply angled shoulder. About eighteen, strong, spiral ridges carrying closely packed, small spines; a deep canal between ridges. Up and outward pointing frill on shoulder with triangular hollow spines – about twelve on body whorl – open towards aperture and having fine, beaded striae on their surface. Dentate lip; smooth columella; quite long siphonal canal; umbilicate; small fasciole with lamellated edge. White.

L. dunkeri Kuroda and Habe 1961 Japan and Taiwan, 40mm. Solid with about six whorls and fine, rough spiral cords. Angular shoulders with triangular spines, about ten on body whorl. Finely dentate lip is strongly ridged within; about twelve spiral ridges. Smooth columella; long, open, recurved siphonal canal; umbilicate; strongly lamellate fasciole. White.

L. mawae Griffith and Pidgeon 1834 Japan, 55mm. Sunken spire with sharp, pointed apex. Shoulder carries small frill with blunt, triangular spines; the area between it and suture slightly concave. Up to last whorl suture is formed with shoulder of previous whorls as its edge, but last whorl develops outward and downward, so that suture ceases to exist. Smooth lip; no columella in an adult shell; deep, quite long, curved siphonal canal. White.

L. pagodus A. Adams 1853 Japan, 30mm. High spire; about six whorls with very deep suture. Convex above shoulder which bears long, pointed, triangular, hollow spines, open towards aperture; second row of smaller spines below; about four rough cords below that. Lip faintly plicate within; smooth columella; moderate, recurved siphonal canal; umbilicate fasciole with long spines – the ends of earlier canals. White or pale brown with purple stains in aperture; columella white.

L. lischkeanus Dunker 1882 Japan, 40mm. Rather similar to *L. japonicus*, but more finely sculptured with spinous, spiral cords and curved, triangular spines on shoulders; about eighteen on body whorl, and with two rows of smaller spines near base of siphonal canal. Finely dentate lip; smooth columella; moderately long, open, curved siphonal canal; umbilicate; lamellate fasciole. White.

L. pilsbryi Hirase 1908 Japan, 22mm. Spire almost flat with sharp apex. Whorls with sharp shoulder carrying flat, slightly up-pointing, triangular spines. In adults last two whorls break away and suture ceases. Fine spiral threads. Smooth lip; rather short, curved siphonal canal; wide, deep umbilicus; spinous fasciole. White.

L. idoleum Jonas 1847 Japan and Taiwan, 40mm. Solid with about seven, rather inflated whorls, constricted at suture. Fine, distinct, slightly scabrous, spiral ridges; low, uneven, oblique axial ribs. Finely dentate lip; smooth columella; open siphonal canal is a little recurved; umbilicus may be very wide or narrow and rather shallow; very rough fasciole. White.

L. kiranus Kuroda 1959 Japan and Singapore, 30mm. Rather high spire and angular whorls. Spiral ridges with irregular, very small spines; larger, triangular spines on shoulders; a row of small ones on siphonal canal. Axially ribbed, about nine ribs on body whorl. Dentate lip ridged within; smooth columella; almost straight siphonal canal; narrow, shallow umbilicus; lamellate fasciole. Cream.

L. gyratus Hinds 1844 west Pacific, 45mm. High spire; inflated whorls; constricted sutures. Spiral cords close-set, finely beaded. Axially uneven rather than ribbed. Strong, beaded keel projects from shoulders of whorls Lirate inside lip; wide, rather shallow umbilicus; strong fasciole. White.

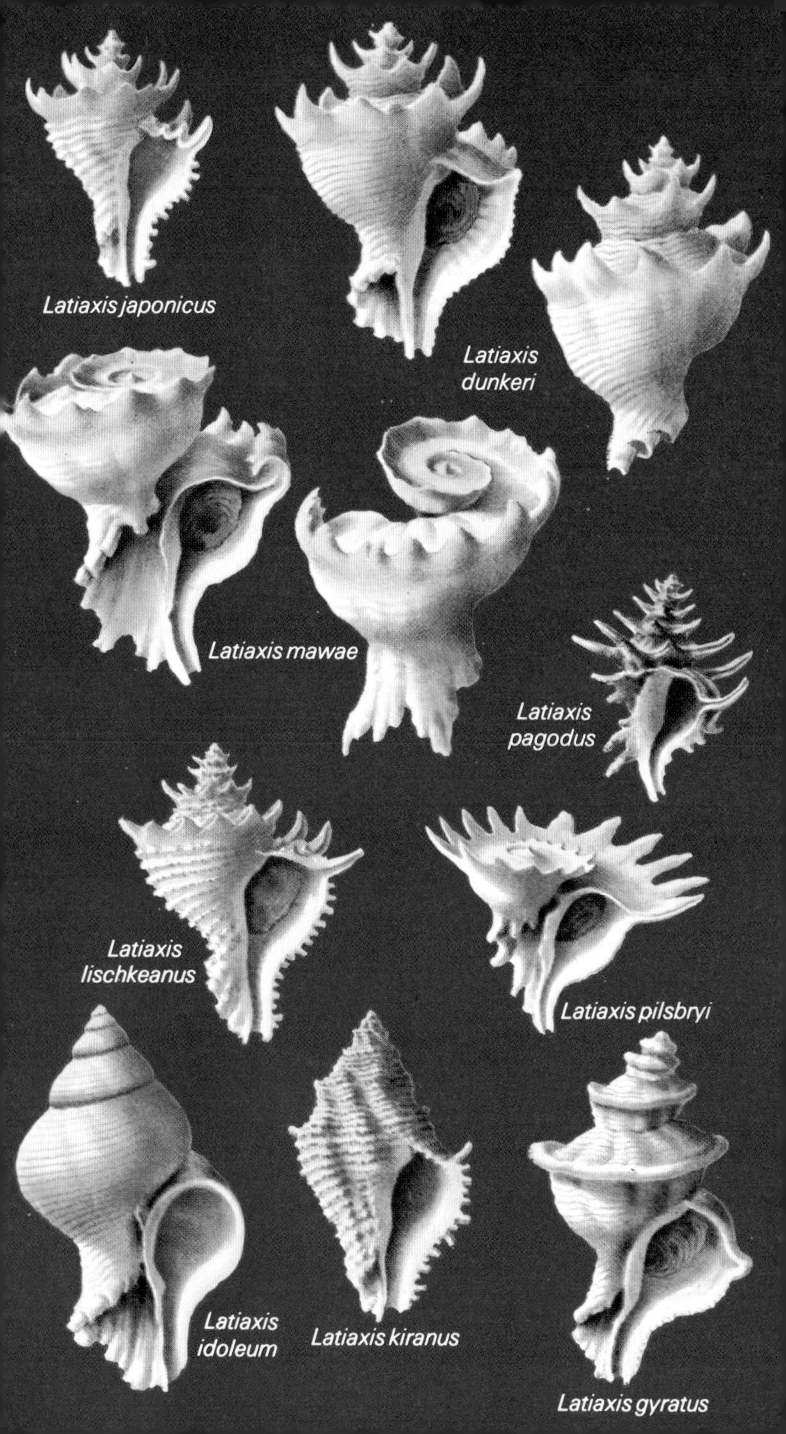
Latiaxis japonicus
Latiaxis dunkeri
Latiaxis mawae
Latiaxis pagodus
Latiaxis lischkeanus
Latiaxis pilsbryi
Latiaxis idoleum
Latiaxis kiranus
Latiaxis gyratus

Superfamily: Buccinacea

Family: Pyrenidae

The dove shells. Generally small and colourful. Live in coral or sand; tropical or semi-tropical. Most are carnivorous. Synonym Columbellidae.

Pyrene testudinaria Link 1807 Pacific, to Singapore, 25mm. Solid, pyriform with deep suture. Corded spiral lines below shoulder, stronger towards base. Nine teeth on lip; four columellar teeth anteriorly. White; brown-black spots or streaks or brown with white spots or streaks; aperture blue-white.

P. varians Sowerby 1832 Pacific, 10mm. Short spire. Smooth, or strong axial ribs on early whorls and posterior of body whorl; spiral striae. Toothed lip; bifid columellar tooth, five teeth on edge. White, cream or brown; sometimes dark brown, wavy, axial lines and rows of V-shaped marks; aperture white or blue-white; purple marks on lip and canal.

P. rustica L.1758 Mediterranean, west Atlantic, 30mm. Variable. Moderate spire. Spiral striae, obsolete in middle of body whorl. Thickened, strongly toothed lip indented centrally; five columellar teeth anteriorly. White or blue-white, more or less mottled purple, purple-brown, or red.

P. flava Brugière 1789 Indo-Pacific, 25mm. Solid; deep suture. Ten, strong, spiral ridges anteriorly on body whorl and in short, slightly recurved siphonal canal. Bevelled lip has axial rib with three, strong and two, weak teeth. Columella smooth but with obsolete nodules on edge of narrow columellar shield. Colour variable, usually dark or pale brown; dark brown, zig-zag, axial lines or spiral bands cut by white spots on shoulder, or with irregular mottling; aperture white.

P. philippinarum Recluz Philippines, Malaysia, 25mm. Concave spire. Angular shoulder, narrowing to slightly recurved and twisted siphonal canal. Bevelled lip has twelve, weak teeth. Smooth columella. White or cream; strong, wavy, axial, red-brown or purple-black marks; aperture white.

P. ocellata Link 1807 Indo-Pacific, 20mm. Solid; short spire; angled shoulder; narrowing to anterior. Spirally corded anteriorly. Lip turned in with nine internal teeth on axial ridge. Columella with row of small teeth near edge of narrow shield. Dark brown to black; white or yellowish spots or zigzag stripes; lip mauve; outer edge of columella brown.

P. splendidula Sowerby Indo-Pacific, 30mm. Low spire. Inflated body whorl, narrowing anteriorly to short siphonal canal. Anterior with spiral ridges crossing columella. Large brown and white blotches; aperture white; thick periostracum (illustrated).

Strombina maculosa Sowerby 1832 north tropical west America, 25 mm. High spire; nodulose shoulder. Spiral ridges on lower body whorl. Thickened, bevelled lip with six teeth. Smooth columella. White maculated with brown, white showing mainly as tent marks.

Microcithara harpiformis Sowerby 1832 Central America, 18mm. Solid; short spire. Axial ribs, fourteen on body whorl, form blunt points on angular shoulders. Narrow aperture. Finely dentate lip, thickened, especially on inside, extends out and up to deep, narrow posterior canal, and has sharp ridge on edge opposite columella; latter has obsolete fold. Dark chocolate; white spots; suture, knobs on spire, apex white; inside lip white, may have some brown dots; columella as body whorl.

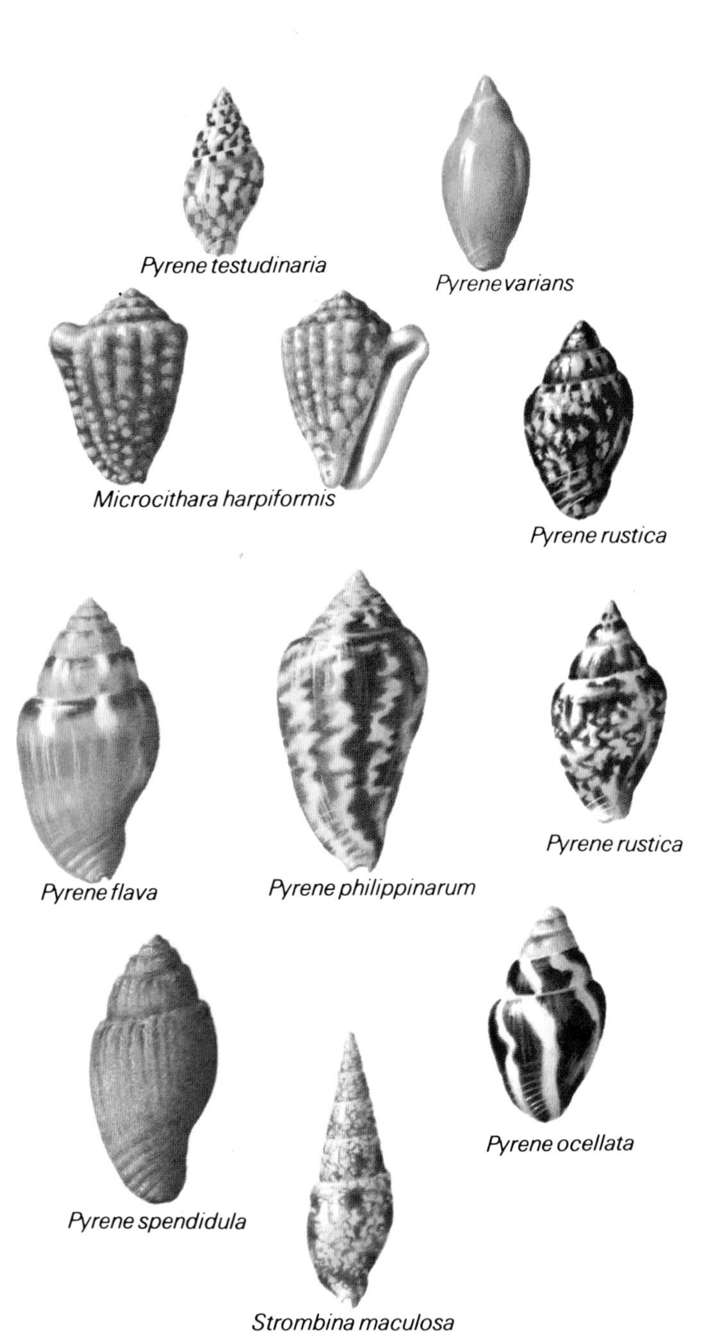
Pyrene testudinaria
Pyrene varians
Microcithara harpiformis
Pyrene rustica
Pyrene flava
Pyrene philippinarum
Pyrene rustica
Pyrene spendidula
Pyrene ocellata
Strombina maculosa

Family: Nassariidae

Dog whelks or basket shells live mostly in tropical or semi-tropical shallows. Carnivorous, scavengers; active, especially at night.

Bullia grayi **Reeve 1846** South Africa, 75mm. Smooth; deep suture; strong, flattened ridge below, then incised line and short, sharp shoulder. Body whorl has nine, faint, incised lines; obsolete, axial striae. Livid; grey band on shoulder; early whorls blue-grey; columella white; interior brown.

Niotha gemmulata **Lamarck** Philippines, Japan, 30mm. Stubby; expanded body whorl; channelled suture; cancellate; nodulose. Dentate lip, ridged within; granulose columella and shield; short, recurved canal; fasciole. Cream; light brown and blue-grey clouding; aperture white; interior purple.

Zeuxis olivaceus **Bruguière 1789** Pacific and China Seas, 45mm. High spire. Early whorls obliquely ribbed, later smooth bar growth striae and ten ridges anteriorly on body whorl. Thickened, bevelled lip, dentate within; shallow, well-developed siphonal canal; short anal canal; outer edge of columella with twelve blunt teeth. Dark brown sometimes with a yellow band; aperture purple-white; interior purple.

Plicarcularia pullus **L. 1758** west Pacific, China Seas, 20mm. Solid; short spire; humped dorsum. Axial ribs on columellar side, other smooth. Early whorls ribbed; incised line below suture. Heavily callous parietal shield forming flat 'pad' with lip; latter thickened, dentate. Prominent columellar tooth bordering short, deep anal canal, three or four teeth anteriorly; short, deep siphonal canal. Brown or green; usually yellow band above suture, dark spot on hump; columella, callus white; interior purple-brown.

Nassarius arcularius arcularius **L. 1758** central and west Pacific, 30mm. Globose; short spire; flat shoulders. Prominent axial ribs, disappear on last third of body whorl, except on shoulder where ends remain as nodules. Lip ridged internally; posterior columellar tooth, two plicae anteriorly. Rugose with expanded callus over parietal area. White or cream; sometimes brown spots between nodules on shoulder of body whorl; aperture white.

N. arcularius plicatus **Röding 1798** Indian Ocean, 30mm. Differs from above in having spiral, incised lines overall.

N. coronatus **Bruguière 1789** Indo-west Pacific, 30mm. Moderate spire; sharply angled shoulders with blunt, heavy nodules, twelve on body whorl. Finely dentate lip, ridged within; columella has posterior ridge, four, small teeth anteriorly. Callous parietal shield; short siphonal and anal canals. White with brown band, or green-brown; knobs, lip, aperture, callus white.

Tarazeuxis reeveanus **Dunker 1847** Indo-Pacific, 20mm. Smooth; slightly expanded. Lip toothed within; columella with ridge posteriorly, rugose anteriorly; short, deep anal and siphonal canals. Opaque blue-white-green; white spots and dashes; central band of white and brown; ten, red-brown lines on body, a few on earlier whorls; aperture white; interior purple-brown, two white stripes.

Family: Buccinidae

Found worldwide, these shells are carnivorous, living mostly on bivalves.

Northia northiae **Griffith and Pidgeon 1834** tropical west America, 50 mm. High spire; sharply angled shoulders. Early whorls with spiral cords and axial riblets, disappearing on penultimate whorl. Body whorl smooth bar fine growth lines and six, spiral ridges anteriorly. Slightly thickened lip

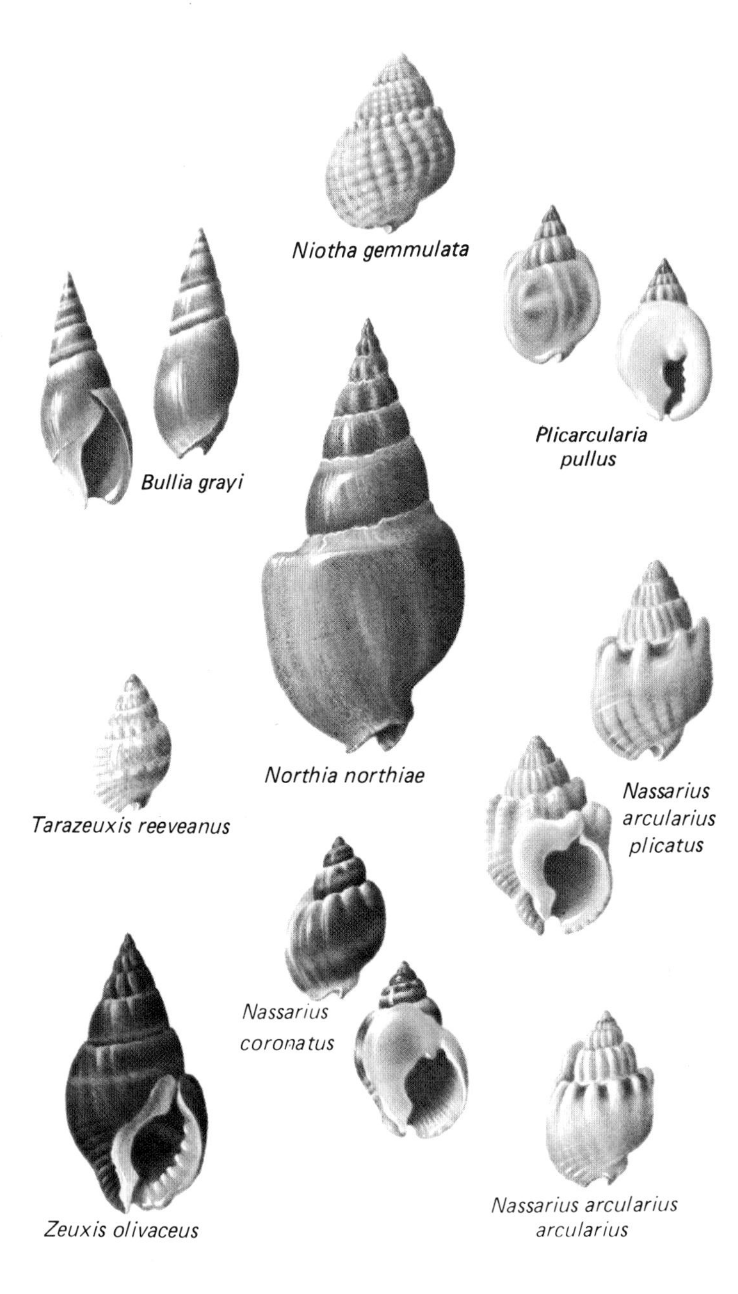
Niotha gemmulata
Bullia grayi
Plicarcularia pullus
Northia northiae
Tarazeuxis reeveanus
Nassarius arcularius plicatus
Nassarius coronatus
Zeuxis olivaceus
Nassarius arcularius arcularius

with blunt knob on shoulder; finely dentate, ridged within. Columella with ridge posteriorly; short canal; broad fasciole. Olive brown; early whorls darker; lip, canal, edged inside red-brown; columellar shield flesh; interior white; ridges brown.

Hindsia magnifica Lischke Japan, 45mm. High spire; slightly inflated; restricted suture; spiral ridges; axial ribs; may have earlier varices. Lip with swollen varix, ridged internally. Columella ridged posteriorly, pustulate anteriorly; recurved siphonal canal; weak fasciole. White; brown clouding, banding.

Cantharus erythrostomus Reeve 1846 Indian Ocean, 50mm. Moderate spire; angular whorls. Strong spiral ridges; axial ribs strong on shoulder, obsolete anteriorly. Crenulate lip, ridged within; ridges cross posterior of columella. Pale orange; darker ridges, shoulders; inside lip orange; columella paler; interior white.

C. undosus L. 1758 China Seas and Pacific, 40mm. Solid; spiral ridges and axial ribs, obsolete on penultimate whorl, rounded and heavy on body whorl. Lip with varix, bevelled ridges within; columella smooth on inner side, unevenly plicate on outer; weak fasciole. White, blue-white or pale yellow; ridges dark brown or purple; aperture edge pale orange; thick periostracum.

C. fumosus Dillwyn 1817. As *C. undosus* but axial ribs overall. Cream or yellow-tan; ribs brown; pale band on body whorl; interior white.

C. elegans Griffith and Pidgeon 1834 tropical west America, 45mm. High spire and angular whorls. Well-separated, strong, spiral ridges; spiral threads and axial striae between; axial ribs on early whorls and upper half of body whorl, nodulose on intersections. Lip weakly ridged within; columella weakly plicate. Dark brown maculated with some white; interior blue-white.

C. ringens Reeve 1846 tropical west America, 30mm. Solid; moderate, sharp spire; concave betwen suture and shoulder. Spiral ridges and cords, and axial ribs. Finely dentate lip, with strong varix, ridged within, indentation posteriorly. Well-developed anal and siphonal canals; columella ridged posteriorly, pustulate anteriorly. Dark brown; aperture blue-white, darker interior.

Pisania pusio L. 1758 West Indies and south Florida, 45mm. Smooth; slightly swollen; high spire. Weakly toothed lip is spirally ridged. Smooth columella, strong ridge posteriorly. Narrow, rugose parietal shield; moderate siphonal canal. Purple-brown; narrow, spiral, broken, dark brown bands; aperture edge orange-red.

P. ignea Gmelin 1791 Pacific and China Seas, 35mm. High spire. Early whorls have axial riblets, later whorls smooth bar faint lines anteriorly. Lip a little expanded, obsoletely fluted, teeth anteriorly; smooth columella. Cream or pale orange; brown axial flames; band on body whorl; interior blue-white.

Buccinulum littorinoides Reeve 1846 New Zealand, 35mm. High spire. Early whorls have obsolete axial ribs; fine growth lines. Bevelled lip, ridged within; smooth columella. Grey; darker spiral lines; aperture white.

Phos senticosus L. 1758 Indo-Pacific, 40mm. High spire; early varices; square shoulders; constricted suture. Spiral ridges, threads between; axial ribs, short sharp riblets at intersections, axial striae between. Lip ridged within; irregularly plicate columella; strong fasciole. Cream or pink-brown; darker band on body whorl, suture; aperture white; interior white or violet.

P. veraguensis Hinds 1843 tropical west America, 25mm. High spire and constricted suture. Spiral ridges, cords between; axial ribs, nodulose at inter-

Pisania pusio

Cantharus erythrostomus

Hindsia magnifica

Cantharus undosus

Buccinulum littorinoides

Cantharus fumosus

Phos senticosus

Cantharus ringens

Cantharus elegans

Pisania ignea

Phos veraguensis

sections. Lip ridged within; columella has ridge anteriorly; small fasciole. Pale brown; darker brown bands; interior white-cream.

Buccinum undatum **L. 1758** north Atlantic and the Mediterranean, 160mm. The European Whelk has a shell which is very variable in shape, sculpture and colour. More or less solid with usually moderate spire of about seven, slightly inflated whorls. Spirally ridged and with oblique axial ribs becoming obsolescent on the body whorl. Outer lip a little expanded and S-shaped; smooth columella; short siphonal canal; weak fasciole. Dirty white, grey or cream, sometimes with brown band at the suture and on the body whorl; interior cream or white.

Siphonalis signum **Reeve 1846** Japan, 60mm. Moderate or short spire; about six whorls. Spirally ridged; rather angular shoulders with flattened nodules, about twelve on body whorl and less on early whorls. Straight sides from shoulder to suture. Fluted lip is ridged within; very thin columellar wall, callous posteriorly; produced, recurved and open siphonal canal; strong fasciole. Colour very variable; white or pale yellow to light or dark brown, with axial, dark brown flame markings, blotches and spiral lines; aperture edge white; interior grey-brown.

Penion adustus **Philippi 1845** north New Zealand, 125mm. Solid with moderate to high spire; about six whorls; spiral ridges and cords between. Weak axial ribs; nodulose at the angular shoulder. Outer lip slightly fluted and lirate within. Columella smooth, with a strong callus posteriorly; extended, open, slightly twisted and curved siphonal canal; a small fasciole. Light red-brown to grey; aperture white.

P. mandarinus **Duclos 1831** south New Zealand and Cook Strait, 125mm. Lacks the angular shoulders of its northern counterpart, *P. adustus*, the whorls being a little inflated. Spiral sculpturing very similar. Dirty grey-white to brown; aperture white.

Hemifusus ternatana **Gmelin 1798** north China Seas, 200mm. Long, rather slender, with a high spire of about seven, more or less angular whorls, slightly constricted at the suture. Spirally ridged and corded and axially finely striated. Also ribbed on early whorls with small nodules on the shoulders tending to become obsolete on later whorls, but growing stronger again on the last half of the body whorl. Weakly fluted and ridged lip, smooth within; columella smooth anteriorly, weakly callous posteriorly over the body sculpture; long, open and slightly twisted siphonal canal. White to cream; aperture flesh-cream. This shell has a thick, strong, dark brown periostracum, some of which is shown in the illustration, still adhering. The animal is used for food.

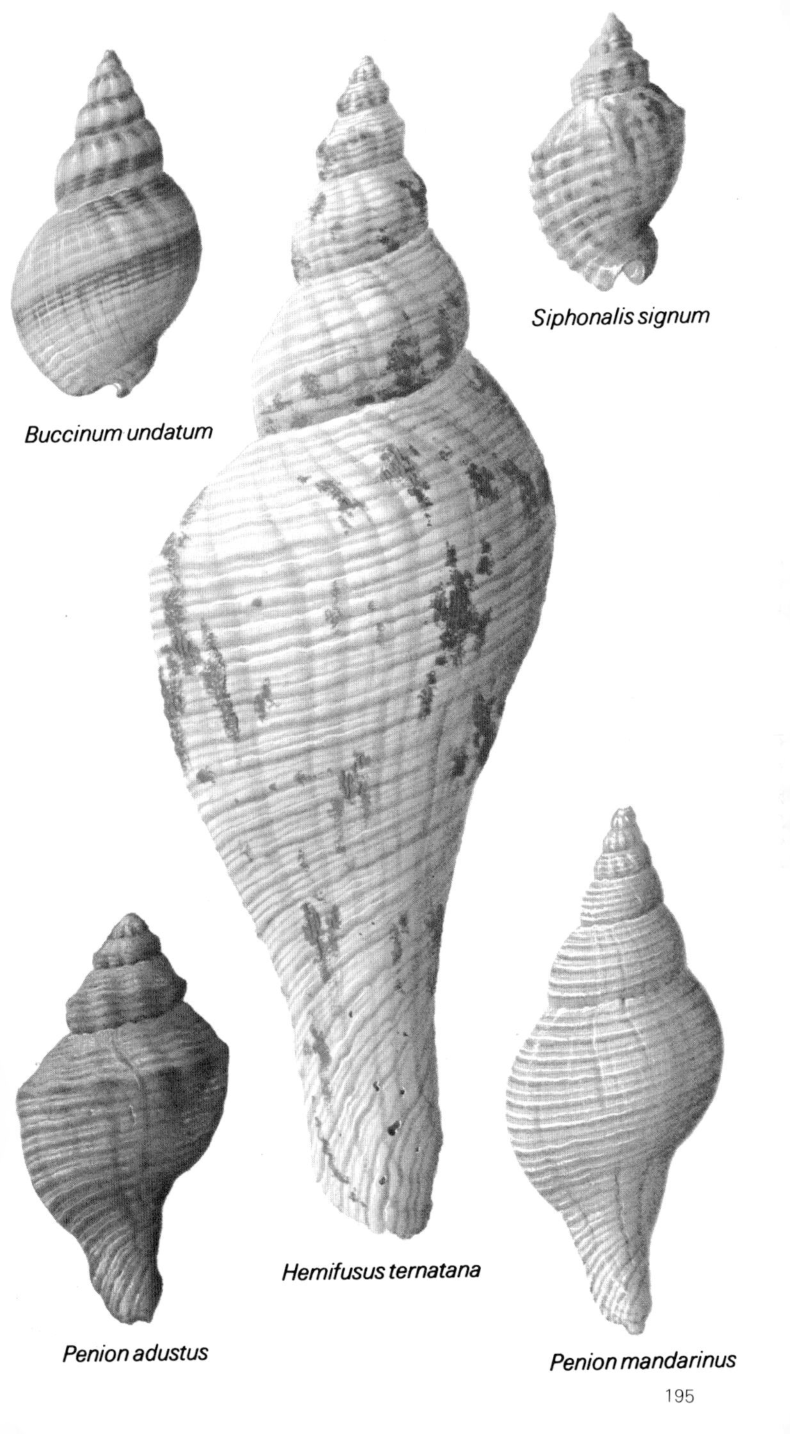
Siphonalis signum

Buccinum undatum

Hemifusus ternatana

Penion adustus

Penion mandarinus

Babylonia lutosa Lamarck 1822 east Asia, 60mm. Rather high spire and about six slightly inflated whorls and sharply rounded shoulders. Axial growth lines and microscopic axial and spiral striae. Smooth lip is a little expanded. Smooth columella, heavily callous, especially posteriorly. Rather weak anal canal; short, deep siphonal canal; strong fasciole; deep umbilicus. White with pale fawn clouding on broad spiral bands; aperture white.

B. japonica Reeve 1842 Japan and Taiwan, 75mm. A shorter spire than *B. lutosa*, and not such a solid shell. More smoothly rounded shoulders and more moderate sculpturing. Smooth lip; less callous columella; shallower siphonal canal; smaller umbilicus; weaker fasciole. Creamy white with a row of V-shaped, liver-coloured blotches below the suture and at the widest part of the body whorl, profusely spotted elsewhere with the same colour; post-nuclear whorls have a mauve tinge.

B. formosae Sowerby 1866 Taiwan, 50mm. Moderate spire, slightly constricted at the suture. The whorls have a sharply angular shoulder and are flat from shoulder to suture. Smooth lip and columella, the latter heavily callous posteriorly. Deep, open umbilicus; strong fasciole; weakened anal canal and short, deep siphonal canal. Cream with a broad band of wide, <-shaped marks below the suture and two rows of squarish blotches below this, all dark liver-brown; aperture white, the brown showing through on the edge of the lip; the post nuclear whorls have a mauve tinge.

B. zeylanica Bruguière 1789 India and Sri Lanka, 75mm. Rather like *B. japonica* in shape but less robust with less inflated whorls. White with large brown blotches and spots generally smaller in centre of the last whorl and above the suture. Inside edge of fasciole ridge, and apex purple.

B. areolata Link 1807 south-east Asia, 50mm. Rather similar in shape to *B. formosae*, but slightly shorter spire, shoulders a little more rounded, narrower umbilicus and weaker fasciole. Very smooth and shiny. White; a row of dark liver-coloured oblongs below the suture, then a row of rather square blotches followed by a row of more or less triangular blotches; blotches tend to have a golden lining; aperture white; post-nuclear whorls mauve.

B. canaliculata Schumacher 1817 Arabian Sea, 65mm. Solid, heavy, with short spire and channelled at suture. Whorls somewhat inflated; smooth except for some coarse growth lines. Smooth lip; heavily callous columella; weak anal canal; deep, short siphonal canal. Fasciole is broad and strong but shiny, and umbilicus is very narrow and shallow. White-cream with pale liver-coloured blotches, streaks and spots, the blotches forming a discontinuous band low on the body whorl; aperture white.

B. spirata L. 1758 Indian Ocean, 75mm. Also solid and heavy, but less coarse than *B. canaliculata* and with a high spire. The suture more deeply and widely channelled and the edge of the channel much sharper. Smooth lip and heavily callous columella. Anal canal is more developed than in others of the genus. Short, deep siphonal canal; wide fasciole; shallow umbilicus. White, heavily marked with light brown blotches, oblique streaks and spots; aperture white; fasciole mostly orange-brown; nuclear and post-nuclear whorls purple, becoming lighter, and the brown growing stronger anteriorly.

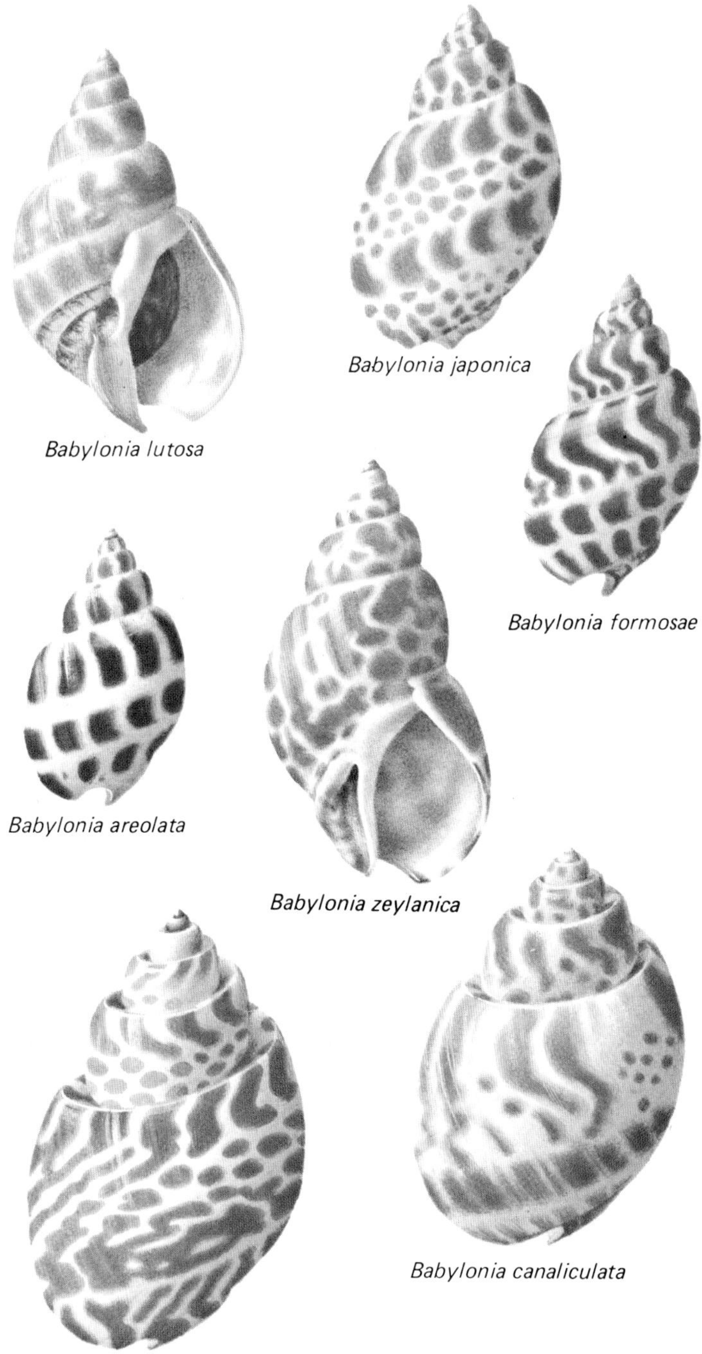

Babylonia lutosa

Babylonia japonica

Babylonia formosae

Babylonia areolata

Babylonia zeylanica

Babylonia spirata

Babylonia canaliculata

Family: Fasciolariidae

The tulip shells are found worldwide and, like Buccinidae, are carnivorous.

Fasciolaria hunteria Perry 1811 south-west USA, 100mm. Inflated; smooth bar fine growth lines and weak ridges at base of body whorl and on canal. Smooth lip, columella. White; three, broken, spiral, green bands; five, purple, spiral lines on body whorl, two on early whorls; purple bands inside lip.

F. trapezium L. 1758 Indo-Pacific, 200mm. Solid; heavy; moderate spire; constricted suture. Shoulders with slightly pointed, strong knobs. Incised spiral lines, mainly paired; lip with seven pairs of teeth where lines meet edge; strong ridges internally. Columella smooth posteriorly, three plaits anteriorly; extended siphonal canal; weak fasciole. Flesh; incised lines, teeth, interior cords, fasciole, brown; columella pale coffee; dark periostracum.

F. filamentosa Röding 1798 Indo-Pacific, 150mm. High spire; round shoulders, low on whorls; constricted suture. Pairs of incised spiral lines with ridges between and obsolete axial ribs, ten on body whorl. Lip has pairs of small denticles at end of incised lines, spirally ridged within. Columella callous anteriorly and with three plaits; extended, open, slightly twisted siphonal canal; weak fasciole. Blue-white profusely clouded, especially on ribs, with red-brown; teeth brown; aperture flesh; internal ridges, columella red-brown.

F. salmo Wood 1828 tropical west America, 125mm. Solid; heavy; short spire; fine spiral ridges. Angular shoulders have blunt knobs, nine on body whorl; deep suture. Finely denticulate lip, ridged within; columella has ridge edging anal canal, and two anterior plaits; long, open, slightly curved siphonal canal; fasciole. Yellow-flesh; columella, lip edge salmon pink; interior white; end of siphonal canal purple; strong, brown periostracum (illustrated).

Latirus polygonus Gmelin 1791 Indo-Pacific, 70mm. Fusiform; angulate; constricted suture; spiral ridges; axial ribs, seven on body whorl; nodulose at shoulder and lower on body whorl where one ridge is strong. Finely dentate lip is ridged within; columella has about four small plicae anteriorly; extended, open siphonal canal; weak fasciole. Cream to pale yellow-brown; axial, broken, dark brown stripes on ribs; aperture pale brown at edge; ridges, interior white.

Peristernia philberti Recluz 1844 South China Sea, 30mm. Moderate spire; angular; constricted suture. Spiral ridges, channels between. Axial ribs, ten on body whorl, nodulose on shoulder. Finely dentate lip, ridged within, blunt tooth anteriorly. Smooth columella, posteriorly callous, narrowly umbilicate. Red-brown; ridges white; aperture violet; strong fasciole dark brown; interior white.

P. incarnata Kiener 1840 Indo-Pacific, 30mm. Solid with moderate spire, channelled sutures and broad, flat, slightly oblique axial ribs, twelve on body whorl. Fine spiral cords. Finely dentate lip, ridged within, blunt tooth anteriorly. Columella has strong ridge posteriorly and large, flat, rectangular tooth anteriorly. Well-developed anal canal; short, deep, open siphonal canal. Ribs orange-red; interstices dark brown; aperture pinky mauve; fasciole flesh.

Opeatostoma pseudodon Burrow 1815 tropical west America, 45mm. Moderate spire and very angular, flattened shoulders. Strong, flat spiral ridges with weaker cords between. Lip has obsolete spiral ridges internally, long, strong, pointed spine anteriorly. Columella has ridge posteriorly and six plicae anteriorly. Short siphonal canal; fasciole. White; ridges dark brown-black, showing within lip; interior white; strong, brown periostracum (illustrated).

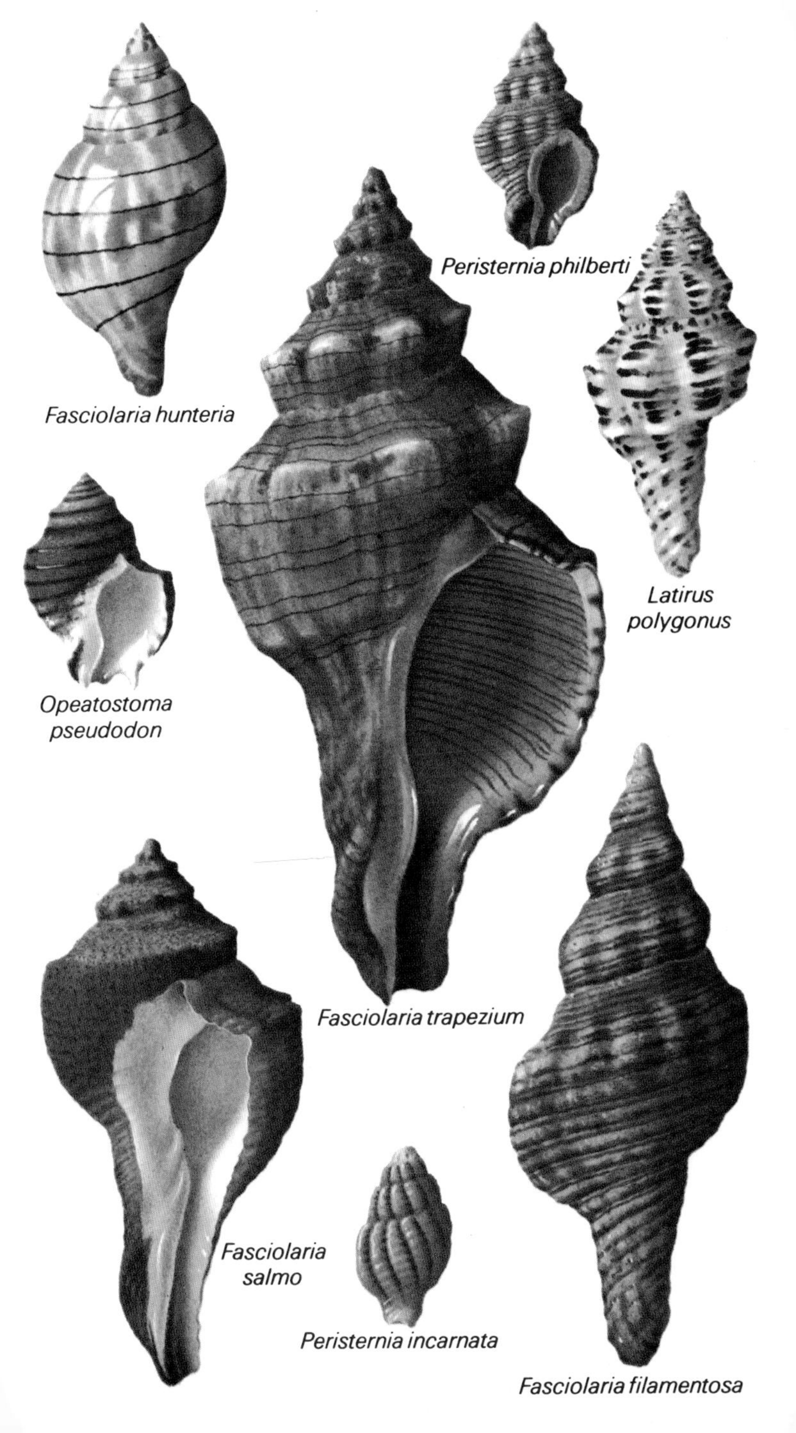
Fasciolaria hunteria
Peristernia philberti
Latirus polygonus
Opeatostoma pseudodon
Fasciolaria trapezium
Fasciolaria salmo
Peristernia incarnata
Fasciolaria filamentosa

Genus: *Fusinus*

There are over fifty species of spindle shells. They are quite distinctive with a long spire and a long, generally straight siphonal canal and spiral ribbing. In the small subgenera *Aptyxis* and *Barborofusus,* however, the canal may be relatively short. The columella is smooth and most species are white, often with some brown marks or clouding. They are found in tropical and temperate waters worldwide. As the rest of the family, they are carnivorous. They live on sandy bottoms and are usually found in pairs.

Fusinus colus L. 1758 Indo-west Pacific, 200mm. Fusiform and rather angular with high spire. Spiral ridges and cords; fine axial striae; early whorls axially ribbed; on last two or three whorls these become obsolete and remain only as nodules on the shoulders. Finely dentate lip ridged within. Thinly callous columella, ridges showing through; long, slender and slightly wavy siphonal canal. White or cream with a few, axial, brown streaks; brown between the ribs and nodules, more conspicuous on earlier whorls; a few spots at the suture; siphonal canal brown growing dark towards the tip; aperture white, edged faintly with brown.

F. dupetitthouarsi Kiener 1840 tropical west America, up to 250mm. Rather similar to *F. colus* but less slender, rounder shoulders, shorter, broader and more open siphonal canal, bigger aperture. White; faint brown marks.

F. longicaudus Lamarck Japan, 100mm. Narrow spire and long, narrow siphonal canal; spirally ridged and axially ribbed on early whorls. Fluted lip; columella smooth anteriorly but only thinly callous posteriorly where the ridges show. White with faint shading between ribs; darker between ribs on early whorls.

Family: Melongenidae

Genus: *Busycon*

The fulgur whelks are found off the east coast of North America. It seems, however, that one species is also found off the east African coast (see below). It includes in addition to those species illustrated and described below: *Busycon perversum* L. which may be sinistral or dextral; *B. carica* Gmelin ; *B. spiratum* Lamarck ; and *B. coarctatum* Sowerby The latter three species are all dextral.

B. contrarium Conrad south-east North America, up to 400mm. Sinistral with low spire. Angular shoulder with more or less developed, triangular knobs. Spirally striated, except in middle of the body whorl where it is almost smooth. Smooth lip and columella; long and open siphonal canal. Grey with irregular, axial, dark brown lines; two grey-brown bands on body whorl; siphonal canal grey-brown; aperture with brown staining; columella thin, callous posteriorly allowing shell colour to show through, flesh-coloured anteriorly The illustration is of a juvenile.

A sinister busycon (probably also *B. contrarium*) is found off east Africa, and a specimen 160mm is illustrated. Its shape and sculpture are the same, but it is coloured pale yellow-brown with axial purple-brown lines strongest on the spire; aperture white.

B. canaliculatum L. 1758 east North America, 180mm. Rather short spire with angular shoulders and a deep, wide channel at the suture. Fine, obsolete, spiral cords; strong ridge at the shoulder which is nodulose on early whorls. Smooth lip; columella only weakly callous; extended and open siphonal canal. Light brown-grey; shoulder ridge nodules cream-white, with brown interstices; inside edge of lip light brown.

Busycon contrarium juvenile

Busycon spec.

Busycon canaliculatum

Fusinus colus *Fusinus dupetitthouarsi* *Fusinus longicaudus*

Melongena patula **Broderip and Sowerby 1829** west tropical America, 250mm. Short, sharp, concave spire; angular, nodulose shoulders; sometimes spinous, sometimes almost smooth on body whorl. Spirally corded above the shoulder and at base of body whorl. Smooth lip; callous, smooth columella; extended siphonal canal; weak fasciole. Chestnut brown banded with yellow, white or tan; lip edged interiorly with browns merging into white then blue-white; columella cream-peach; fasciole dark brown; and tip of siphonal canal purple.

M. melongena **L. 1758** Caribbean, 100mm. Solid, heavy, with short spire and channelled suture. Axial growth striae; two spiral rows of short sharp spines in the widest part of the rather inflated body whorl, and one near the base. Uneven lip; heavily callous columella; strong fasciole; short, open siphonal and anal canals. Dark chocolate brown; canals and suture white; more or less narrow white bands below the two upper rows of spines; white axial hairlines and tips to spines; aperture and fasciole white.

M. corona **Gmelin 1791** south-east America and Mexico, 100mm. Variable. Generally with a moderate spire and angular whorls. Spirally corded and axially ribbed, ribs becoming obsolete on body whorl. One or two rows of spines on shoulder, hollow, more or less numerous, and vertical to horizontal, sometimes a row at base of body whorl. Smooth columella; strong fasciole; short, open siphonal canal. White; spiral bands of purple-brown or blue-black; spines white; lip stained brown internally; interior, columella white.

M. morio **L. 1758** west Africa, Brazil and West Indies, 175mm. Rather elongate; angular shoulders; spiral ridges. Early whorls ribbed, ribs becoming obsolete on later whorls but shoulders more or less nodulose. Lip ridged internally; columella weakly callous anteriorly and barely at all posteriorly. Long, open siphonal canal; weak fasciole. Chocolate brown; one moderately broad yellow-white band and a number of narrow ones; aperture brown; columella dirty white.

M. galeodes **Lamarck** Red Sea, Indian Ocean to China and Philippines, 65mm. Short sharp spire; about six whorls with short frilly spines at the suture. Spirally corded; a row of about eight, heavy, sharp spines at the shoulder of the body whorl, and may have two other rows of spines on body whorl, centre one usually small, one nearer base larger. Finely dentate lip; callous, smooth columella; narrowly umbilicate; strong fasciole; moderate siphonal canal; short, fairly deep anal canal. Grey or pale brown, ridges darker; inside of lip white with purple-brown banding, deeper within; columella cream.

Volema cochlidium **L. 1758** Indian Ocean and South China Sea, 150mm. Solid; heavy; angular whorls; concave from suture to shoulder. Spirally corded; early whorls axially ribbed; body and sometimes penultimate whorls with about eight, strong, heavy, flattened spines. Lip weakly ridged within; smooth columella; moderately short, open siphonal canal; very narrow, shallow umbilicus; short, strong fasciole. Red brown; aperture peach with brown edging. It has a thick periostracum (illustrated).

Syrinx aruanus **L. 1758** north Australia, up to 600mm (about 2 ft). Probably the world's largest gastropod. Heavy with angular, rounded shoulders. Spirally ridged and axially threaded; early whorls axially ribbed; later whorls nodulose at the shoulder. Some ridges on body whorl larger than others. Finely ridged lip; smooth columella; extended, open, straight siphonal canal. Umbilicus is a narrow slit. Straw-coloured. The protoconch of some five-and-a-half whorls with constricted sutures and axial ribs is normally lost in adult shells. Illustration shows a juvenile with protoconch still attached.

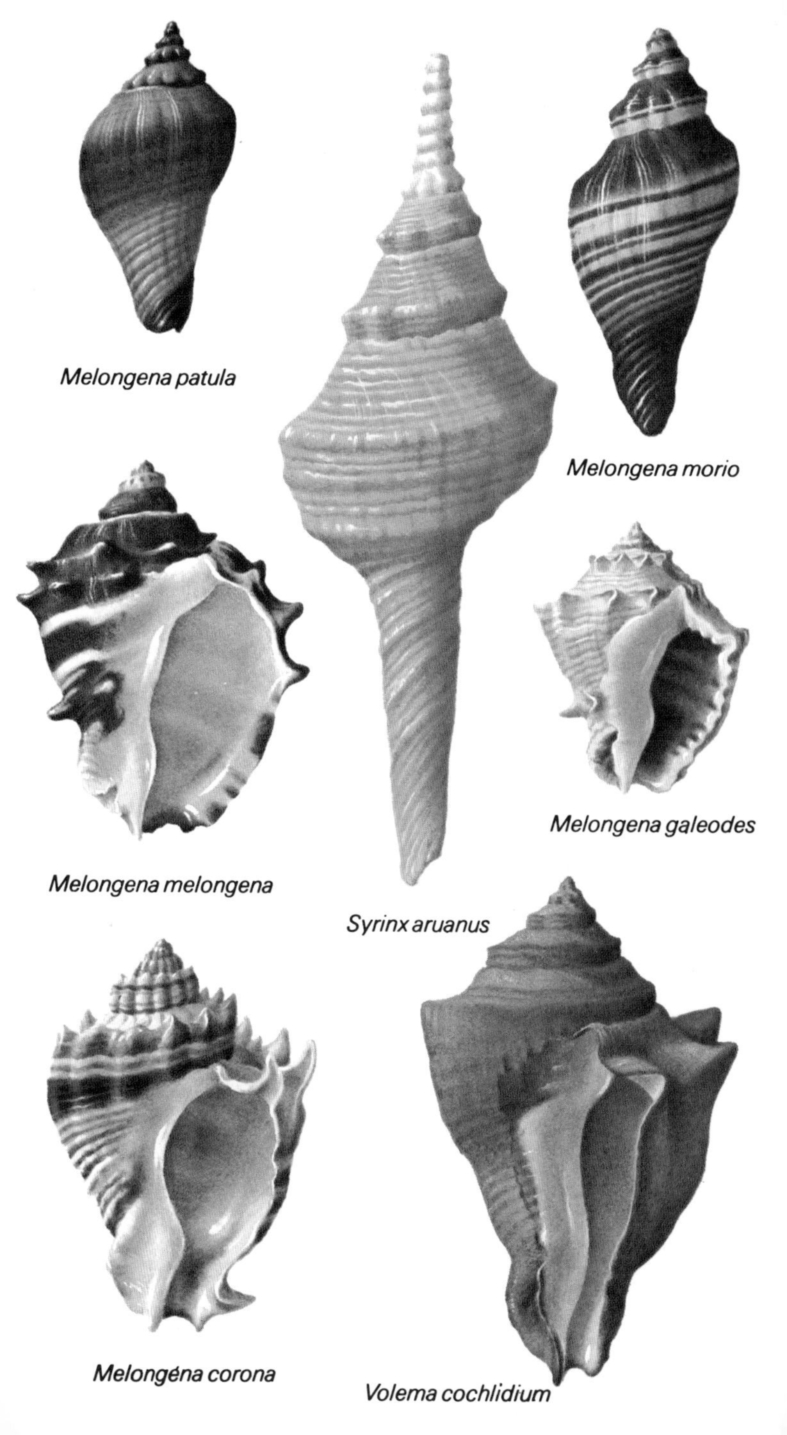

Melongena patula

Melongena morio

Melongena melongena

Syrinx aruanus

Melongena galeodes

Melongéna corona

Volema cochlidium

Superfamily: Volutacea
Family: Olividae

Variable in size, generally cylindrical, with short spire, a siphonal notch, the columella with folds anteriorly, and a fasciole. They are sand dwellers in all tropical and warm seas. Carnivorous, they feed at night and capture their prey by enfolding it in their large 'foot'. They have no periostracum or operculum, and their shells are very hard, shiny and usually brightly coloured.

Oliva sayana Ravenel 1834 west Atlantic, 80mm. Rather extended spire; narrowly and deeply channelled suture; columella has numerous plaits. Fawn-grey with brown zigzag marks in two, generally spiral bands, and small, yellow tent marks; brown and yellow flecks at the suture; inner edge of lip brown shading into white and then violet; columella white. Variable in pattern and range of colours. The form *O.s. citrina* Johnson 1911 is a golden yellow.

O. reticularis Lamarck 1811 West Indies and Florida Keys, 60mm. Rather long spire, narrowing sharply towards the anterior; channelled sutures. White, reticulated with pale or dark pinky brown; brown specks at suture; aperture white. *O.r. bifasciata* Küster 1878 has two red-brown bands, a narrow one in the middle of the body whorl, and a broad one anteriorly. Other forms include dark brown, pale yellow, white and red-brown, as well as variations in shape.

O. caribaeensis Dall and Simpson 1901 Caribbean, 40mm. More cylindrical than the above. Low spire and channelled suture. Colour rather like that of *O. sayana*. Dark flecks below the suture; aperture purple. The form *trujilloi* Clench 1938 is rather narrow, red-brown, and has a light aperture.

O. flammulata Lamarck 1811 west Africa, 35mm. Similar in shape and colour to *O. reticularis*. High spire and channelled suture. Usually pink-grey with red-brown, zigzag, axial markings and white or pale yellow tent marks; darker specks below suture; aperture white or with a purple tinge. Brown and yellow forms exist.

O. spicata Röding 1798 tropical west America, 60mm. Fairly high spire; a little concave, widest at the shoulder and narrowing anteriorly; deep narrow channel at the suture. Very variable in colour. General form is pale yellow-grey; spotted or reticulated with brown or red-brown, denser at the suture; aperture white. Colour forms vary from white to very dark red-brown. The common form *venulata* Lamarck 1811 is reticulated brown over a yellow-grey background.

O. porphyria L. 1758 tropical west America, 100mm. The largest of the genus, it has a low spire with a sharply pointed apex. The outline is a smooth curve from suture to fasciole, somewhat expanded at the shoulder. Narrow, deep channel at the suture; lip a little concave in the middle; columella strongly spirally ridged along its full length, except where it lacks a callus at the posterior end; strong fasciole. Violet-grey, covered profusely with tent marks of varying size outlined in dark brown-red; two indistinct bands of the same colour; aperture and columella flesh; base of columella, outside edge of lip and siphonal notch violet; apex and early whorls also violet above the suture.

O. incrassata Lightfoot 1786 tropical west America, 90mm. Solid, heavy and angular; very short callous spire; sharp apex; channelled suture. Ventricose, with a very angular shoulder for the genus, thickened before the lip. Callous columella with rather weak plaits anteriorly; heavy fasciole. Pale grey or yellow-grey; darker grey spots and dark brown, broken, axial bands of

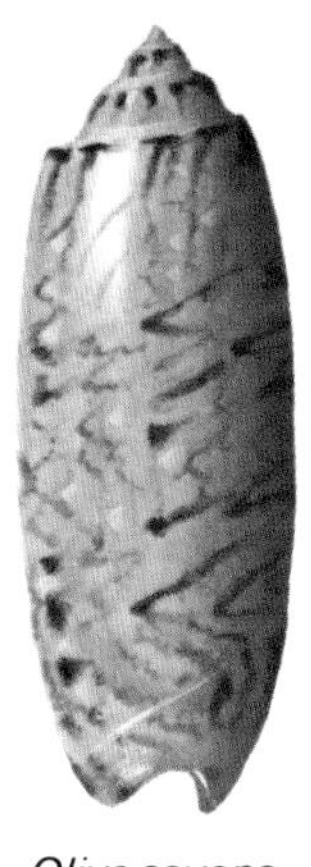

Oliva sayana

Oliva sayana form *citrina*

Oliva reticularis bifasciata

Oliva caribaeensis form *trujilloi*

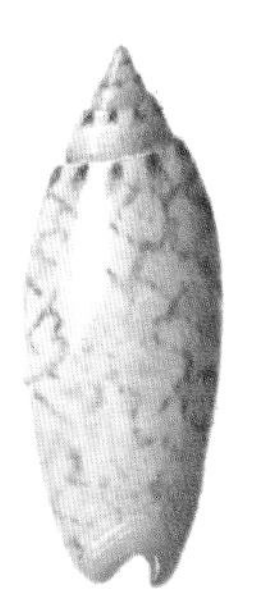

Oliva reticularis

Oliva porphyria

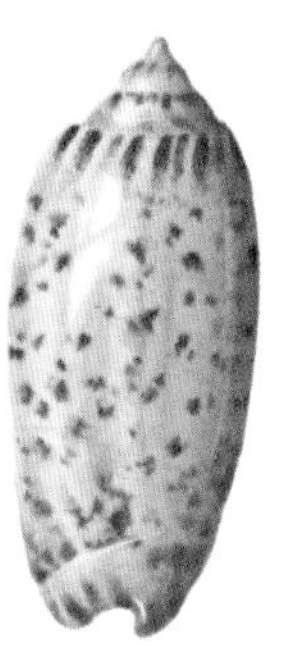

Oliva spicata form *fuscata*

Oliva incrassata

Oliva spicata form *venulata*

Oliva flammulata

horizontal dashes, and sporadic spots and dashes; aperture white with pale pink or flesh tinge, especially on the columella. Gold, white and black forms are known.

Oliva splendidula Sowerby 1825 Pacific coasts of Panama and Mexico, 50mm. Short spire with pointed apex; channelled suture; cylindrical. Markings comparatively constant. Pink-grey with small flesh-coloured tent marks and dark red-brown spots; overall two, uneven, broad, broken bands of light brown; dark red-brown spots and fine lines below the suture; lip white; interior with a yellow tinge; columella and apex pale purple.

O. peruviana Lamarck 1811 Peru and Chile, 50mm. Very variable in both colour and shape. The typical form is short-spired; narrowly canaliculate suture; somewhat inflated posteriorly. Columella with plaited callus on anterior half; no callus posteriorly. Blue-grey to flesh-coloured, profusely spotted with red-brown axial dashes and spots; aperture white with faint blue tinge; upper plaits of columella and fasciole cream, lower white. Varieties include all dark brown to white, sometimes with axial streaks; and variations in shape include a flatter spire with an angular shoulder and heavy callus at the posterior end of the columella, known as form *coniformis* Philippi 1848 (illustrated).

O. tigrina Lamarck 1811 Indian Ocean and west Pacific, 60mm. Short spire, sharp apex. Shallow sutural groove; slightly inflated body whorl. Columella plaited anteriorly. Typically white, well-spotted with blue-grey and some dark brown spots; aperture blue-white; base of columella with a red tinge. Sometimes dark brown almost black in form *fallax* Johnson 1910 (illustrated); or dark brown with white band or bands.

O. rufula Duclos 1835 Philippines and Moluccas, 35mm. Very short spire; channelled suture; cylindrical, narrowing slightly. Thickened lip is very slightly flared. There is a callus on either side of the sutural channel where it ends on the body whorl. Columella callous and plaited anteriorly. Milk-chocolate, coarsely reticulated with darker chocolate and grey diagonal stripes, which become dark brown, almost black, on the edge of the lip; interior white; base of columella pale pink.

O. bulbosa Röding 1798 Indian Ocean, 50mm. Short spire with pointed apex; inflated or bulbous body whorl. Lip more or less thickened before the edge. Columella with callus, sometimes very thin posteriorly and plaited anteriorly; a short sharp diagonal ridge at the top of the fasciole. Many colour varieties. Typical is brown with axial, wavy, dark brown lines. A common form *inflata* Lamarck 1811 (not illustrated) is pale blue-grey, profusely spotted with darker blue-grey; form *tuberosa* Röding 1798 is like *inflata* but with three broken bands of orange-brown; *fabagina* Lamarck 1811 with irregular brown blotches; form *bicingulata* Lamarck 1811 with two dark brown bands. There are other named and unnamed varieties.

O. tricolor Lamarck 1811 Indian Ocean and South China Sea, 60mm. Short, callous spire and deeply channelled suture. Cylindrical, slightly inflated centrally. Heavily callous columella is plaited anteriorly. Cream profusely spotted with blue-green and gold spots, the blue-green showing particularly in two bands below the suture and in the middle of the body whorl; lip edge and spire tessellated with dark brown-black and gold; inside lip white; fasciole salmon pink, red blotch at base. The form *philantha* Duclos 1835 is a light form, with pale yellow, green and blue, spots, dashes and obscure bands.

Oliva tigrina form *fallax*

Oliva splendidula

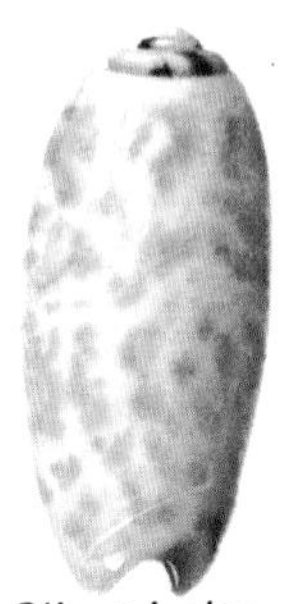
Oliva tricolor form *philantha*

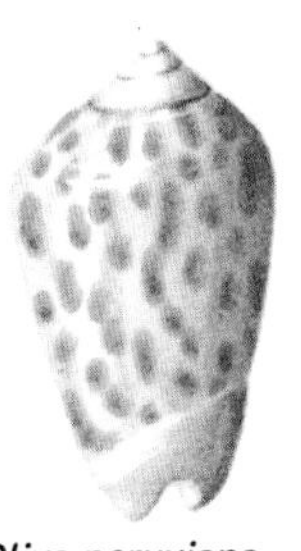
Oliva peruviana form *coniformis*

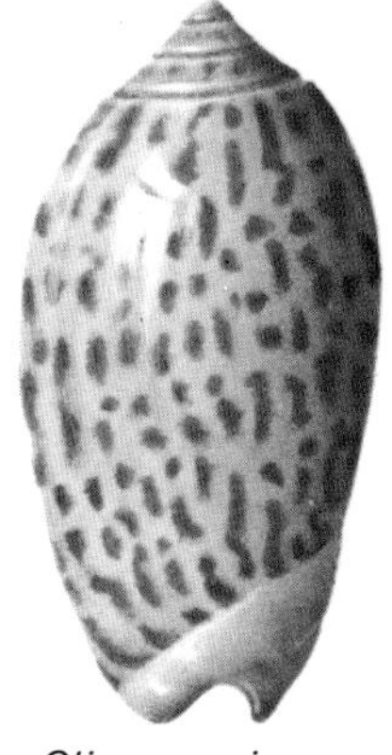
Oliva peruviana

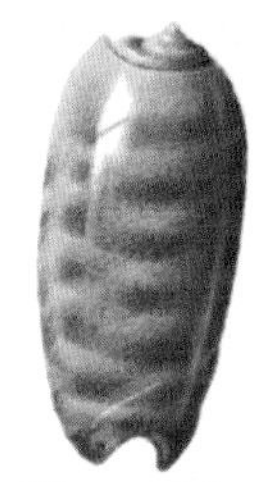
Oliva rufula

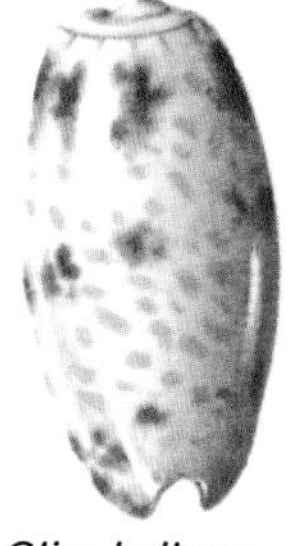
Oliva bulbosa form *tuberosa*

Oliva bulbosa form *bicingulata*

Oliva bulbosa form *fabagina*

Oliva elegans Lamarck 1811 Indo-Pacific, 40mm. Cylindrical, chubby, with very short spire, sharp apex and narrow, deeply channelled suture. The lip is extended posteriorly to level with the apex, and the posterior end of the columella is extended to the same level. Columella plicated its full length. Pale yellow-green with a network of discontinuous oblique olive-green lines; two bands of darker marks on the shoulder and the middle of the body whorl; columella white with a pink tinge at the base; inside lip and interior pale blue-white. Very variable in colour, from very dark to the golden form *flava* Marrat 1871, in which the aperture lacks the blue colour.

O. episcopalis Lamarck 1811 Indo-Pacific, 60mm. High, callous spire; narrow, deep suture; slightly swollen centrally. Columella ridged posteriorly and plicate anteriorly. White, profusely spotted – except below the suture – with blue-grey and golden yellow, often together giving an overall green effect; the spots tend to join up in lines and are more dense in two faint bands; inside edge of lip and columella white; fasciole with yellow tinge; interior deep violet. There are a number of named forms of varieties in colour and shape.

O. vidua Röding 1798 Indo-Pacific, 60mm. Very short or flat spire; a little inflated posteriorly; deep narrow suture ending in raised callous projection as in *O. elegans* above. Columellar callus thin near posterior end, ridged throughout. This species has many named colour forms, the typical being black with a white aperture. It is also well known by its synonym *O. maura* Lamarck 1811. Form *albofasciata* Dautzenburg 1927 is brown with two darker broken bands; form *aurata* Röding 1798 is orange, gold or golden brown, without pattern; form *cinnemonea* Menke 1830 is cinnamon with darker, axial, wavy lines; other forms include grey backgrounds and dark, zigzag, axial lines and spiral bands.

O. reticulata Röding 1798 Indo-Pacific, 35mm. Fairly low spire; deep, narrow suture, callous near its end as in *O. vidua*, but not as produced posteriorly. A little inflated; columellar callus weak posteriorly, obsoletely ridged centrally. Cream-white background, heavily covered with a network of dark grey-brown, two darker brown bands; aperture white; columella, plaits, and extreme edge of outer lip, red, giving it the name by which it was long known of *O. sanguinolenta* Lamarck 1811. The form *azona* Dautzenberg 1927 lacks the bands, and *evania* Duclos 1835 (illustrated) is paler and has a dense pattern. There are other named colour forms.

O. mustellina Lamarck 1811 India to Japan, 40mm. Rather narrow with moderate to short spire and channelled, deep, relatively wide suture. Shoulder rather square and high on the whorls, giving a somewhat oblong appearance. Columella weakly callous but plaited its full length, with a callus posteriorly bordering the end of the suture. Mustard yellow with an irregular network of purple-brown; columella blue-white; interior deep violet.

O. multiplicata Reeve 1850 Taiwan, 40mm. A high, very slightly convex spire, widest a little below the suture which is narrowly and deeply channelled. Narrows considerably towards the anterior end. Columella with many plaits as its name implies. Yellow-cream, heavily marked with pale purple-brown; darker purple-brown blotches at the anterior end, and with a few below the suture; the yellow-cream background shows as tent marks as in *O. porphyria* and many of the Conidae; columella and interior white.

Oliva elegans

Oliva elegans form *flava*

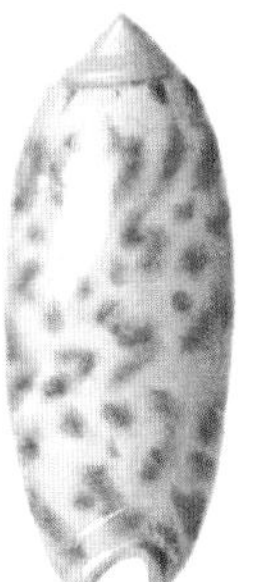

Oliva episcopalis

Oliva vidua form *albofasciata*

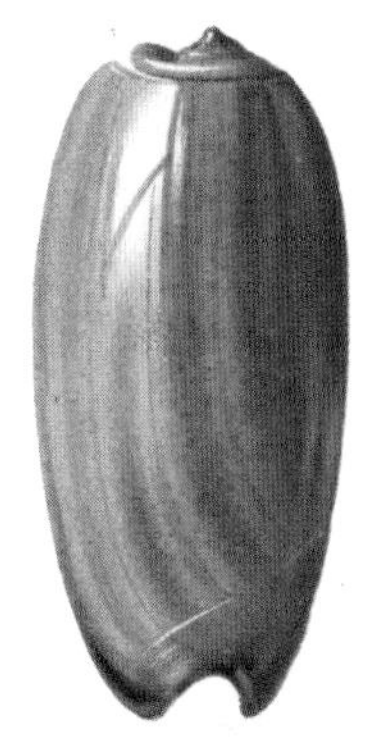

Oliva vidua form *cinnemonea*

Oliva vidua form *aurata*

Oliva reticulata form *evania*

Oliva mustellina

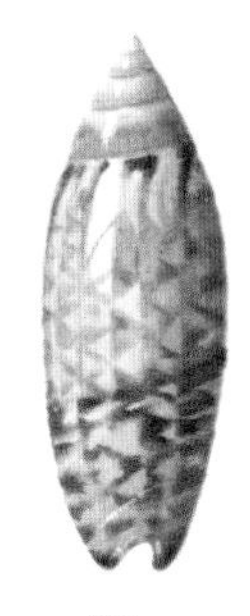

Oliva multiplicata

O. oliva **L. 1758** Indo-Pacific, 30mm. Perhaps better known still by its synonym *O. ispidula* L. 1758. Short spire and narrow sutural groove. Columellar callus on the anterior two-thirds, which is unevenly plaited; rather narrow aperture. Colour and pattern are extremely variable; white to black, through browns, chestnut, yellow, grey; with spots, zigzag lines, blotches, or without any markings; the columella is white and the interior usually brown, in some forms white or pink. Three forms are illustrated including the all-black *oriola* Lamarck 1811.

O. australis **Duclos 1835** Australia and New Guinea, 20mm. Moderate to high spire and narrow sutural groove. Columella a little concave posteriorly; uneven plaits covering the whole columella. White, covered with fine, wavy, axial lines of cream-grey to grey; a row of purple-brown dashes at the suture; columella and aperture white.

O. caldania **Duclos 1835** west and north Australia and Indonesia, 22mm. Similar to *O. australis* except that it has a much shorter spire. It may be a form of *O. australis.*

O. sidelia **Duclos 1835** Indo-Pacific, 20mm. Narrow with a short spire which is covered by a callus except for the suture of the last whorl. The suture is deep and wide. Columella coarsely plaited along its full length. White with brown markings, generally tent-shaped. Columella and aperture white, sometimes with a violet tinge. The form *volvaroides* Duclos 1835 (illustrated) is all one colour, either medium or dark brown or white.

O. carneola **Gmelin 1791** Indo-Pacific, 25mm. Short, almost flat spire; covered with a callus except for the deep sutural groove of the last whorl. Generally rather swollen and with a slightly angular shoulder. There are many colour varieties and patterns but the shell is basically white, banded with orange, red or purple-red; many varieties are named.

O. paxillus **Reeve 1850** Guam and Fiji, 27mm. High spire with narrow, deep sutural channel. Body whorl expanded at the shoulder and narrowing quite sharply towards the anterior. Columella with rather coarse plaits overall. Ivory white with faint, grey-brown network below the shoulder; purple-brown, comma-shaped marks below the suture; aperture white; inside of lip sometimes with two or three, spiral, purple lines. A form from the Hawaiian Islands, named *sandwichiensis* Pease 1860 has a shorter spire, is less angular at the shoulder and has darker markings with two, obsolete, spiral bands.

O. tessellata **Lamarck 1811** west Pacific and east Indian Oceans, 35mm. Short spire; rather callous with wide, shallow sutural channel. Body whorl a little expanded centrally; strongly plaited columella. Creamy yellow with rather evenly placed, purple-brown spots. Edge of lip and anterior end of fasciole white; aperture and columella rich, very deep violet.

O. buloui **Sowerby 1887** New Britain, 30mm. Moderately high, callous spire. Body whorl a little expanded at the shoulder. Columella with coarse plaits. Spire and area below the suture, and anterior edge, pink; elsewhere apricot with brown-red streaks becoming more frequent and darker anteriorly.

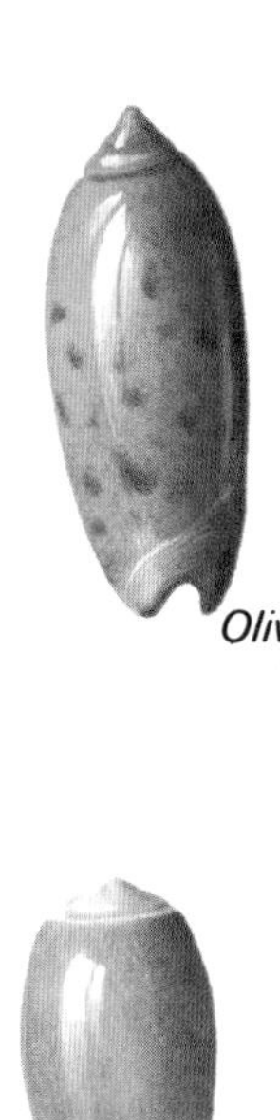

Oliva oliva

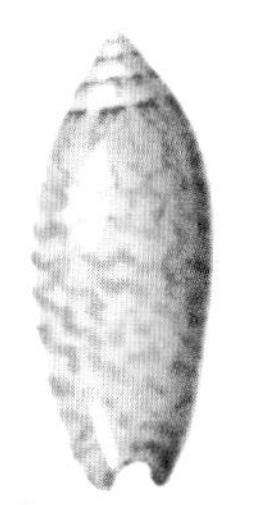

Oliva caldania

Oliva paxillus form *sandwichiensis*

Oliva sidelia form *volvaroides*

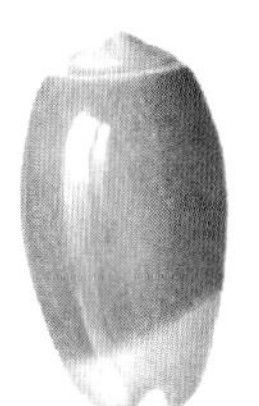

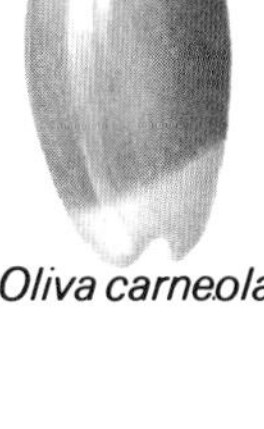

Oliva carneola

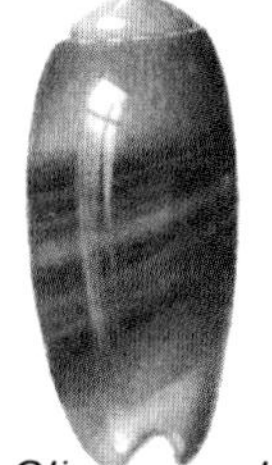

Oliva carneola

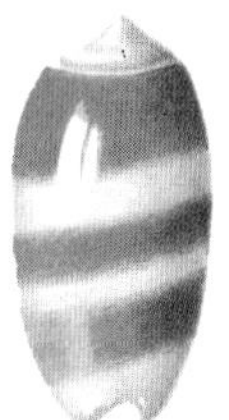

Oliva carneola

Oliva australis

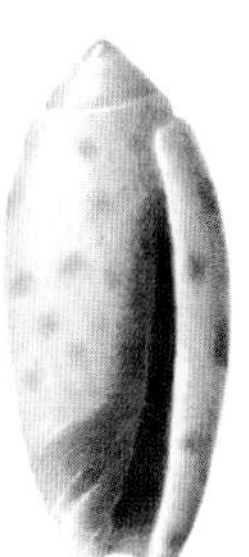

Oliva paxillus

Oliva tessellata

Oliva oliva

Oliva oliva form *oriola*

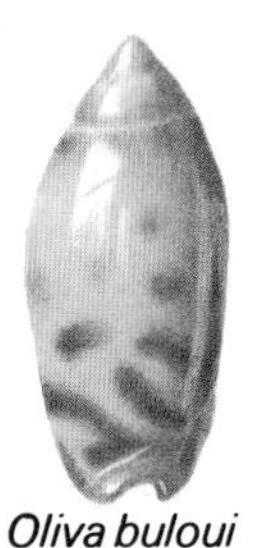

Oliva buloui

O. miniacea Röding 1798 Indo-Pacific, 90mm. Heavy, solid, handsome, with moderate to short spire, deep, channelled suture and slightly inflated body whorl. Its colour is very variable except for its characteristic rich orange interior. Typically it is a cream-yellow with irregular, wavy, brown-grey, axial lines, and two, broad, broken bands of dark chocolate, one below the suture and one in the middle of the body whorl; inner edge of lip brown with a cream band between it and the rich orange interior; columella white; fasciole band apricot; the axial streaks may be purple, brown, green or blue.

O. tremulina Lamarck 1811 Indo-Pacific, 90mm. This may be a form of *O. miniacea*, from which it differs only in having a white interior. The colour varieties are otherwise like those of *O. miniacea*. The form *oldi* Zeigler 1969 is pale grey with darker grey, close, wavy, axial lines, with dark bands at the shoulder and in the middle of the body whorl; columella with an orange tinge.

O. textilina Lamarck 1811 Indo-Pacific, 85mm. Short spire and deep, narrow, channelled suture. Top of columella with a strong callus rising above the suture on the penultimate whorl and broadening towards the end of the channel. White, more or less heavily marked with light or dark grey; some irregular, wavy, axial, grey-brown lines; usually a dark brown band below the suture and in the middle of the body whorl. An albino form is *O.t. albina* Melville and Standen 1897. The illustrated specimen is somewhere between this and the typical form.

O. ponderosa Duclos 1835 Indian Ocean, 85mm. Solid, heavy, with a short spire and a callous area at the top of the plaited columella rising up as in *O. textilina*. White with a network of faint pale brown on much of the body whorl; streaks and spots of purple-brown below the suture, on the middle of the body whorl and the fasciole; aperture and columella pure white.

O. lignaria Marrat 1868 east Indian Ocean to Taiwan and north Australia, 65mm. Long known by its synonym *O. ornata* Marrat 1867. Short spire is heavily callous; such of the sutural channel not covered is narrow and deep. Pale grey-cream, with dots and dashes of brown and light blue-grey, and sparse dashes of dark brown on two broken bands; aperture pale violet; columella tinged with red anteriorly. The golden or orange form *cryptospira* Ford 1891 often has this colour pattern showing faintly through the overlying colour.

O. annulata Gmelin 1791 Indo-Pacific, 60mm. High spire and narrow, deep sutural channel. A little thickened before the edge of the lip and rather sharply shouldered for the genus; sometimes with a protruding ring around the centre of the body whorl. The type form is an uncommon near white or cream colour, while the more common form *amethystina* Röding 1798 is flesh-coloured with pale purple spots and darker ones just below the suture; apex pale yellow. Form *intricata* Dautzenberg 1927 has a network of fine red-brown lines and sparse dark brown spots; aperture rich orange. The white form *mantichora* Duclos 1835 is similar but the posterior edge of the last whorl is more or less sharply angled. Variable like so many of the genus.

O. bulbiformis Duclos 1835 Indo-Pacific, 35mm. Short and rather bulbous. Short spire usually completely covered by a callus. Body whorl inflated especially posteriorly. Mustard or grey-yellow, more or less heavily reticulated with grey-brown and usually with two darker bands; one colour variety is green-grey; columella may have a red tinge; interior usually violet-brown or chocolate brown.

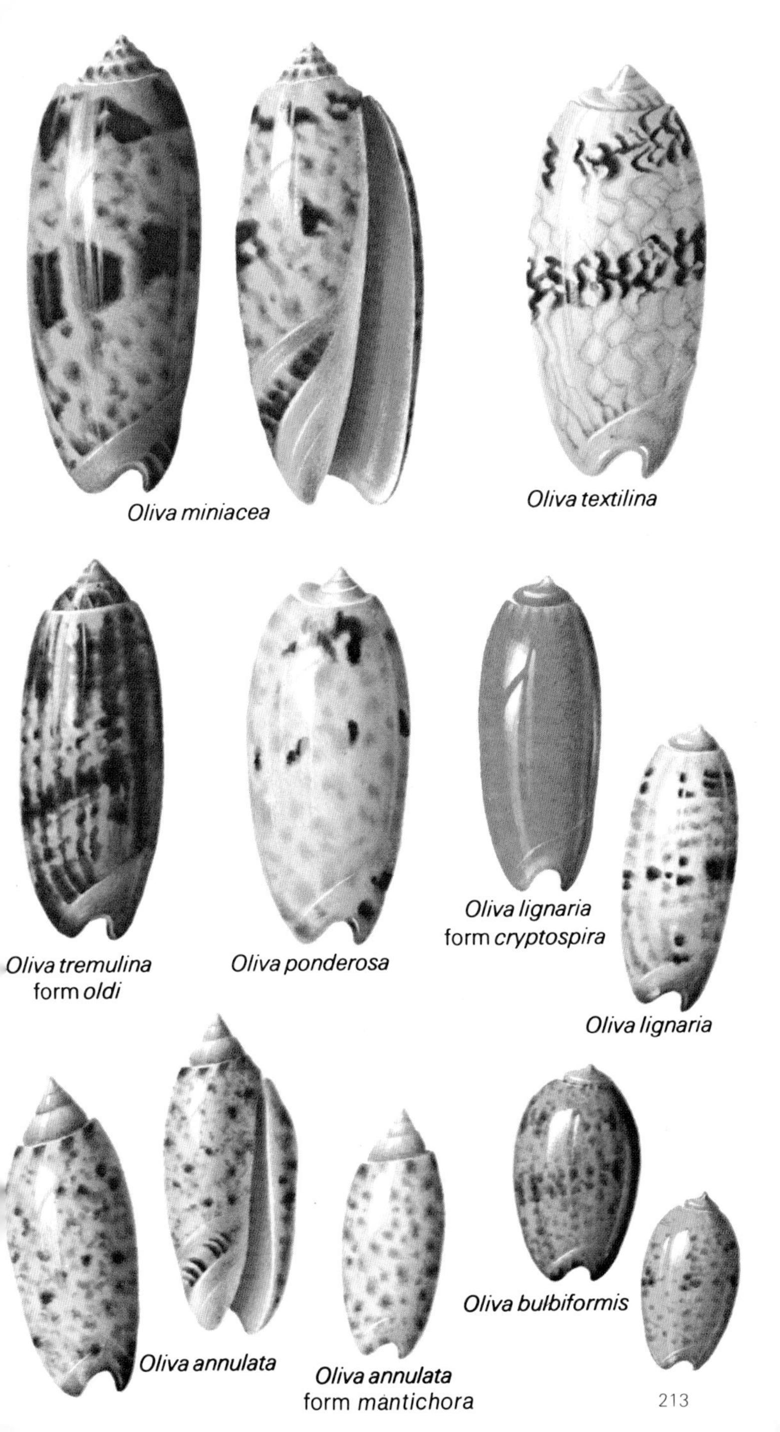
Oliva miniacea
Oliva textilina
Oliva tremulina
form *oldi*
Oliva ponderosa
Oliva lignaria
form *cryptospira*
Oliva lignaria
Oliva annulata
Oliva annulata
form *mantichora*
Oliva bulbiformis

Ancilla lienardii **Benardi 1821** Brazil, 30mm. Chubby with moderate spire, swollen body whorl, concave columella, and deep, open, twisted umbilicus. Golden orange; aperture and columella white; narrow white band immediately above fasciole.

A. albocallosa **Lischke** Japan, 45mm. High, callous spire; inflated body whorl; wide aperture; no umbilicus. Fawn; brown band below suture and on fasciole, both edged white; aperture, columella, inner side of callus pink.

A. tankervillei **Swainson 1825** West Indies to Brazil, 60mm. High spire and inflated, with deep narrow twisted umbilicus. Yellow; pink band below suture and above fasciole; aperture and columella white.

A. cingulata **Sowerby 1830** east Australia, 90mm. Shiny; high spire; rounded apex; restricted suture; fine growth striae; folds on fasciole disappear into aperture. Fawn; white band above and below suture; narrow brown line on suture edge on last half of body whorl; penultimate whorl rich red-brown; antepenultimate paler; earlier whorls pink-white; interior fawn; columella, inner edge of siphonal notch white; fasciole has two red-brown bands, light band between.

A. australis **Sowerby 1830** New Zealand, 25mm. Moderate spire; biangulate body whorl; suture entirely covered with transparent callus. Brown; broad, dark blue-brown band edged with white covering most of body whorl; interior dark brown; columella white; fasciole dark brown.

A. mucronata **Sowerby 1830** New Zealand, 45mm. Solid, stubby, shiny; short, callous spire; wide aperture. Columella with central ridge, twisted anteriorly. Pink to brown; apex paler; fasciole darker; faint, pale line on shoulder.

Olivancillaria urceus **Röding 1798** Brazil to Uruguay, 40mm. Solid and squat. Almost flat, broad, callous spire with sharp, pointed apex and wide, deep suture. Swollen just below suture then narrowing sharply. Aperture wide, especially anteriorly. Columella with large, heavily callous area posteriorly, plaited anteriorly and fasciole is very broad and callous. Lip extends beyond where it is attached to body whorl at end of sutural canal. White; profusely axially streaked with fawn; inside lip fawn; columella white; posterior half of callous area fawn; fasciole fawn-edged and clouded with brown.

O. vesica auricularia **Lamarck 1810** Brazil to Argentina, 50mm. Solid, coarse with very expanded body whorl. Very short spire, entirely covered with thick callus. Deeply channelled suture on dorsal side of body whorl, elsewhere covered by callus. Fine growth striae; lip extending further than its juncture with body whorl; smooth columella, plaited anteriorly; very broad, callous, fawn fasciole. Grey-white; columella and callous area white.

O. gibbosa **Born** Sri Lanka, 50mm. Short spire and inflated body whorl; penultimate whorl callous almost to shallowly channelled suture. Columella smooth posteriorly, plaited over lower two-thirds, and curved convexly. Fasciole strong and raised, giving the effect of a 'dent' above it. Dark brown with light spots and axial dashes; white band low on body whorl; horn-coloured band on fasciole; aperture, columella and callus white; creamy touches inside lip and base of columella. Sometimes pale yellow-green, heavily maculated with light brown.

Agaronia testacea **Lamarck 1811** tropical west America, 50mm. Slender, high, pointed spire; deeply channelled suture; wide aperture. Columella weakly callous posteriorly, strongly plaited and twisted anteriorly. Pale grey-green with uneven axial rows of grey-brown dots, dark below the suture; lip

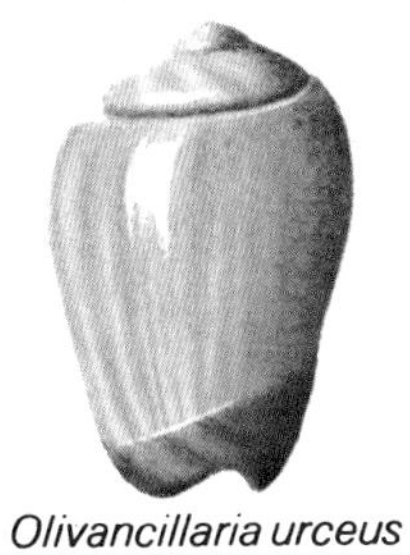
Olivancillaria urceus

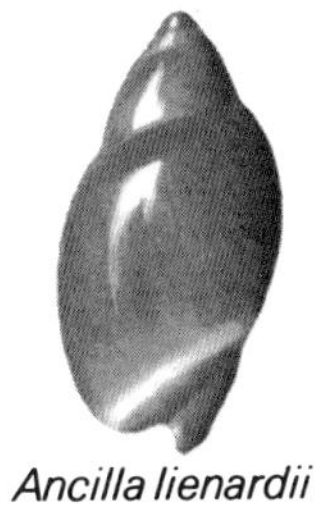
Ancilla lienardii

Ancilla albocallosa

Ancilla tankervillei

Ancilla cingulata

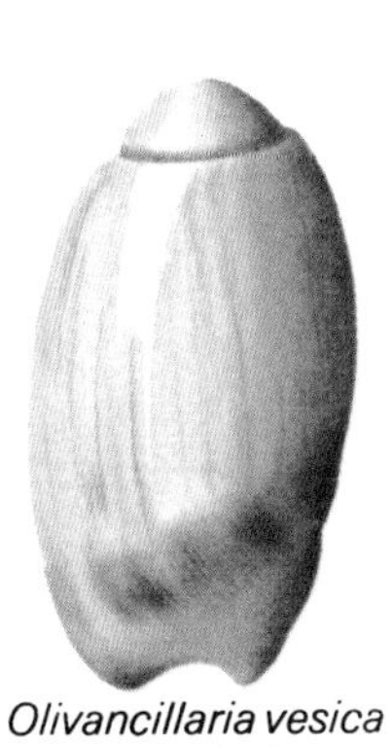
Olivancillaria vesica auricularia

Olivancillaria gibbosa

Ancilla australis

Ancilla mucronata

Agaronia testacea

blue-grey lined with brown; columella white; fasciole clouded with dark brown, yellow and green.

Family: Vasidae

A small family, most are moderately large shells. There are some twenty-five species, usually solid and heavy, spirally ridged, with a plaited columella. They are tropical, most living in shallow water on the reefs.

Vasum turbinellus L. 1758 Indo-Pacific, 85mm. Moderate to short spire. Indistinct suture with two rows of more or less long blunt spines below it, the upper one on the shoulder being the larger; a discontinuous cord below these forms a row of small blunt nodules, then a rough ridge, two more rows of smaller blunt spines, the upper again the larger, and finally the fasciole ridge. The surface is generally malleate. The columella has three slightly oblique ridges with fine thin plaits between. Narrow, patterned parietal area; lip has coarse blunt teeth, some paired; open, short siphonal canal. White with dark brown maculations; aperture white; teeth dark brown; columella cream with brown marks round the edge.

V. rhinoceros Gmelin 1791 Kenya and Zanzibar, 85mm. Rather short spire. Spirally ridged and corded with two rows of blunt spines/nodules, one pair below the suture and one pair above the fasciole, the posterior one in each pair is the larger, and overall finely lamellated axial striae. Thickened lip has teeth, generally in pairs, and one larger tooth posteriorly. Columella has three plaits; strong parietal callus; short siphonal canal; umbilicus may be open or sealed by the parietal callus. White maculated with brown; aperture white, some brown streaks within; columella cream, heavily clouded, especially on the parietal wall, with brown.

V. capitellum L. 1758 West Indies, 65mm. Fusiform with high spire and angulate. Fine spiral striae and coarse spiral ridges, about nine on body whorl; fine axial lamellate striae. Strongly crenulate lip is toothed within, some teeth bifid. Columella has three strong plaits and two obsolete ridges posteriorly; fasciole strong. White with brown clouding; aperture cream.

V. ceramicum L. 1758 Indo-Pacific, 140mm. Fusiform, solid and heavy with high spire. The sculpturing is the same as in *V. turbinellus*, though the shell is narrower for its length, and the two small plaits on the columella are not always present. White heavily maculated with dark brown-black; aperture white; teeth almost black: columella with dark brown marks on parietal wall.

V. tubiferum Anton 1839 Philippines, 115mm. Very similar in shape and sculpture to *V. turbinellus,* but bigger spines on the two rows, and overall larger and stronger. Narrowly open umbilicus. White maculated with brown, not black; lip and teeth white; columella white with a pink tinge and a dark brown-purple blotch.

Tudicula armigera A. Adams 1855 Queensland, Australia, 75mm. Delicate with rather short spire, large body whorl and long, narrow siphonal canal – rather like some of the Muricidae. Finely spirally corded with long, sharp, hollow spines on the shoulders – about eight on the body whorl – which show just above the suture on earlier whorls. Three cords on the body whorl are larger than the others and have sparse, short spines. There is a row of about five, long spines round the posterior end of the long siphonal canal and another of short ones below it. Finely dentate lip is ridged; columella has four plaits and a broad, partly free-standing parietal shield. White, cream or pale brown with brown clouding; aperture white.

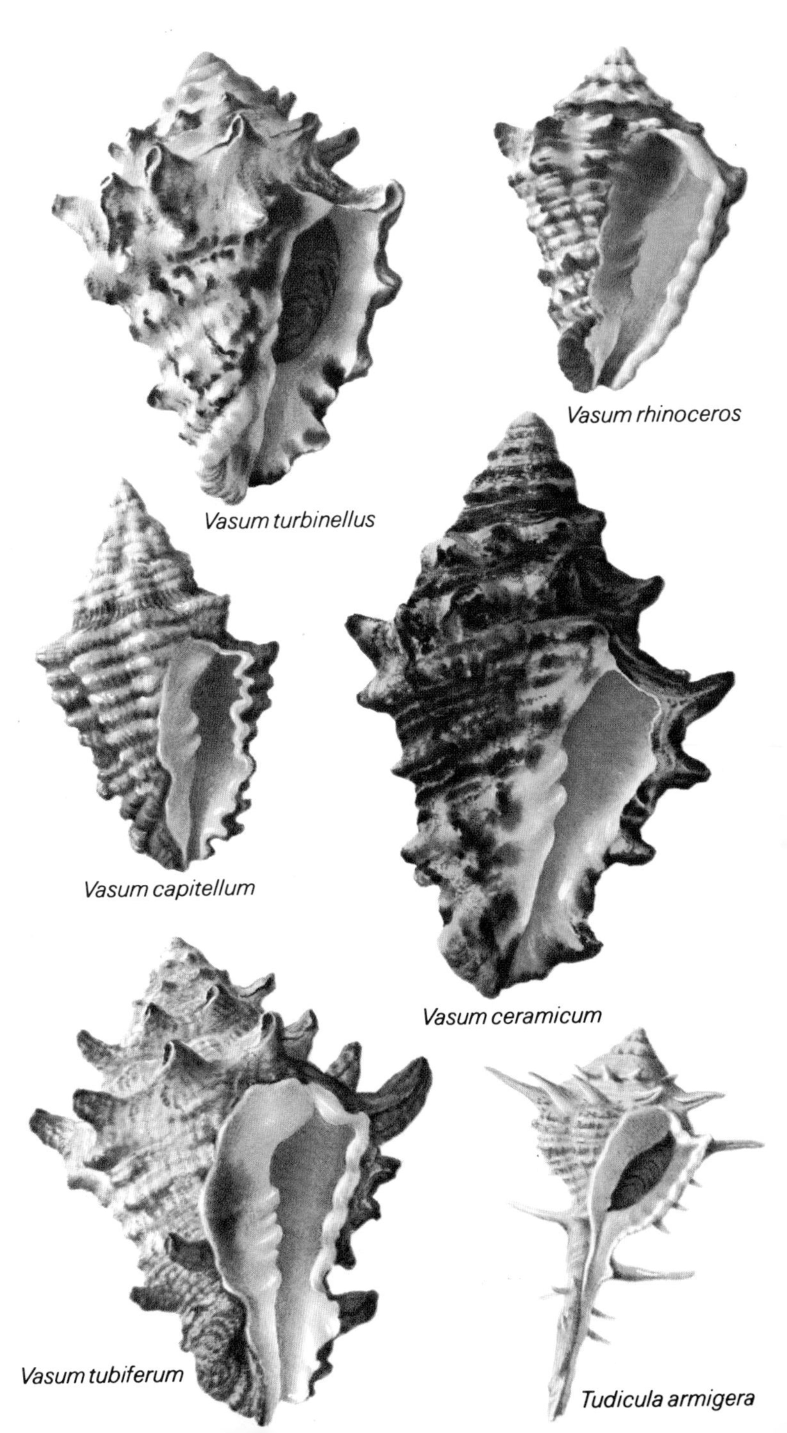

Vasum turbinellus

Vasum rhinoceros

Vasum capitellum

Vasum ceramicum

Vasum tubiferum

Tudicula armigera

Family: Turbinellidae

Better known by their synonym Xancidae, as the generic name *Xancus* is better known than that of *Turbinella*. This family and the Vasidae are considered by some to be subdivisions of the same family. They are generally large, heavy shells with plaited columellas. They are found only in a few areas.

Turbinella angulatus **Lightfoot 1786** Caribbean, 350mm. Very heavy and solid. Fusiform with high spire; about seven angular whorls; well-defined suture. Spirally corded, cords well-separated and obsolete on the middle of the body whorl. Strong, blunt knobs on each whorl—about eight on bi-angulate body whorl—which point slightly posteriorly and extend anteriorly into weak, wide ribs; growth lines are prominent from the suture to the shoulder and towards the lip. Wide aperture; simple outer lip; concave columella has three, large plaits; callous parietal area; extended and straight siphonal canal; open, rather narrow umbilicus. White; outer lip and interior white; columella and parietal callus light orange-brown. This shell has a strong, brown periostracum and traces of this still adhering can be seen in the illustration.

T. pyrum **L. 1758** Bay of Bengal, Sri Lanka, 170mm. Very heavy. Moderate spire of about six whorls with adpressed sutures. Inflated body whorl; malleate surface; corded at the posterior end and on the short siphonal canal. Simple outer lip; columella with four plaits, the posterior one the largest, others decreasing towards the anterior; callous parietal wall. White; edge of lip, columella and parietal callus pale peach. This shell also has a thick heavy periostracum. It is a sacred object of the Hindus associated with Vishnu, and is frequently seen in Hindu and Buddhist temples, in sacred pictures, and so on. It is often carved with encircling grooves, and is drilled for use as a horn, and cut and carved into ornaments. Rare sinistral specimens are very valuable and most highly prized by Hindus.

T. laevigatus **Anton 1839** (not illustrated) Brazil, 120mm. Solid with bulbous body whorl. Spire tapering to heavy nucleus. Moderately elongated siphonal canal; columella with three strong plaits. White with a strong, brown periostracum. This is an uncommon shell even in its very restricted range.

Turbinella angulatus

Turbinella pyrum

Family: Harpidae

A small family of two genera and some twelve species. Among them, to me, are the most beautiful of all shells. They have short spires, an inflated body whorl, a greater or lesser number of axial folds or ribs, wide aperture, no umbilicus and no operculum. Columella has no plaits and there is no periostracum. They live in and among coral, are active creatures and carnivorous. Most live in the Indo-Pacific region. Many of the harps are known as well or better by synonyms which have been given.

Harpa major Röding 1798 Indo-Pacific from East Africa to Hawaii, 100 mm. Moderate to low spire with sharp apex. Body whorl has about twelve ribs, some wide, some narrow, with small points on the sub-angulate shoulder; finely striate between ribs; earlier whorls callous; simple, blunt lip; large, callous parietal area; strong fasciole. White; ribs with transverse bands of shades of pale pink-brown; interstices scalloped axially with the same colours; penultimate whorl rosy pink over a purple background below the shoulder and yellow above to the pink apex; inside lip with red-brown marks at the end of the darker exterior bands, and inner aperture with golden brown clouding; outer side of columella and parietal area dark chocolate brown almost divided by a yellow wedge and with lighter brown areas on the anterior, inner half of the columella. Synonym *H. conoidalis* Lamarck 1843.

H. articularis Lamarck 1822 Philippines and Indonesia to north Australia and Fiji, 95mm. Moderate spire; sharp apex. Body whorl has about fourteen ribs, generally rather narrow, ending in small sharp points on the shoulder; finely axially striate; penultimate whorl callous below the shoulder. Simple lip; strong fasciole. Parietal area heavily glazed. Pink-flesh to tan with typical *Harpa* markings between the ribs. Bands of dark brown to black show on the ribs. Columella and parietal area dark brown, the ribs showing through in a lighter brown. Inside lip white, and interior light brown.

H. amouretta Röding 1798 Pacific, 50mm. More slender than others of the family and with a higher spire. About thirteen, axial ribs, wider towards the aperture and finely pointed at the shoulder; finely axially striate between the ribs. Simple lip; parietal wall not very callous; long, rather narrow fasciole. White; ribs shiny and crossed by pairs of dark brown lines enclosing yellow-brown or grey-brown; the interstices are dull with profuse, brown, axial scalloping; the outside pattern shows obscurely through the outer lip; the columella is fawn anteriorly with a brown blotch in the middle and a brown spot near the anterior end, the posterior half continuing the pattern on the body whorl. Though this is the typical form, much less slender and more globose species are not uncommon, and the overall colour can be more or less dark. The form *crassa* Krauss 1848 is heavier, coarser and more angulate; the ribs especially being heavier and sometimes more numerous. It is often lighter coloured.

H. crenata Swainson 1822 (not illustrated) west tropical America, 90mm. Rather similar to *H. doris* but more inflated; lower spire; spines on the shoulder smaller but with two or sometimes three other rows of more or less obsolete spines below the shoulder. The markings are similar but it is generally lighter in colour and the central band is purple-brown instead of orange-red.

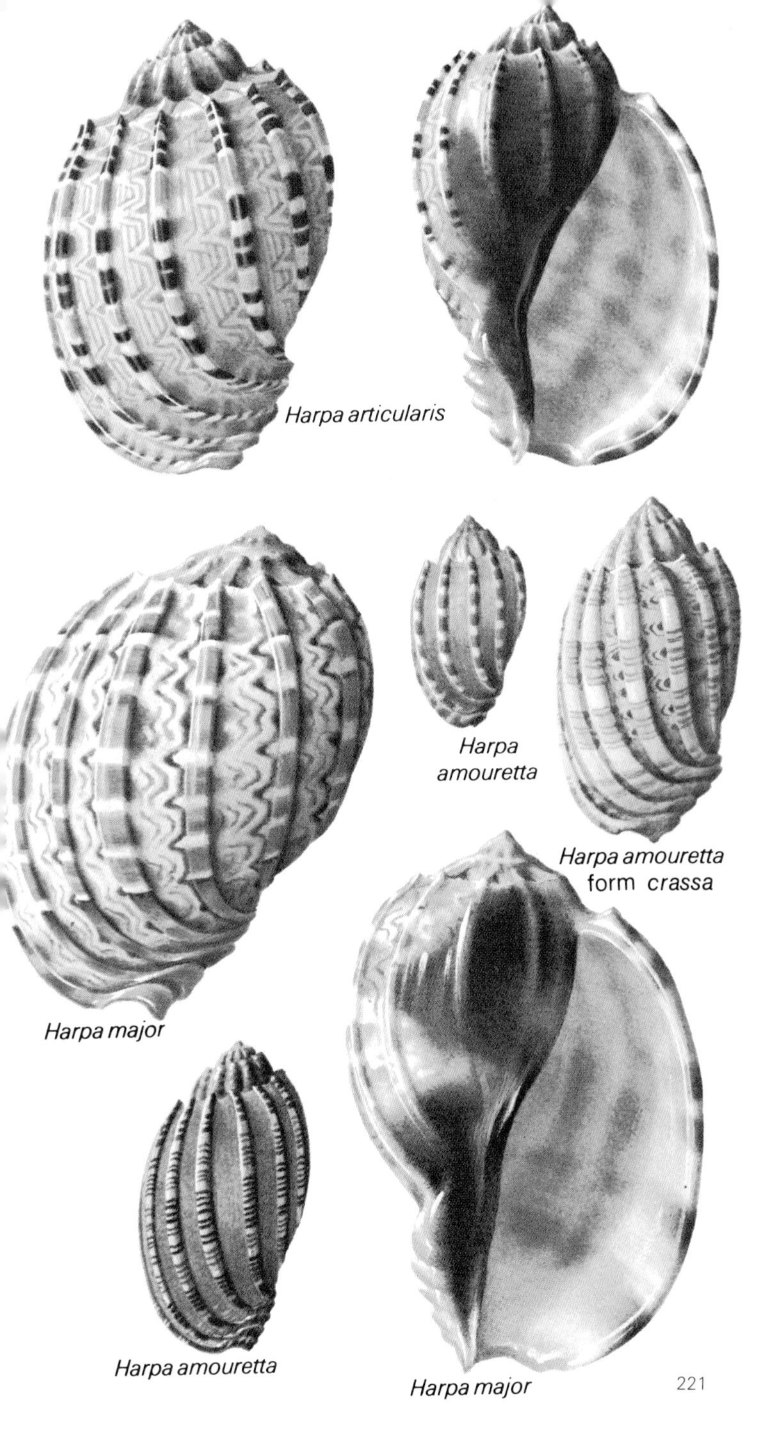
Harpa articularis
Harpa amouretta
Harpa amouretta form crassa
Harpa major
Harpa amouretta
Harpa major

H. davidis **Röding 1798** Bay of Bengal, 90mm. Light, moderate spire. Expanded body whorl with about eleven ribs, generally bigger towards aperture. Small, sharp spines on shoulder. Smooth lip; twisted fasciole. Fawn-pink; spiral bands of lighter pink scalloping and V-marks; ribs with single, red-brown lines; posterior of columella, parietal area dark brown, divided by >-shaped pale band; small, brown mark near anterior of columella.

H. kajiyamai **Rehder 1973** south Philippines, 70mm. Moderate spire; rather elongate. Twelve to seventeen ribs on body whorl bearing short, sharp spines on shoulder. A little inflated below shoulder. Early whorls with axial riblets and spiral cords. Penultimate whorl callous below shoulder. Concave between shoulder and suture with axial striae. Lip dentate anteriorly, obsolètely so posteriorly. Smooth, slightly concave columella; strong, heavily glazed fasciole. Flesh-pink with mauve-pink banding and white, mauve, pink and chestnut scalloping, the colours brighter and more vivid than in other Harpidae; blotches of red between alternate pairs of ribs spirally on periphery; red-brown lines in pairs or threes on ribs, which form dark brown marks on inside of lip; large mauve-brown blotch just above the fasciole on columella and parietal area, and small ones anteriorly and posteriorly; in the interior the outside colours show through the tan-white.

H. doris **Röding 1798** tropical west Africa, 75mm. Body whorl more elongate and less inflated than in other species. About thirteen, low ribs. Short spines on the shoulder. Fine axial striae. Lip with teeth which are obsolete posteriorly. Rather dull colouring; pink-brown to tan, edge of ribs with thin, broken, dark brown line; narrow chestnut bands carrying white >-marks end in teeth on lip; a more or less indistinct orange-red band on widest part of the body whorl; aperture pink-fawn with interrupted, darker, spiral bands; columella pale yellow anteriorly; posteriorly and on parietal area the basic pattern shows through the light glaze; large, dark chestnut blotch above fasciole running on to columella, a smaller one on anterior, outer edge of columella and a third smaller still where lip joins body whorl. Synonym *H. rosea* Lamarck 1822.

H. harpa **L. 1758** east Africa to Tonga, 75mm. Ovate; moderate spire; square shoulder. Twelve axial ribs, bigger near aperture. Short, sharp spines on shoulder. Fine axial striae between ribs. Short, sharp teeth on posterior two-thirds of lip. Strong, twisted fasciole. Flesh-pink; ribs crossed by dark brown lines in twos, threes or fours; between ribs, axial scalloping in pink, white and red-brown, rows of dashes and V-shaped marks, the various colours and patterns forming bands only broken by the ribs; band on periphery of red-brown blotches between alternate pairs of ribs; lip with brown marks where groups of lines end on last rib; columella fawn with brown mark centrally, smaller blotch posteriorly on parietal wall and one smaller still at base of columella. Synonym *H. nobilis* Röding 1798.

H. costata **L. 1758** the Imperial Harp, Mauritius, 80mm. Almost flat spire with sharp, pointed apex. Body whorl with thirty to forty, close-set, axial ribs which project in short, sharp points on the shoulder, then cross a broad, shallow channel to the suture; posteriorly the ribs curve parallel with the lip and cross the fasciole forming lamellae. Simple lip; wide aperture. Columella with a faint ridge continuing the fasciole ridge into the interior. White, spirally banded with fawn and pink-brown; aperture white with yellow clouding, especially anteriorly; columella yellow with a large, triangular, dark red-brown area centrally; a lighter brown streak from it running towards the apex and with brown streaky blotches on the parietal area. Synonym *H. imperialis* Lamarck 1822.

Harpa davidis
Harpa kajiyamai
Harpa doris
Harpa harpa
Harpa costata

Family: Mitridae

The mitres are a family of some ten genera and many species. They are large and small, generally fusiform and elongate, smooth or spirally ridged and some axially ridged or cancellate. Aperture is generally narrow, columella with plaits and siphonal canal has a notch. The animals live in sand or among coral and seaweed in intertidal and shallow areas. They are carnivorous and scavengers, and inhabit tropical and temperate seas worldwide.

Mitra papalis L. 1758 Indo-Pacific, 125mm. High spire. Punctate, spiral grooves becoming obsolete on the last two whorls which are coronated at the suture. Columella with about five plaits. White with dark red spots, generally rather square and in spiral rows; coronations white; aperture cream.

M. imperialis Röding 1798 Indo-Pacific, 65mm. Punctate, spiral grooves and small, blunt coronations at the suture. Columella with about six plaits; dentate lip. Yellow-brown with dark brown and white clouding; aperture orange-brown.

M. stictica Link 1807 Indo-Pacific, 75mm. High spire. Sharply angular, coronated shoulders; two rows of deep punctures on early whorls which disappear on the penultimate whorl. Slightly waisted body whorl with three spiral grooves anteriorly. Lip edge is parallel with the axis of the shell for most of its length and curves sharply through 90° to end at the siphonal notch; finely dentate lip; columella with about four plaits. White, with generally rather square red spots in spiral rows; aperture cream.

M. ambigua Swainson 1829 Indo-Pacific, 75mm. Rather broad for its length; narrowly channelled suture. Spirally grooved; grooves obsolete on middle of body whorl. Lip coarsely dentate within; columella with about five plaits. Light red-brown with a wide cream band below the suture; aperture white.

M. cardinalis Gmelin 1791 Indo-Pacific, 75mm. High spire and simple suture. Spiral rows of small punctures, about twenty-five rows on body whorl. Lip with tiny blunt teeth; columella with about five plaits. White, with spiral rows of brown dashes and a few blotches of the same colour on the spire; aperture cream-white.

M. coronata Lamarck 1811 Pacific, 30mm. High spire, coronated suture and cancellate sculpture. Finely toothed lip; columella with about five plaits. Dark brown; coronations and a single spiral ridge below the suture cream; aperture white.

Neocancilla papilio Link 1807 Indo-Pacific, 60mm. Spiral ridges a little elevated posteriorly, two or three smaller ridges between each of these; axial grooves divide the ridges to give the overall effect of scales. Columella with about five plaits; finely dentate lip. Creamy white with sparse purple-brown and red-brown spots, but profuse in two areas to form red-brown spiral bands; aperture cream-fawn.

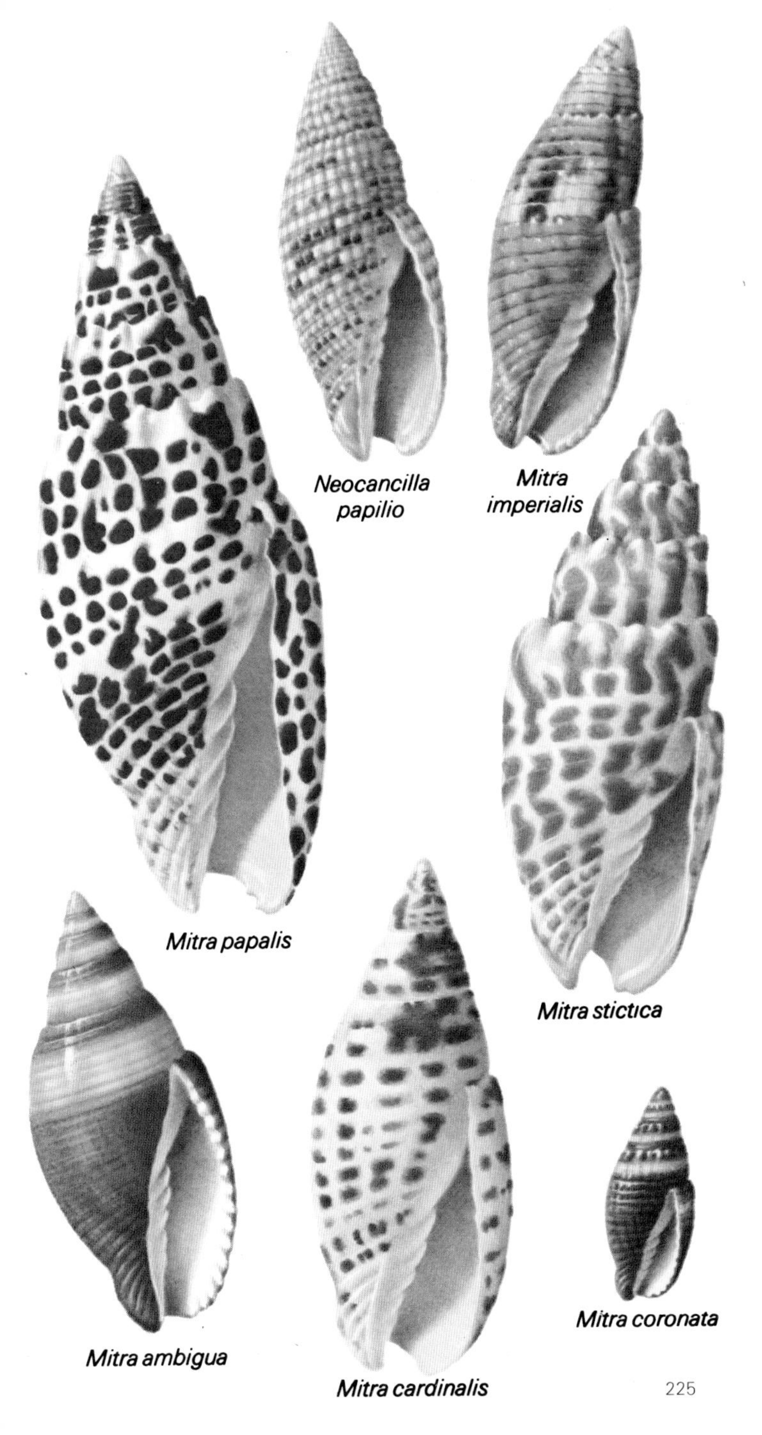
Neocancilla papilio
Mitra imperialis
Mitra papalis
Mitra stictica
Mitra ambigua
Mitra cardinalis
Mitra coronata

Mitra fraga **Quoy and Gaimard 1833** Indo-Pacific, 30mm. Stoutly fusiform; sculptured with spiral ridges divided by shallow channels in which there are axial lirae. Near the lip, the body whorl has axial grooves breaking the ridges into rows of nodules. Lip toothed within; columella with about five plaits; short, open siphonal canal. Dark wine colour with pale orange spots on the ridges and nodules; aperture pale brown.

M. lugubris **Swainson 1821** south Indo-Pacific, 40mm. Fusiform with high spire, spiral ridges and axial grooves. Obsolete coronations at the suture and shoulder. Finely dentate lip is flared a little anteriorly; columella with about five plaits. Red-brown with a white band from suture to shoulder, spire therefore being white with a little brown showing above the suture; aperture white; columella brown posteriorly, white anteriorly including plaits.

M. mitra **L. 1758** Indo-Pacific, 140mm. Largest of the mitres, it has a high spire and adpressed suture. Fine, punctate spiral lines becoming obsolete and finally disappearing altogether on the last two whorls. Lip with short, sharp spines anteriorly; columella with about four plaits. Narrow columellar shield over fasciole area. White with spiral rows of varying sized, generally rather oblong, red spots; spots becoming large irregular blotches below the suture; aperture creamy yellow.

M. eremitarum **Röding 1798** Malaysia and Philippines across the Pacific, 85mm. High spire; coarse edging to adpressed suture. Spiral grooves, pustulate on early whorls; axial striae. Dentate lip; columella with about five plaits. Siphonal canal is short and a little twisted. Light yellow-cream with irregular axial streaks of dark and light browns.

M. incompta **Lightfoot 1786** Indo-Pacific, 100mm. Slender with a very high spire; adpressed suture and very finely coronated. Obsolete axial riblets; spiral punctate grooves. Finely dentate lip, sometimes recurved; columella with about five plaits; siphonal canal short and rather straight, a little recurved. Cream or orange-cream with dark brown axial streaks and a pale band low on body whorl; aperture creamy tan.

M. puncticulata **Lamarck 1811** Indo-Pacific, 50mm. Stubby with moderate spire. Suture with small coronations; small axial ribs are crossed by incised, deeply punctate, spiral grooves. Finely dentate lip; columella with about four plaits. Short siphonal canal is a little twisted. White with two broad bands, one below the suture and one anteriorly of light orange-brown carrying dark brown axial streaks, and a red-brown thread in the middle of the central white band; aperture cream.

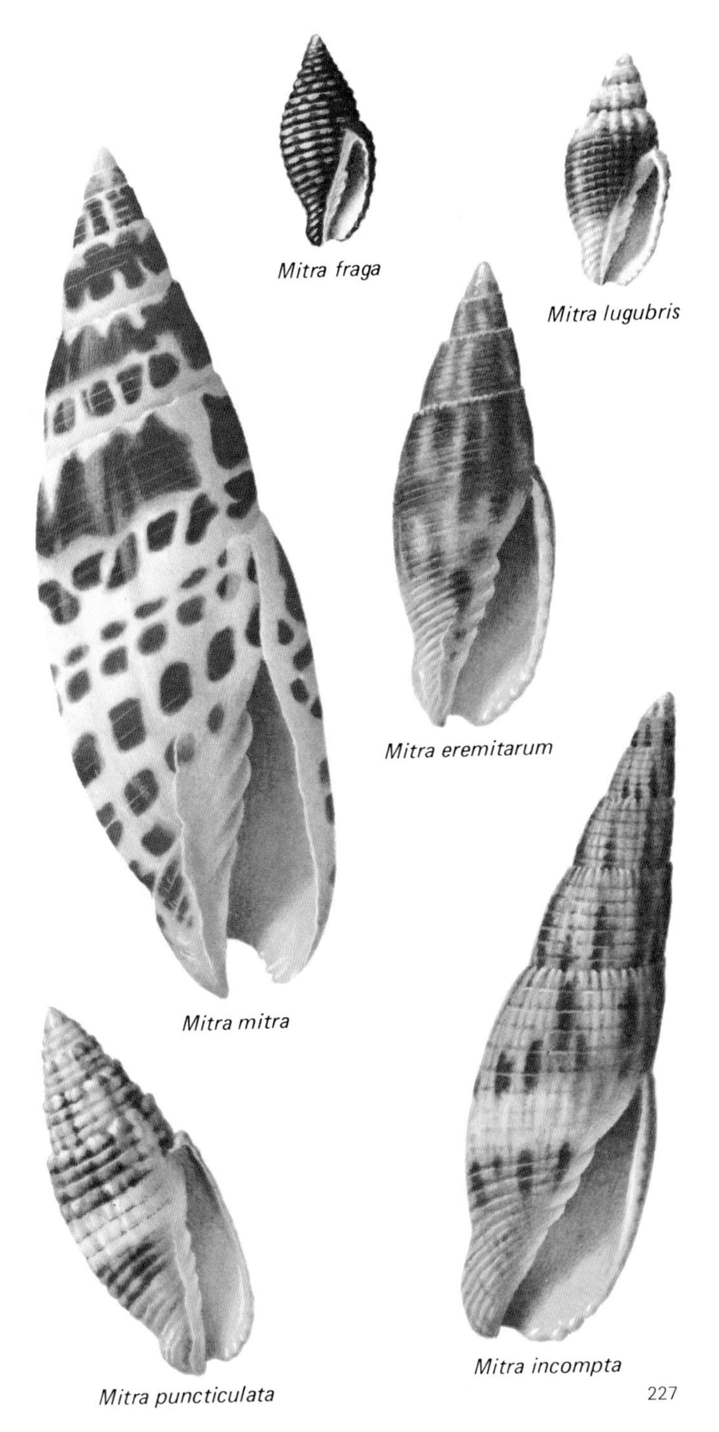
Mitra fraga

Mitra lugubris

Mitra eremitarum

Mitra mitra

Mitra incompta

Mitra puncticulata

M. chrysostoma Broderip 1836 Indo-Pacific, 50mm. Rather broad with moderately high spire. Adpressed, coarse sutures; axial riblets and spiral grooves on early whorls; centre of body whorl almost smooth, grooves stronger anteriorly. Thickened lip; about five plaits on columella. White with broad, broken, brown band below the suture and another anteriorly; some splashes of the same colour across the central white band and as axial streaks on the spire; aperture dirty white.

M. isabella Swainson 1831 Japan, 100mm. Long and slender with a high spire. Spiral ridges crossed by axial threads. Narrow aperture; finely dentate lip; columella with about five plaits. Long and curved siphonal canal with quite a strong fasciole. White profusely clouded and streaked with pale yellow-brown; aperture pinky-cream.

M. floridula Sowerby 1874 south Indo-Pacific, 50mm. Moderately high spire. Adpressed, coronated suture; fine, spiral, punctate grooves becoming obsolete on the middle of the body whorl. Finely dentate lip; columella with about six plaits. Red or orange-brown, with white blotches in a band below the suture, a row of sparse white dots on the middle of the body whorl, and white dashes and dots anteriorly; aperture white.

M. nubila Gmelin 1791 Pacific, 70mm. Solid with moderately high, slightly concave spire. Whorls a little inflated; spiral punctate grooves. Dentate lip; columella with about five plaits. White with red-brown blotches in two indistinct bands; paler clouding and small chalky white spots, partly edged with dark brown; aperture and apex white.

M. cucumerina Lamarck 1811 Indo-Pacific, 25mm. Stubby with a moderate spire. Spiral ridges divided by channels. Obsolete, broad, blunt axial ribs. Dentate lip; columella with about four plaits. Red with a broad, central, white band and white flecks elsewhere; aperture white.

Strigatella paupercula L. 1758 Indo-Pacific, 30mm. Chubby with short spire and inflated at the shoulder of the body whorl, it looks like a member of the Columbellidae. Smooth except anteriorly where it has fine spiral threads. Lip slightly concave posteriorly; about five columellar plaits. Chocolate brown to black with wavy white axial streaks; lip and plaits white; interior deep brown.

S. litterata Lamarck 1811 Indo-Pacific, 30mm. Short, fat with low spire and spiral punctate lines. Lip thickened and turned in, blunt tooth posteriorly; columella with about five plaits. Creamy white with brown, wavy, axial streaks and blotches, generally in three spiral rows; aperture blue-white.

S. scutulata Gmelin 1791 Indo-Pacific, 50mm. Moderately high spire. Spiral cords, obsolete on the body whorl, and fine axial striae. Sinuate lip; columella with about four plaits. Dark brown-black; a narrow yellow band below the suture and white spots and short axial streaks; aperture blue-white.

Mitra chrysostoma

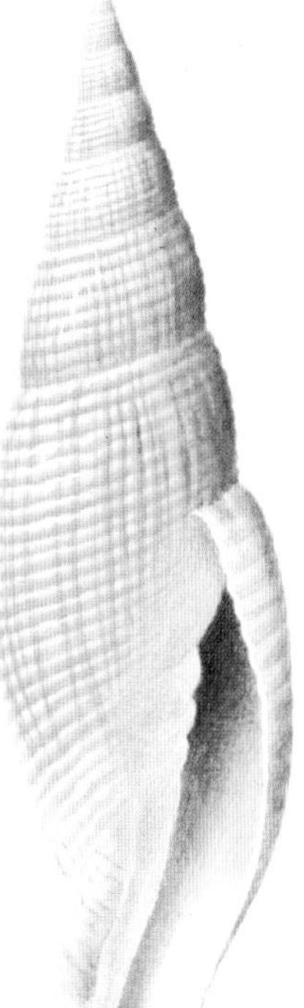

Mitra isabella

Mitra floridula

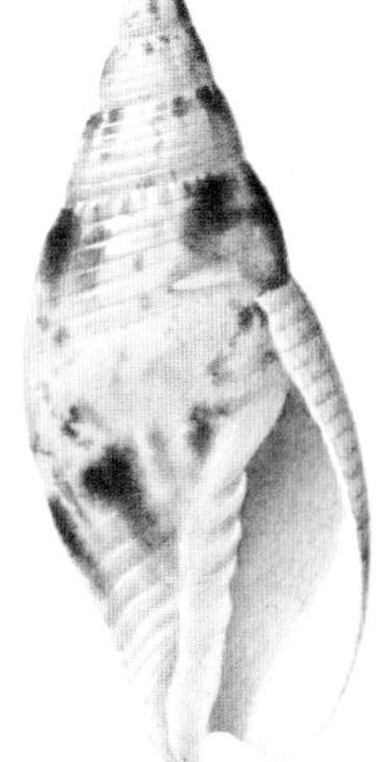

Mitra nubila

Mitra cucumerina

Strigatella paupercula

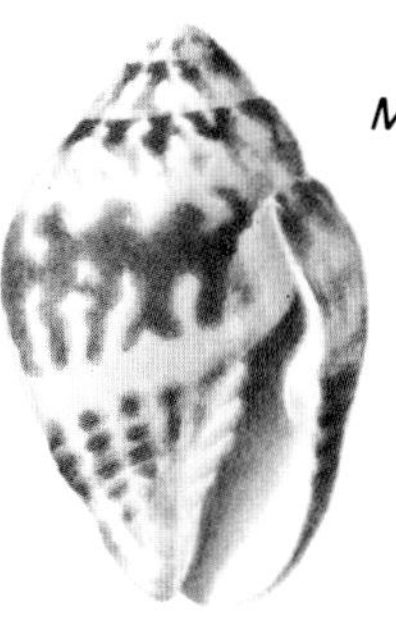

Strigatella litterata

Strigatella scutulata

Swainsonia variegata Gmelin 1791 Pacific, 45mm. Solid, high-spired and slightly shouldered with axial ribs between shoulder and suture. Incised and punctate spiral lines, some fourteen on body whorl. Dentate lip; columella with about six plaits. Creamy grey with white blotches edged on the aperture side with wavy dark brown streaks; two obscure bands of pale and dark brown; aperture cream.

Neocancilla granatina Lamarck 1811 Indo-Pacific, 65mm. High spire. Sculpture the same as *N. papilio* (see page 224). Columella has five plaits. Shiny white with spots and rather sparse, transverse dashes of magenta, rose, and orange, more profuse on two bands, one in the centre and one at the anterior end of the body whorl; aperture cream to light tan.

N. antoniae H. Adams 1870 Indo-Pacific, 40mm. Slightly inflated with nodulose spiral ridges, threads and axial incised lines. Lip dentate within; columella with about five folds. White with red-brown dashes on the rows of nodules; aperture pinky white.

Cancilla filaris L. 1771 Indo-Pacific, 30mm. Fusiform with high spire and finely dentate lip. Columella with about four plaits. Surface somewhat cancellate; evenly spaced spiral cords crossed by vertical striae. The first cord is further below the suture than the distance between the other cords. White; spiral cords red; aperture white; cords showing through faintly but clearly at the edge.

C. praestantissima Röding 1798 Indo-Pacific, 40mm. Very high spire. Sculpture similar to *C. filaris* but rather finer; lacking the extra width from suture to first cord. Colour similar.

Imbricaria conularis Lamarck 1811 Indo-Pacific, 25mm. Shaped and looks like a cone shell, except for the five or so columellar plaits. Moderate spire with high, sharp apex, sometimes with spiral lines of minute punctures. Lip uneven rather than dentate. Opaque white; chalky white squares; brown-red spiral lines and a central purple band which also carries the white squares; lip white; interior purple-brown; columella opaque and chalky white; apex purple-brown.

I. olivaeformis Swainson 1821 Pacific, 20mm. Olive-shaped, hence its name. Low rounded spire with sharp, pointed apex; smooth or with obsolete spiral punctate lines. Smooth lip; columella with about five plaits. Green-yellow; apex and tip of columella purple.

I. punctata Swainson 1821 south Indo-Pacific, 20mm. Low, almost flat spire with sharp, pointed apex. Rounded shoulder narrowing to the suture and spiral, incised, punctate lines. Lip with very small, fine teeth; columella with about six plaits. Pale orange-yellow; paler on the spire and towards the lip; aperture pale cream-white.

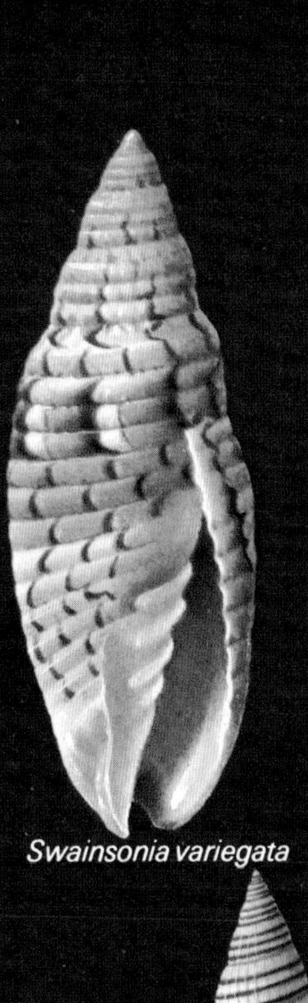
Swainsonia variegata

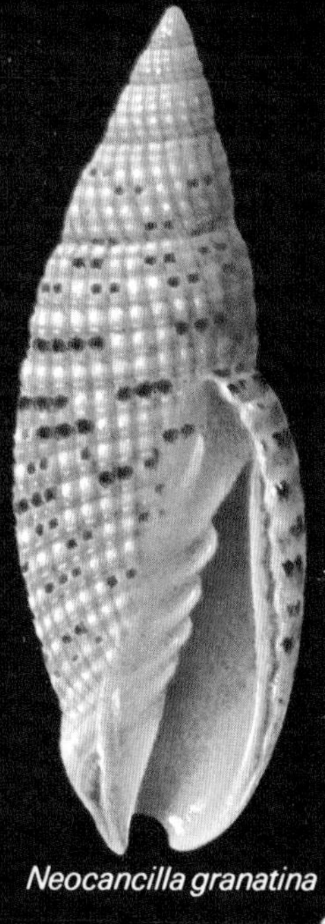

Neocancilla granatina

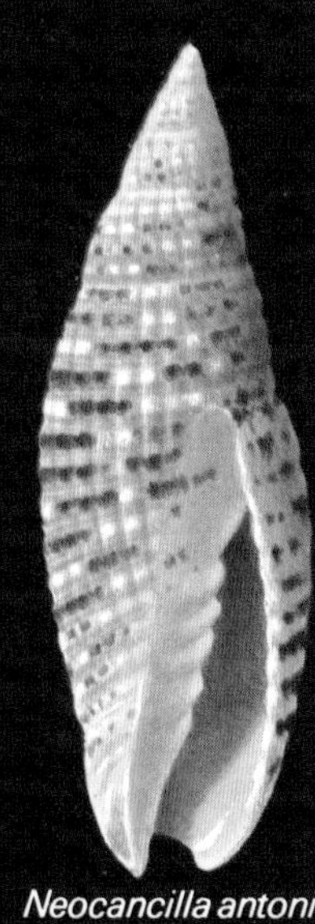
Neocancilla antoniae

Cancilla filaris

Cancilla praestantissima

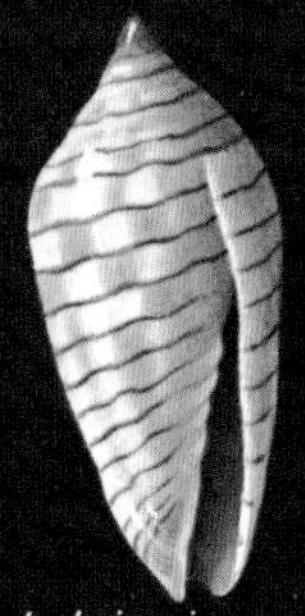
Imbricaria conularis

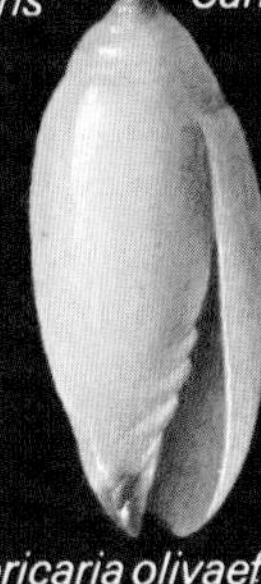
Imbricaria olivaeformis

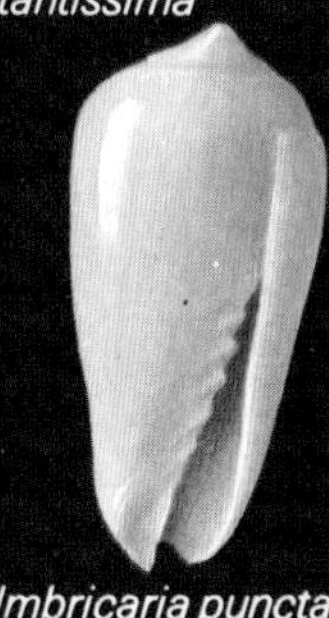
Imbricaria punctata

Pterygia dactylus L. 1767 Indo-Pacific, 50mm. Solid and heavy with short spire and adpressed suture. Swollen at the rounded shoulders. Incised spiral lines and bevelled lip. Columella has about seven strong plaits; small fasciole. White, heavily clouded with brown, on which three or four indistinct bands are discernible; aperture white; columella with sparse brown marks.

P. crenulata Gmelin 1791 Indo-Pacific, 40mm. Solid with short spire. Narrower than the above. Suture has a narrow ledge; rounded shoulder; finely reticulated. Lip faintly toothed within and columella has about eight plaits. White with light brown maculations; interior white.

P. nucea Gmelin 1791 Indo-Pacific, 60mm. Solid, heavy, with moderate spire and adpressed suture. Fine axial striae and faint spiral grooves. Lip has tiny nodules; columella has about five strong plaits; small but strong fasciole. White with three spiral rows of irregular dark brown dashes and spots and about six to eight rows of very small, lighter brown dots; nodules on the lip pale brown; aperture white.

Vexillum taeniatum Lamarck 1811 Indo-Pacific, 75mm. Narrow, fusiform, with high spire and sharply rounded shoulders below the suture. Axially ribbed, some ten ribs on body whorl; coarse spiral cords rather weak on middle of body whorl. Inside of lip lirate; columella with about five plaits; extended, open and recurved siphonal canal. White with three orange-red bands, one below the suture, one centrally on body whorl, and one anteriorly, all edged with strong black lines; with or without a narrow red line through the posterior white area; the bands show on the lip; aperture cream.

V. formosense Sowerby 1890 western Pacific, 55mm. Narrow, fusiform, with high spire and slightly constricted suture. Axial riblets, about twenty on body whorl, becoming obsolete on the last half of the body whorl. Spiral cords overall, but rather weak on middle of body whorl. Lirate inside lip; columella with about five plaits; open, slightly extended and recurved siphonal canal; small fasciole. Very dark brown with a narrow, white, spiral band, sometimes carrying a red thread; inside lip edged with dark purple-brown; columella brown; columellar plaits cream; interior violet.

V. caffrum L. 1758 Indo-Pacific, 50mm. Fusiform, and wider and heavier than the two preceding species. High spire and adpressed suture. Small crowded axial riblets become obsolete on middle of body whorl; spiral cords anteriorly on body whorl. Lirate lip is a little concave and inverted; columella has about four plaits; siphonal canal is open and a little recurved. Chocolate brown with two narrow yellow-white spiral bands; inside lip edge dark brown; columella brown; columellar plaits cream; interior cream.

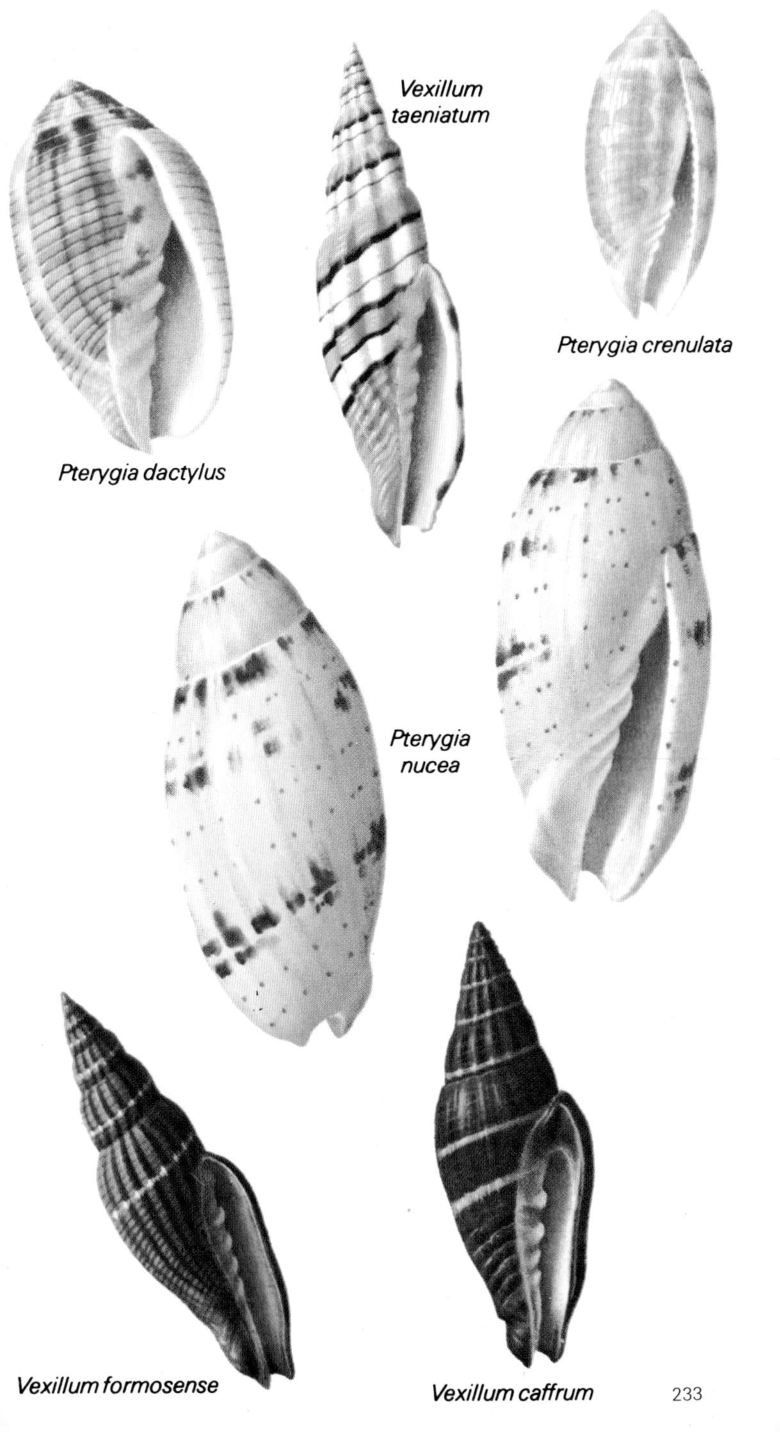
Vexillum taeniatum
Pterygia crenulata
Pterygia dactylus
Pterygia nucea
Vexillum formosense
Vexillum caffrum

Vexillum sanguisugum L. 1758 Indo-Pacific, 42mm. Fusiform with high spire and about six whorls. Axial riblets are broken into nodules by fine, incised, spiral grooves. Very slightly concave lip; narrow aperture; columella with about four plaits. White; spiral rows of shiny white nodules divided into three groups of five by two rows of scarlet nodules; spiral grooves purple; aperture edged with purple-brown; columella brown; interior white; fasciole red-brown.

V. melongena Lamarck 1811 Singapore to Fiji, 50mm. Narrow and fusiform with high spire and about eight whorls. Strong axial ribs and weak spiral ridges which do not cross the axial ribs. Narrow aperture; lirate lip; columella with about four folds; extended siphonal canal. Grey with a broad white band on the shoulder which is edged with purple and has a scarlet thread through the middle; two, narrow, brown bands anteriorly; aperture blue-white edged with purple-brown marks; columella pale brown; columellar folds white.

V. stainforthi Reeve 1841 west Pacific and China Seas, 50mm. Narrow and fusiform with high spire and about eight whorls. Adpressed suture; widely spaced axial ribs, about ten on body whorl; fine spiral striae cross the ribs. Slightly concave lip; columella with about four plaits; small fasciole; extended, slightly recurved siphonal canal. Cream; five scarlet bands which only appear on the axial ribs; inside and outside of lip with purple-brown marks where the bands end; columella brown anteriorly, rest of aperture white; apex and end of siphonal canal purple-brown.

V. exasperatum Gmelin 1791 Indo-Pacific, 25mm. High spire; about seven whorls; body whorl bi-angulate. Cancellate with finely nodulose axial ribs, the ribs protruding slightly at the shoulder. Lip with nodulose striae; columella with four plaits; small fasciole. White; a broad brown band between the upper and lower shoulders and one between the anterior shoulder and the fasciole, only show on the axial ribs; aperture white.

V. plicarium L. 1758 east Indian Ocean and south-west Pacific, 50mm. Broadly fusiform with a high spire, about eight whorls and adpressed suture. Concave from suture to sharp shoulder. Fine axial striae, and spiral cords anteriorly. Strong axial ribs, about ten on body whorl, end in blunt knobs on the shoulder. Columella with about four plaits. White with a broad central blue-grey or brown band edged between the ribs with red-brown or black lines, and with a similar line below the shoulder and anteriorly; inside lip shows the banding; columella white; apex purple.

V. vulpecula L. 1758 Pacific, 55mm. Solid and fusiform with high spire, about seven whorls and narrowly channelled suture. Coarse axial ribs becoming obsolete near the aperture; narrow spiral grooves, and spiral cords anteriorly; the ribs are particularly strong at the angular shoulder. Slightly concave lip, lirate within; columella with about four plaits. Colour variable; cream, yellow or orange with three black or red-brown bands, one at the suture, one centrally and one anteriorly; siphonal canal and fasciole of the same colour; lip carries the colour band; interior white; columella white, with the black or red colour anteriorly.

Pusia microzonias Lamarck 1811 Indonesia to Polynesia, 25mm. Fusiform with about six whorls. Coarse, rounded axial ribs replaced by fine, nodulose spiral cords anteriorly. Lirate inside lip; columella with about four folds. Dark red-brown, with a row of white spots on the ribs joined by a white line on the middle of the body whorl; a pale spiral line below the ribs; aperture and columella white with brown showing through the lip.

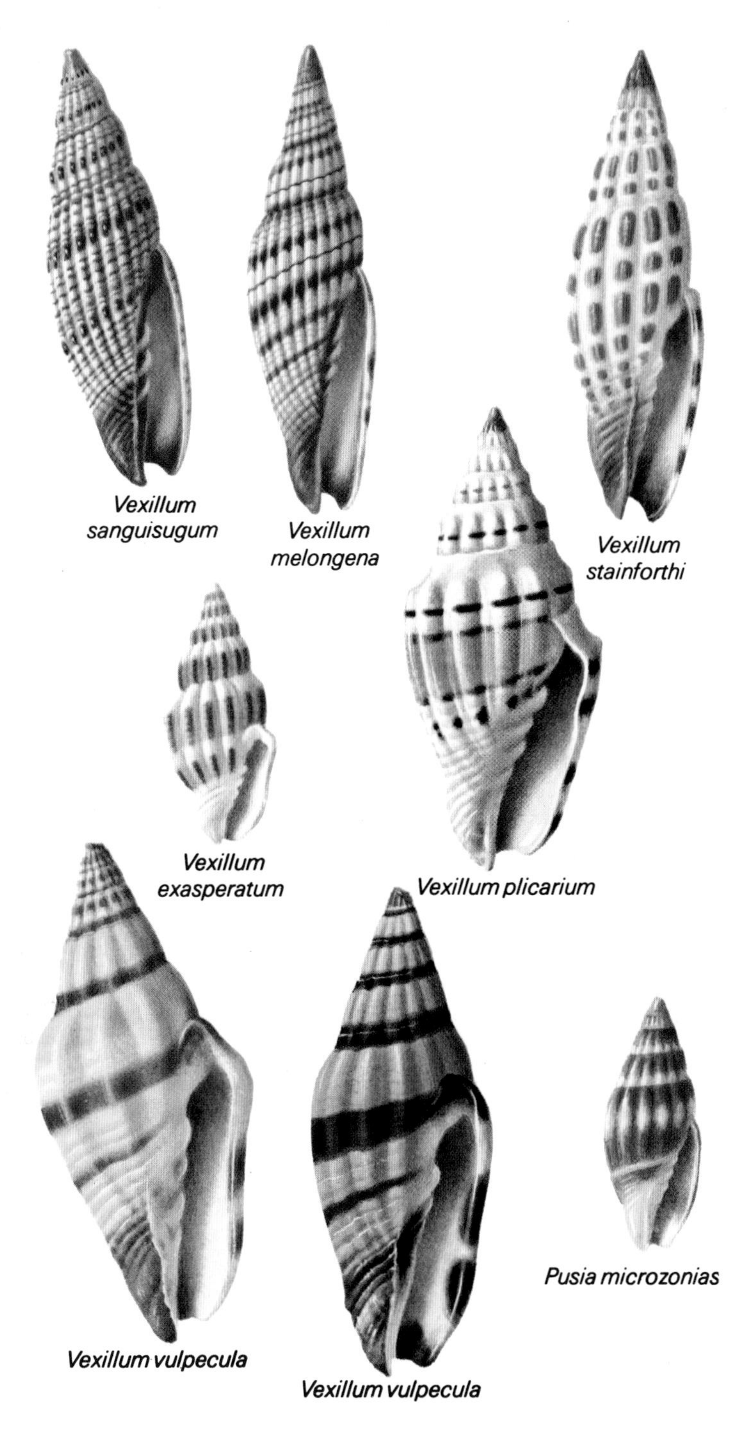

Vexillum sanguisugum
Vexillum melongena
Vexillum stainforthi
Vexillum exasperatum
Vexillum plicarium
Vexillum vulpecula
Vexillum vulpecula
Pusia microzonias

Family: Volutidae

Some 200 species belong to this family and vary from large to fairly small. They are colourful, most having a reasonably simple sculpture. Most are carnivorous and live in sand in both shallow and deep water. Although found worldwide, Australia has far more than its fair share of species.

Ericusa sericata **Thornley 1951** east Australia, 125mm. Solid and rather narrow. Moderate spire with large rounded protoconch forming the apex. About two-and-a-half whorls, nucleus eccentric; subsequent whorls about three-and-a-half. Lip slightly thickened and flared anteriorly. Columella is somewhat extended posteriorly and a little recurved, with four plaits. Siphonal notch is wide and rather shallow. Cream-tan, heavily covered with orange-brown, leaving tent marks of the base colour; short orange-brown axial streaks below the suture; aperture cream-grey.

Livonia mammilla **Sowerby 1844** south-east Australia from Queensland to south Australia, and north Tasmania, 300mm. Large and ovate. Short spire has very large, semi-globular protoconch of one whorl and three subsequent convex whorls. Very inflated body whorl; wide aperture; simple lip is extended posteriorly to a point on the penultimate whorl near the suture. Curved columella has three plaits and siphonal notch is wide but not deep. Weak fasciole and narrow callous parietal area. Cream or pale yellow-cream; two broad bands of sparse, dark brown-red, generally axial and oblique lines forming separate patterns; inside of lip, area at posterior end of lip and columella apricot; interior flesh-coloured.

Fulgoraria mentiens **Fulton 1940** south Japan, 215mm. Fusiform; moderate protoconch of about two whorls, with eccentric nucleus; about five subsequent whorls. About sixteen axial ribs on penultimate whorl, becoming obsolete on body whorl, as also do the spiral striae which are strong on early whorls. Simple, slightly recurved lip; extended columella has about four plaits, the strongest anteriorly with a wide gap between it and the next; very shallow siphonal notch. Pale flesh, with three spiral bands, one below the suture, one centrally, and one anteriorly, of wavy, generally axial, dark red-brown lines, two of which show on earlier whorls; edge of lip white; interior pinky flesh.

F. rupestris **Gmelin 1791** Taiwan, 130mm. Fusiform and solid with moderate spire. Large protoconch has about two whorls with eccentric nucleus, and about three-and-a-half subsequent whorls. About fourteen axial ribs on penultimate whorl becoming obsolete on body whorl. Somewhat angular shoulders and strong, spiral, incised lines. Outer lip is a little thickened, bevelled and slightly crenulated anteriorly. Extended columella is convex centrally with a ridge bearing about nine plaits. No siphonal notch. Creamy white to very pale creamy tan; brown wavy axial lines; aperture white; interior with a pale brown tinge.

F. delicata **Fulton 1940** south-east Japan, 55mm. Small, narrow and fusiform with relatively high spire. Small protoconch has about two whorls, eccentric nucleus and about four subsequent whorls. About fifteen axial ribs on penultimate whorl becoming obsolete on body whorl. Constricted suture; very fine spiral striae; simple lip; columella has about two plaits; no siphonal notch. Creamy white.

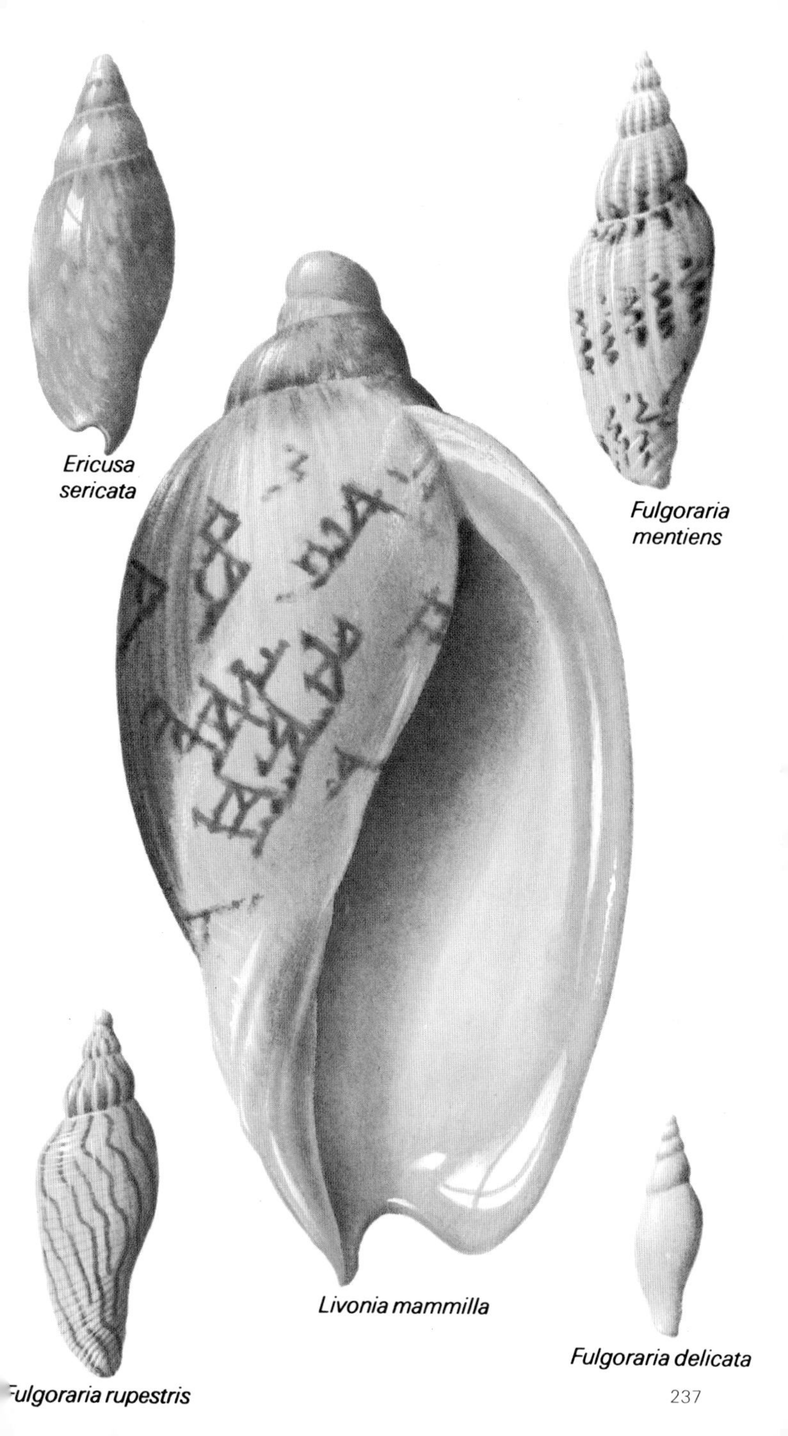

Ericusa sericata

Fulgoraria mentiens

Livonia mammilla

Fulgoraria delicata

Fulgoraria rupestris

Voluta ebraea L. 1758 north Brazil, 150mm. Rounded protoconch; angular shoulders carry short spines at end of low ridges which disappear anteriorly. Uneven suture; thickened, sharply recurved lip. Recurved columella with five strong plaits and very weak plaits posteriorly. Narrow, callous parietal area; deep, narrow siphonal notch. Cream to tan; fine axial and spiral red-brown lines, mainly in two bands; faint purple marks on bands and strong fasciole; lip with twelve purple-brown spots; aperture, columella pale peach.

V. musica L. 1758 north-east South America and east Caribbean, 90mm. Blunt protoconch; shoulders carry heavy, pointed, laterally compressed knobs at end of low, wide ribs becoming obsolete anteriorly. Uneven suture; long aperture, wider anteriorly; thickened, recurved lip; straight columella, recurved anteriorly, five strong plaits anteriorly, strong lirae posteriorly. Parietal area narrow posteriorly, wider over part of strong fasciole; deep, narrow siphonal notch. Ivory or cream; spiral red-brown markings, especially in two bands; aperture cream. Variable; knobs more or less strong, colour more or less heavy.

Lyria mitraeformis Lamarck 1811 south Australia, 55mm. Small, inflated, blunt protoconch; indented suture. Axial ribs, sixteen on body whorl. Aperture wide anteriorly. Concave columella with three plaits, low tooth posteriorly, lirae between. Shallow, narrow siphonal notch; small, partly callous fasciole. Cream heavily mottled with grey-brown spiral dashes, rectangular spots; lip with red-brown hair lines; aperture white; columella and notch with yellow tinge.

L. lyraeformis Swainson 1821 Kenya, 145mm. Bulbous protoconch with spur at apex, convex, restricted at suture and slightly concave below. Axial ribs, eighteen on body whorl. Short aperture, wider anteriorly; thickened, bevelled lip; columella has three plaits anteriorly, lirate centrally; parietal area and part of small strong fasciole callous; small siphonal notch. Creamy tan; three spiral bands of uneven red-brown edged with dark brown lines, broken in interstices of ribs, thin red-brown lines between; aperture pale orange.

L. delessertiana Petit de la Saussaye 1842 Madagascar, Comoro and Seychelle Islands, 55mm. Small, rounded protoconch; close axial ribs, twenty on body. Indented, constricted suture. Narrow, long aperture. Thickened lip with a low ridge carrying three ribs. Concave columella has small plaits almost covering narrow parietal area; deep, narrow siphonal notch; weak fasciole. Pinky white clouded with orange-red; twelve, spiral, thin, red, broken lines; aperture white.

L. cumingii Broderip 1832 west Central America, 35mm. Small, rounded protoconch; low shoulder; blunt, laterally compressed knobs at end of low ribs. Ten ribs on body whorl, obsolete anteriorly, enlarged in middle. Uneven suture; narrow aperture; thickened lip has axial ridge before bevelled edge and small, blunt tooth centrally on inner edge. Concave columella, three plaits anteriorly, three or four weaker posteriorly; narrow parietal area; deep, narrow siphonal notch. Cream; brown clouding, especially in two spiral bands; aperture cream.

Volutocorbis abyssicola H. Adams and Reeve 1848 South Africa, 100mm. Shallow, channelled suture; fine grooves. Thickened, slightly recurved lip; white columellar plaits, strongest anteriorly; wide, callous parietal area; wide, shallow siphonal notch; obsolete fasciole. Pale brown; long, narrow aperture grey-brown.

V. boswellae Rehder 1969 South Africa, 60mm. Rounded spire; strongly

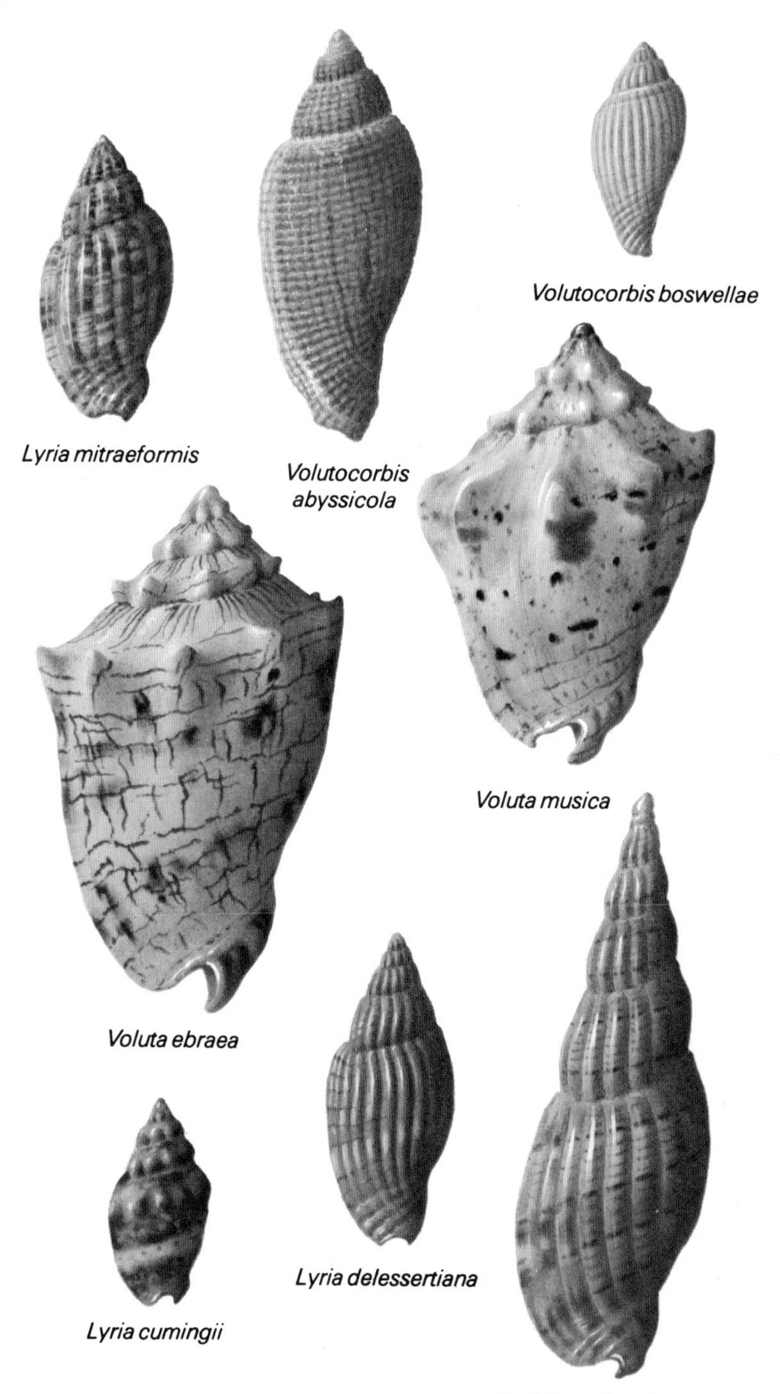
Volutocorbis boswellae
Lyria mitraeformis
Volutocorbis abyssicola
Voluta musica
Voluta ebraea
Lyria delessertiana
Lyria cumingii
Lyria lyraeformis

shouldered; channelled suture; close-set axial ribs. Shallow depression below suture forming two rows of sharp nodules, giving coronated effect at suture. Narrow aperture; thickened, faintly denticulate lip; ten columellar plaits, strongest anteriorly; shallow siphonal notch. Pale yellow-tan; aperture paler. The genus *Cymbium* is found only in the area from Portugal to the Gulf of Guinea in west Africa.

Cymbium pepo **Lightfoot 1786** north-west Africa, 270mm. Large and globose, often more so than in the illustration. Concave spire; the later whorls extend beyond the apex which is usually callous; a wide channel separates the apex from the edge of shoulder; fine axial striae. Wide aperture; simple, flaring lip. Arched columella with three or four plaits; parietal area covered by this callus is very wide; wide, deep siphonal notch; wide, ridged fasciole. Grey-brown; aperture creamy pink; edge of lip and columella darker; plaits white.

C. olla **L. 1758** Portugal to Morocco, 115mm. The smallest of the genus. Very low spire, deeply channelled suture and rounded shoulder. Wide, flaring lip. Columella arched and with two plaits. Wide, very shallow siphonal notch; wide, ridged fasciole; fine axial growth lines; parietal area glazed, but not very wide. Pinky flesh; aperture paler and shiny; fasciole light red-tan.

C. cymbium **L. 1758** Canary Islands to Senegal, west Africa, 155mm. Rather oblong in outline. The spire is flat or slightly concave. Apex has a heavy callus and a wide, flat platform dividing the suture from the sharply angled shoulder, which has a small uneven projecting ridge. Bevelled lip with the anterior half flaring. Arched columella has three plaits; wide, deep siphonal notch; wide, ridged fasciole. The surface is smooth but the whole shell, including the periostracum, is glazed so that grains of sand and small pieces of other matter get covered by the glaze giving a pustulate effect in places. Light brown; aperture and columella creamy brown.

C. glans **Gmelin 1791** Senegal to Gulf of Guinea, west Africa, 325mm. The largest of the genus and the most graceful. Spire sunken and filled with callus, through which the protoconch is seen as a low knob. Sharp shoulder is curved up and out. Simple, flaring lip; arched columella with four plaits; wide fasciole; broad, rather shallow siphonal notch. The whole surface is covered with a thin glaze, often imprisoning many grains of sand. Pale cream-tan, with chocolate edge to the shoulder and streaked on the spire.

C. cucumis **Röding 1798** Senegal, 160mm. Very similar to *C. cymbium* but not so wide from shoulder to protoconch which is much more pronounced, being long and protruding. Flaring, simple lip; columella with three plaits; deep siphonal notch. Light creamy tan; interior darker. Most gastropods when held with the anterior end downward and aperture towards one, have the aperture on the right; these are known as dextral shells. A very few genera have the aperture on the left and are known as sinistral shells. Rarely a normally dextral shell will be found in a sinistral form. I have in my collection a sinistral *C. cucumis* and have used this to illustrate the species and as an example of this abnormality.

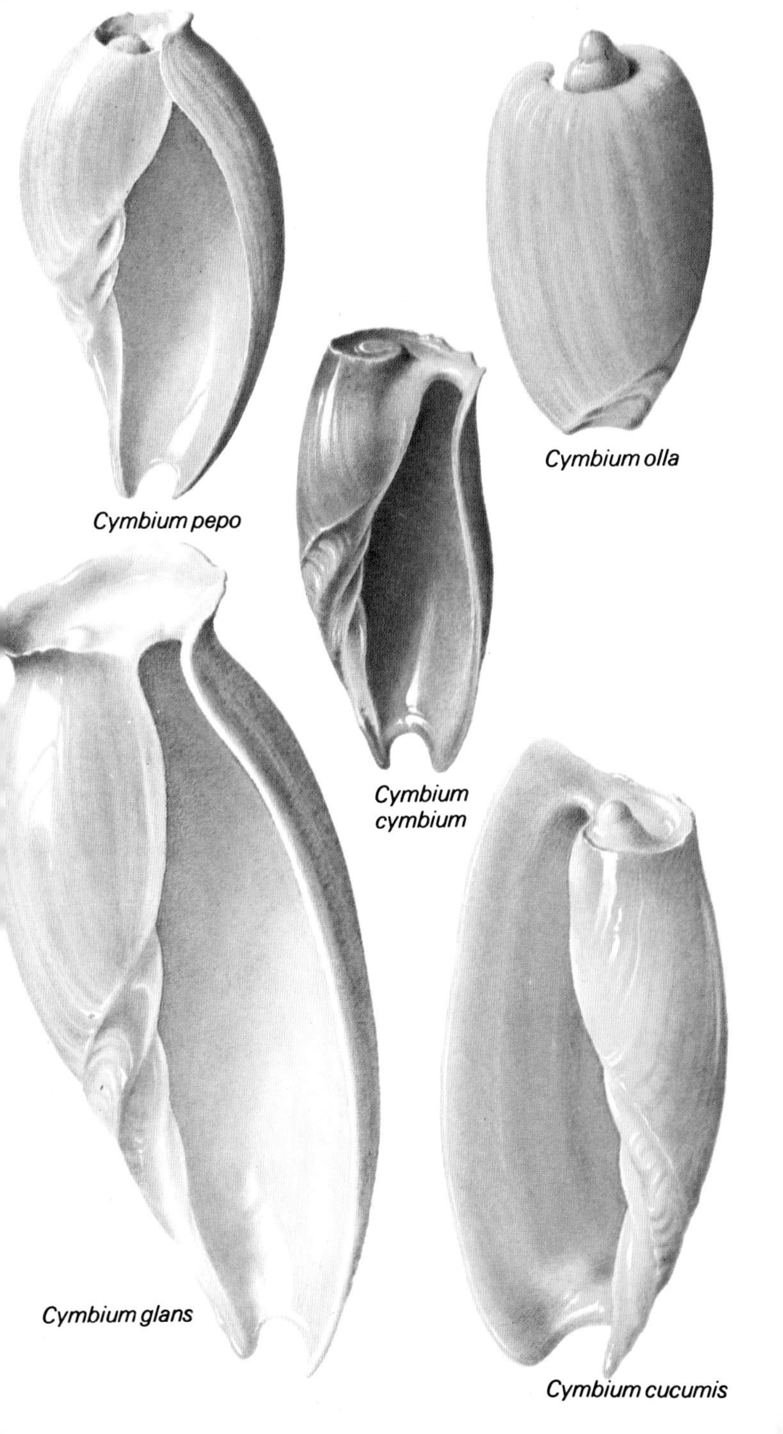
Cymbium pepo
Cymbium olla
Cymbium cymbium
Cymbium glans
Cymbium cucumis

The genus *Melo* is found only in the area of Malaysia to Australia, where they are commonly known as balers, from their use for that purpose.

Melo miltonis **Griffiths and Pidgeon 1834** south-west and south coasts of Australia, 450mm. Rather narrower in relation to its length than others of the genus. Large with rounded, protuberate protoconch of about three whorls; two-and-a-half subsequent whorls. Shoulders have sharp, hollow, even spines sloping inward; deep channel between shoulder and suture. Simple, expanded lip; arched columella with three strong plaits and sometimes a fourth weak one; base of columella curved forwards; strong fasciole; wide but shallow siphonal canal. Creamy white, with purple-brown zigzag axial markings; two or three, broad, spiral bands; inside lip with a few purple marks and purple edging anteriorly; interior creamy brown to cream; columella apricot.

M. umbilicatus **Sowerby 1826** north-east and north coasts of Australia, 400mm. Very inflated and sub-globular with very widely flaring lip. Sunken spire with a large rounded protoconch of about three whorls; about two subsequent whorls. Shoulder with long, sharp, hollow spines which slope inward and in adult specimens almost obscure the spire. Simple lip; arched columella with three, strong plaits; strong fasciole; wide, shallow siphonal notch. Cream or brown-yellow with darker brown, axial zigzags and streaks, mainly forming two, interrupted, spiral bands, narrow and dark near the columella, wide, diffused and pale near the lip; spines dark brown, but paler near the lip; interior creamy tan; columella pale apricot. The illustrated specimen is not fully adult and the apex is only partly obscured.

M. aethiopica **L. 1758** Indonesia and Malaysia, 250mm. Large and heavy. Almost flat spire with large, rounded protoconch of about four whorls; about two-and-a-half subsequent whorls. Rather inflated body whorl; shoulders with rather close-set, short, solid, open spines, separated from the simple suture by a narrow, flat, lamellated area. Almost semicircular outer lip. Columella with three strong plaits and one weak one, straighter than most of the genus, and extending beyond the anterior end of the lip. Broad, shallow siphonal notch; strong, ridged fasciole. Creamy with two rather faint, pale tan spiral bands, and an occasional, small, dark brown streak; aperture cream or peach; columella slightly darker.

M. amphora **Lightfoot 1786** northern half of Australia and south New Guinea, 450mm. Very large and heavy. Almost flat spire, with low rounded protoconch of about four whorls, and three subsequent whorls. Shoulder spines are narrower, longer and more widely spaced than in *M. aethiopica* and disappear on the body whorl of adults. Flaring lip; slightly arched columella with three plaits; wide, shallow siphonal notch; strong fasciole. Creamy brown with white spots and blotches or darker, zigzag, axial lines; rather variable in colour and often with two darker tan spiral bands, aperture peach; columella darker. A thin mid-brown periostracum largely obscures the colour and pattern, as in the illustration.

M. melo **Lightfoot 1786** Strait of Malacca to South China Sea, 275mm. Globose with spire and protoconch hidden beneath the body whorl, the shoulders of which meet at the apex, there being no suture or any spines or coronations. Inflated body whorl; almost semicircular, simple outer lip; arched columella with four plaits. Yellow, or sometimes almost white, with sparse small brown blotches in two spiral bands; aperture cream; periostracum medium brown.

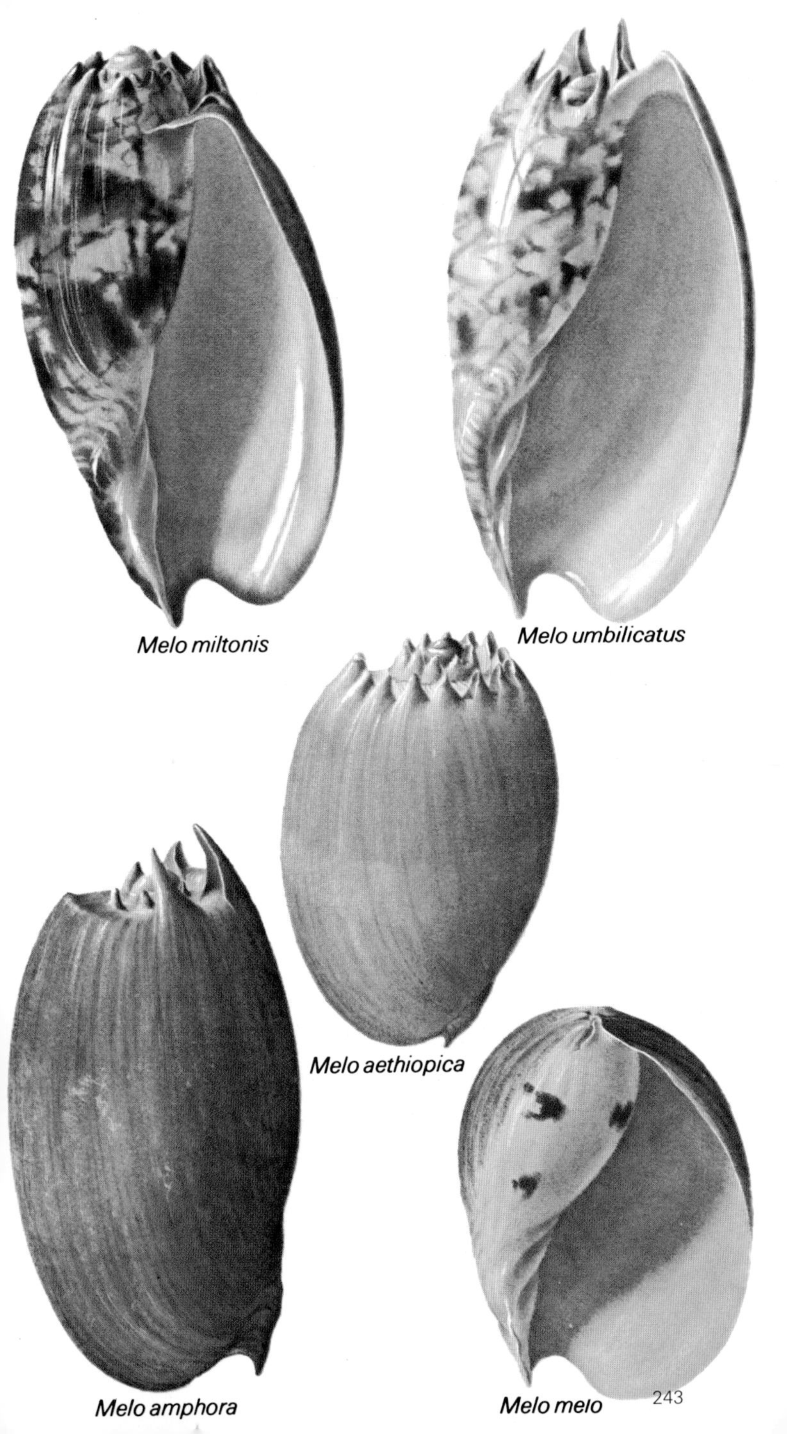
Melo miltonis
Melo umbilicatus
Melo aethiopica
Melo amphora
Melo melo

Cymbiola aulica **Gray 1847** south Philippines, 150mm. Solid with a rather short spire. Large, blunt protoconch of about two-and-a-half smooth whorls; about three subsequent whorls with slightly angular shoulders carrying short laterally compressed spines at the end of ribs, which disappear anteriorly. Rather coarse growth lines; long aperture ending at the shoulder; simple, bevelled lip, expanded anteriorly; slightly concave columella, with four, strong plaits and thickly callous; wide and moderately deep siphonal notch; callous fasciole. Pinky flesh with irregular, generally large, red blotches, mostly in two, broad spiral bands; aperture pinky flesh. This shell is variable in shape and colour; the spines may be knobs or absent altogether, and the red may almost obscure the background colour. A variety mottled with brown and white spots was known formerly as *C. cathcartiae* Reeve 1856. *C. aulica* is rather rare and a collectors' item.

C. flavicans **Gmelin 1791** north-east Australia and south New Guinea, 100mm. Solid and heavy. Short, straight-sided spire with a small rounded protoconch of about three whorls; about four-and-a-half subsequent whorls with rounded shoulders, sometimes with a few, blunt, heavy knobs. Simple lip; almost straight columella with four strong plaits; deep, narrow siphonal notch; strong fasciole. Cream with blue and purple-brown irregular marks below the suture and irregular, wavy, axial splashes of the same colour in two broad bands on the body whorl; inside lip cream with purple-brown marks; columella cream and interior pale blue-grey.

C. imperialis **Lightfoot 1786** Philippines, 250mm. Heavy and somewhat variable, with a low spire but a very large, blunt, rounded protoconch of about four-and-a-half, smooth, shiny whorls; about three subsequent whorls with sharply angled shoulders carrying long, sharp, hollow spines, generally curving a little inwards. Long aperture is wide anteriorly; simple lip is a little thickened; slightly concave columella with four strong plaits; deep, narrow siphonal notch; strong fasciole; callus on parietal area thin but wide; smooth except for growth lines. Protoconch red-brown; base colour fleshy-pink, covered with many wavy axial lines of purple-brown and three, broken spiral bands of the same colour through which the base colour shows in large and small tent marks; aperture and columella pale apricot.
C. imperialis form ***robinsona*** **Burch 1954.** This is a form of the above which lacks the spiral bands.

C. nobilis **Lightfoot 1786** South China Sea, 190mm. Variable. Solid and very heavy, usually having a short spire with a large, blunt, rounded protoconch of about three-and-a-half whorls; about two-and-a-half subsequent whorls. Angular, rounded shoulders are faintly knobbed; coarsely callous suture; wide aperture; thickened, simple lip extends a little posteriorly; slightly concave columella with four plaits; parietal area callous, very heavily in old specimens; a distinct posterior canal and a deep, narrow siphonal notch; strong, partly callous fasciole. Pale flesh colour with axial, zigzag, purple-brown lines and three, broad, broken, spiral bands of the same colour, one above the shoulder, two below. Edge of inside lip marked with the purple-brown of the spiral bands; aperture pale pink; parietal callus white. The colour and pattern are very similar to that of *C. imperialis* and there is a form similar to *C. imperialis* form *robinsona* which lacks the spiral bands. However, it has no subspecific name.

Cymbiola flavicans

Cymbiola aulica

Cymbiola imperialis

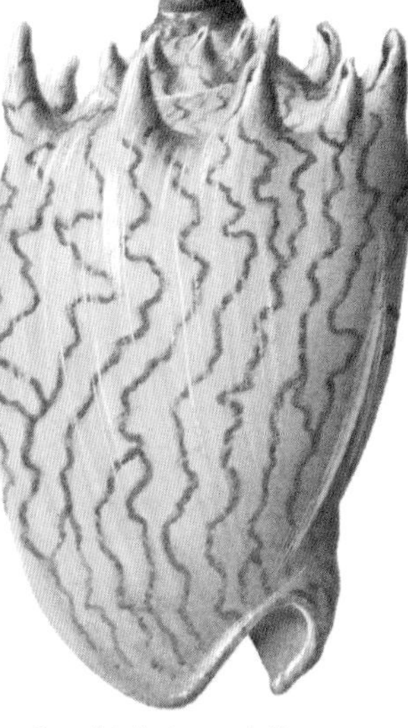

Cymbiola imperialis
form *robinsona*

Cymbiola nobilis

Cymbiolena magnifica Gebauer 1802 east Australia, 300mm. Large but light. Moderately short spire; large, blunt protoconch of three-and-a-half whorls and incised suture; about three-and-a-half subsequent whorls; indented suture; concave between suture and rounded shoulder. Body whorl shoulder may have blunt knobs near aperture. Thickened lip; four columellar plaits; deep, wide siphonal notch; parietal area with opaque callus. Pinky white base showing through heavy, pale brown, zigzag markings as axially flattened tent marks; brown overlay is darker with a purple tinge below suture, in three spiral bands on body whorl and bordering the strong fasciole; inside lip flesh; columella yellow-pink.

Aulicina deshayesi Reeve 1855 New Caledonia, 100mm. Solid. Fairly low spire; large, blunt protoconch of three-and-a-half whorls; faintly nodulose; two-and-a-half subsequent whorls. Shoulders with pointed nodules; a little concave between shoulder and suture. Fine axial growth lines. Thickened, bevelled lip; straight columella with four strong plaits, curved sharply anteriorly; narrow, deep siphonal notch; very strong fasciole; narrow, callous parietal area. White; streaks of red forming spiral bands; aperture apricot.

A. sophiae Gray 1846 north coast of Australia, 75mm. Delicate with low spire and blunt apex. Protoconch of two-and-a-half axially ribbed whorls; two-and-a-half subsequent whorls. Very angular shoulders with short sharp spines, ten on body whorl. Long aperture; simple lip, expanded centrally; four columellar plaits; strong fasciole with sharp central ridge; deep, narrow siphonal notch. Grey-white, heavily mottled with grey-brown, especially on two bands on body whorl, each with two rows of thick black axial dashes, one at each edge of the band; black lines radiate from suture towards shoulder; lip grey-tan; interior grey; columella white; fasciole white crossed by wavy dark brown lines.

A. nivosa Lamarck 1804 Western Australia, 85mm. Variable. Generally moderate to short spire; blunt, rounded protoconch of three whorls with obsolete nodules; two-and-a-half subsequent whorls. A little concave between suture and shoulder which is usually slightly angled. Smooth, long aperture; rounded lip; four columellar plaits; strong fasciole with ridge near posterior margin; deep, narrow siphonal notch. Below shoulder white base shows through brown-grey overlay as small white spots, and two bands of grey with axial, dark brown streaks and dashes; from suture to shoulder cream-white with brown blotches and profuse, dark brown lines; inside lip brown; interior brown-grey; columella pink-orange.

A. vespertilio L. 1758 north Australia, New Guinea, Philippines and Indonesia, 115mm. Variable. Generally solid with short spire. Protoconch of three axially ribbed whorls; three subsequent whorls with short blunt spines extending a short way anteriorly as blunt ribs (sometimes shoulders lack spines). Smooth; long, narrow aperture; slightly thickened, simple lip; four columellar plaits; strong, callous fasciole; narrow, deep siphonal notch. Pale tan; brown, red or olive, zigzag lines creating tent marks; may be almost black.

A. rutila norrisii Gray 1838 east New Guinea to the Solomons, 85mm. Variable. Generally solid with short, blunt spire. Rounded protoconch of three-and-a-half whorls, obsolete nodules; three subsequent whorls with low blunt knobs on shoulder, ten on body whorl which may appear above suture on earlier whorls. Simple, slightly thickened, recurved lip; four columellar plaits; strong fasciole; deep, narrow siphonal notch. Creamy grey; irregular wavy black axial streaks and black dots in three spiral bands. Colour may be cream-pink with red clouding.

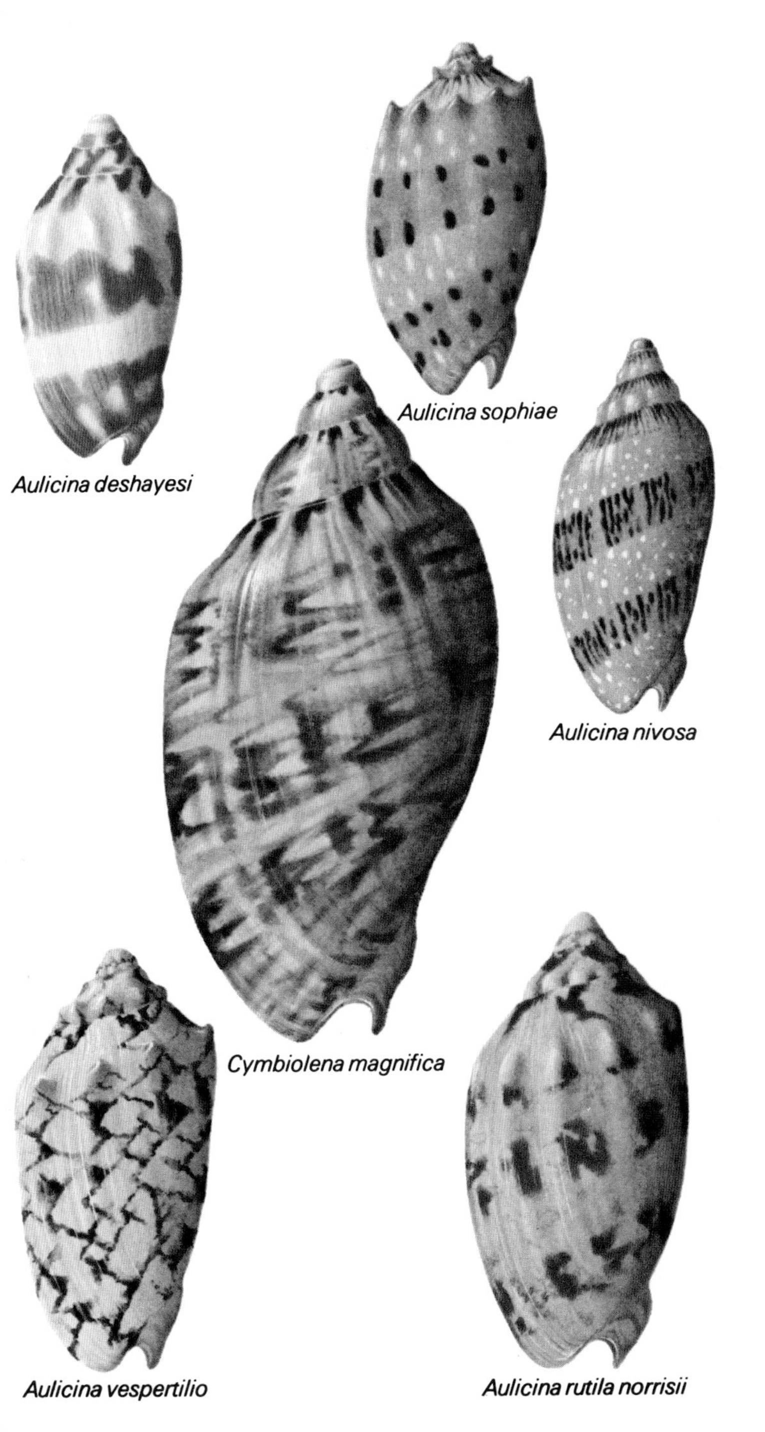

Aulicina deshayesi

Aulicina sophiae

Aulicina nivosa

Cymbiolena magnifica

Aulicina vespertilio

Aulicina rutila norrisii

Callipara bullatiana Weaver and du Pont 1967 South Africa, 70mm. Low spire, indented suture, and long, rather narrow aperture. Simple, slightly thickened lip. Columella indented centrally, with two plaits anteriorly, sometimes a third. Very shallow siphonal notch; weak fasciole. Light brown, speckled with darker brown, thickly below suture and anteriorly, and forming five, indistinct, spiral lines on body whorl; very long aperture; inside lip white; columella and siphonal notch tan. No specimen of the species, the only one in the genus, is known to have been taken alive.

Cymbiolacca wisemani Brazier 1870 Queensland, Australia, 85mm. Moderate spire and blunt protoconch of three ribbed whorls. Axial ribs, posteriorly ending in sharp points on angular shoulders. Smooth with long, narrow aperture. Lip thickened with sharp edge; straight columella with four or five plaits. Narrow, deep siphonal notch; wide fasciole with central ridge. White to pink with pink-brown irregular blotches, generally in four bands; overall red-brown dots or short dashes in pale yellow surroundings; lip with pale yellow-brown clouding; interior, fasciole and columella white.

C. cracenta McMichael 1963 Queensland, Australia, 80mm. Narrow; moderate to short spire; blunt protoconch of three-and-a-half ribbed whorls. Slightly angular shoulders have short, sharp spines, obsolete near long, narrow aperture. Bevelled lip; slightly concave columella with four plaits; deep, narrow siphonal notch; strong fasciole with central ridge. Pink; paler pink streaks and spots above spines and tent marks below; four bands of darker pink blotches and dark brown spots and axial dashes; lip pink; interior pinky grey; columella white.

C. pulchra Sowerby 1825 Queensland, Australia, 90mm. Very variable; typically with short spire and rounded protoconch of three-and-a-half ribbed whorls; ribs ending posteriorly in short, sharp spines on slightly angular shoulders. Long, wide aperture; bevelled lip; slightly concave columella with four plaits; deep, narrow siphonal notch; fasciole with low ridge. Pale red-brown; white tent marks; four darker bands bearing sparse chocolate dots and dashes. The form *woolacottae* (illustrated) is much paler, sometimes white with very pale yellow bands, chocolate dots and white aperture.

Zidona dufresnei Donovan 1823 east South America, 200mm. Variable. Typically with moderately long, narrow spire; protoconch with pointed, claw-like projection of callus covering it. Early whorls convex but body whorl and penultimate whorl (less so) have very wide, rounded shoulders. Simple, slightly thickened lip; slightly concave columella with three plaits; wide fasciole; wide, shallow siphonal notch. Cream; axial, wavy, blue-grey or brown lines; lip, columella and very wide parietal callus apricot; interior cream. Grains of sand may be covered by the overall glaze.

Adelomelon ancilla Lightfoot 1786 south-east South America, 185mm. Long, narrow; high spire; pointed protoconch of two whorls; early whorls sometimes ribbed. Long aperture; simple lip; slightly concave columella with three plaits; wide, deep siphonal notch; strong fasciole partly covered by callus. Cream; sparse, zigzag, axial, brown lines.

A. brasiliana Lamarck 1811 south-east South America, 180mm. Globose with short spire and blunt, rounded protoconch. Body whorl shouldered usually with ten, blunt, axially pinched knobs. Wide aperture; slightly thickened, bevelled lip; concave columella with two strong and one obsolete plaits; wide, rather shallow siphonal notch; wide, strong fasciole; parietal area with thick callus. Flesh to grey-white; brown clouding on knobs near lip; inside lip orange-pink; interior lighter; columella, parietal area rich pink-brown.

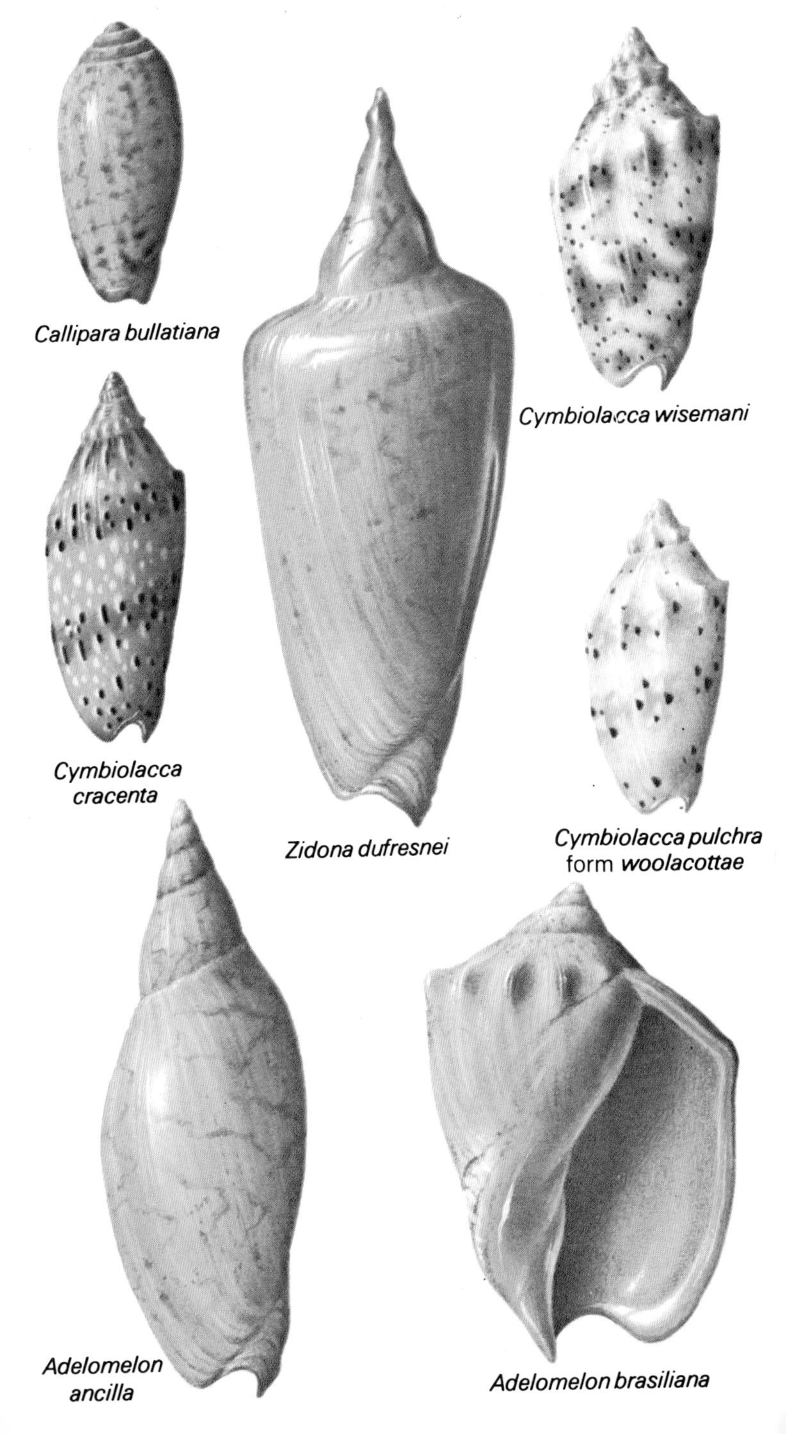

Callipara bullatiana

Cymbiolacca wisemani

Cymbiolacca cracenta

Zidona dufresnei

Cymbiolacca pulchra form *woolacottae*

Adelomelon ancilla

Adelomelon brasiliana

Volutoconus grossi Iredale 1927 north-east Australia, 110mm. Solid, heavy; moderate, bluntly pointed spire. Protoconch of three-and-a-half whorls; fine sharp point at apex; three subsequent whorls. Slightly concave below suture; smooth. Long, narrow aperture; simple, bevelled lip; four columellar plaits, heavy callus posterior to last one. Deep, narrow siphonal notch; moderately strong fasciole. Pink-red; blue-white tent marks; sometimes four, more or less distinct, darker, spiral, red bands; aperture pink.

V. bednalli Brazier 1878 north Australia, 130mm. Solid; moderate, rounded spire. Protoconch of three-and-a-half whorls; fine, sharp point at apex; three subsequent whorls. Slightly concave below suture; smooth; long aperture. Thickened lip; four columellar plaits; deep, narrow siphonal notch; moderately strong fasciole. Ivory or straw; latticed with dark purple-brown or chocolate lines.

Harpulina lapponica L. 1767 south India and north Sri Lanka, 100mm. Solid, ovate; low spire. Prominent, globose protoconch of three whorls; five subsequent whorls. Low axial ribs on early whorls become obsolete on antepenultimate whorl; about ten, low, axially pinched nodules on body whorl become obsolete near lip. Uneven, faintly channelled suture; thickened, bevelled lip; almost straight columella, three strong plaits anteriorly, four or five weak ones behind them. Narrow, deep siphonal notch; narrow anterior canal; callus partly covers fasciole. Cream; three spiral bands of pale brown blotches sometimes darker or absent, and overall spiral rows of dark brown dots and dashes except for a band below suture; aperture white edged with pale yellow; anterior end of columella, callus pale yellow-brown. Variable.

Alcithoe arabica Gmelin 1791 New Zealand, 195mm. Very variable. Moderate, concave spire; blunt, rounded protoconch of two-and-a-half whorls; five subsequent whorls. Usually angular with ribs on early whorls developing into blunt, pointed nodules on body whorl. Thickened lip, sometimes recurved; four strong columellar plaits and sometimes a weak one posteriorly; narrow, deep siphonal notch; parietal area and part of fasciole with callus. Yellow-grey or red-brown; fine, wavy, axial, brown lines and about four, broken, spiral bands of purple-brown, generally axial markings; lip, columella flesh-pink; interior grey-pink.

A. swainsoni Marwick 1926 New Zealand, 225mm. Like *A. arabica* above, of which it may be a form, it is variable. Usually solid with moderate spire. Rounded protoconch of two-and-a-half whorls; five-and-a-half subsequent whorls. Either smooth or with early whorls axially ribbed, ribs becoming obsolete and replaced by low blunt nodules on slightly angular shoulder; coarse, axial growth lines. Thickened recurved lip is projected a little posteriorly; almost straight columella with five strong plaits; deep, narrow siphonal canal; strong fasciole partly covered by shield of callus extending over parietal area. Light or dark red-brown; wavy, axial lines tending to be most obvious in three or four spiral bands; lip and parietal callus metallic pink; long aperture pink-white.

Odontocymbiola magellanica Gmelin 1791 southern South America and the Falklands. Solid but rather light with moderately short spire. Small rounded protoconch of one-and-a-half whorls; three-and-a-half subsequent whorls; faint knobs on rounded shoulder. Body whorl a little inflated. Simple lip; slightly concave columella with three strong plaits; shallow, rather narrow siphonal notch; strong fasciole; lightly callous parietal area. Cream; wavy axial brown lines, forming three spiral bands; aperture pale creamy pink; columella darker.

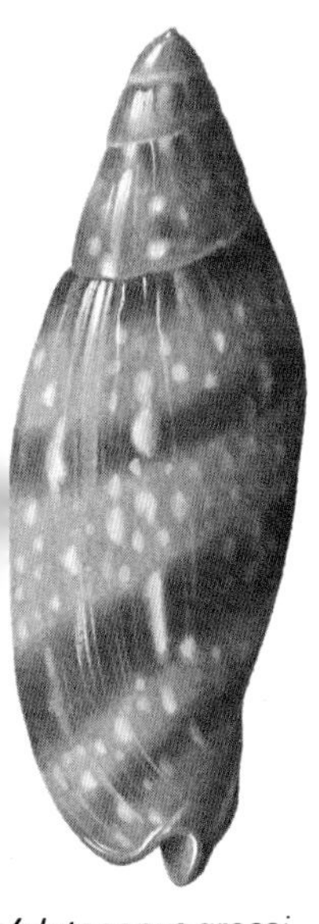

Volutoconus grossi

Alcithoe arabica

Alcithoe swainsoni

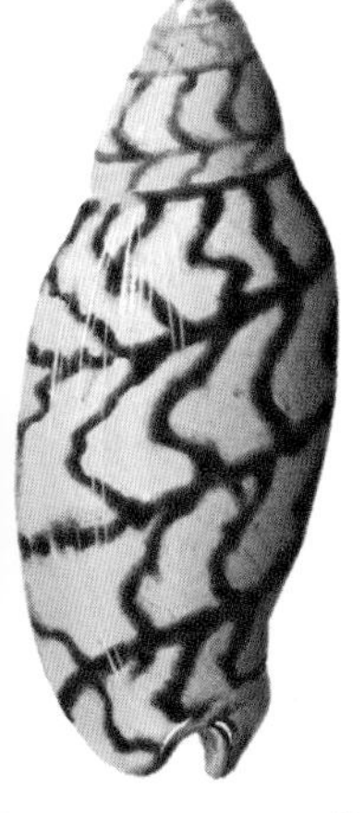

Volutoconus bednalli

Odontocymbiola magellanica

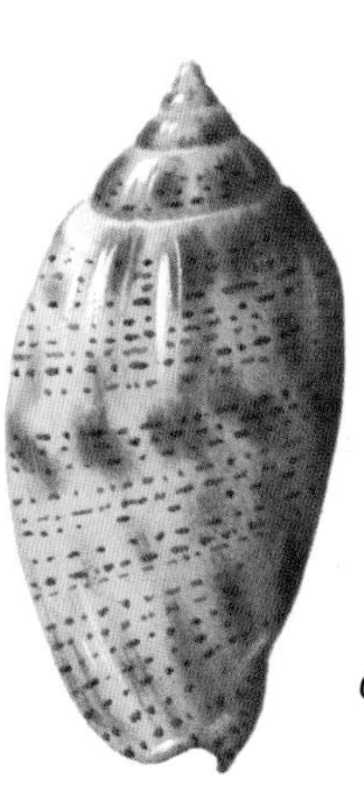

Harpulina lapponica

Amoria grayi Ludbrook 1953 west Australia, 100mm. Large pointed protoconch of four to five whorls. Smooth; very long aperture; bevelled lip, flaring anteriorly. Four columellar plaits; narrow, deep siphonal notch. Apex white; protoconch grey; subsequent whorls grey, axial chestnut lines at suture; lip grey-brown; interior rich brown; sometimes two, pale brown bands.

A. praetexta Reeve 1849 north-west Australia, 70mm. Protoconch of three whorls. Smooth with bevelled lip. Three columellar plaits; narrow, shallow siphonal notch. Protoconch grey-white; subsequent whorls golden brown with fine, white tent marks; chestnut commas below suture; two rows of zigzag lines on body whorl; aperture white; interior pale brown.

A. maculata Swainson 1822 east Australia, 75mm. Protoconch of four-and-a-half whorls. Smooth with bevelled lip. Four columellar plaits; wide, shallow siphonal notch. Cream to brown; four bands of red-brown axial lines; aperture white; interior with brown tinge. Lines may form continuous bands.

A. damonii Gray 1864 north and west Australia, 140mm. Variable. Protoconch of four-and-a-half whorls. Slightly shouldered and smooth with thick ened, sharp-edged lip. Four columellar plaits; wide, deep siphonal notch broad, strong, fasciole with weak ridge. Apex white; protoconch whorls white at suture, opaque grey below; subsequent whorls with closely-set, blue axial streaks crossing on to callus covering suture; wide, brown-red band on penultimate whorl; body whorl cream-grey with light red-brown streaks and with two or three bands of purple-brown marks; lip brown; interior deeper brown; columella white edged with brown next to fasciole, which is white tipped with dark blue. Shell varies from narrow with high spire to inflated and broad with short spire; colour from cream with thin red-brown lines to only red-brown markings.

A. ellioti Sowerby 1864 west Australia, 110mm. Protoconch of four-and-a-half whorls. Faintly shouldered and flaring, thickened, bevelled lip. Four columellar plaits; strong fasciole; narrow, moderately deep siphonal notch. Protoconch opaque, very pale lemon; subsequent whorls rich cream; dark brown spots at suture and thin, brown axial lines, sometimes forming two spiral bands; lip white fading to pale brown interior; columella white.

A. benthalis McMichael 1964 east Australia, 40mm. Protoconch of two, thinly callous whorls; suture with fine channel. Body whorl inflated at shoulder narrowing quickly towards anterior. Thickened, bevelled lip; four columellar plaits; narrow, deep siphonal notch; weak fasciole. Rich cream; yellow-gold clouding in four broken bands obscured by close-set, axial, gold-brown lines with two waves in each; aperture white, internal grey tinge.

A. canaliculata McCoy 1864 Queensland, Australia, 70mm. Protoconch of three-and-a-half whorls; suture deeply channelled on body whorl which is inflated at shoulder. Smooth; bevelled, expanded lip; four columellar plaits; narrow, deep siphonal notch. Shallow water form (illustrated) is white with five spiral bands of well-spaced, axial dashes of pale red-brown; protoconch, aperture white. Deep water form pink; large, red blotches; red axial lines.

Scaphella junonia Lamarck 1804 Florida, Gulf of Mexico, 130mm. Protoconch of two whorls; four-and-a-half subsequent whorls; bi-angulate body whorl. Obsolete axial ribs on first two whorls, last two smooth. Thickened, bevelled lip; four columellar plaits. White to pale yellow; nine spiral rows of rectangular, dark brown blotches, most dividing into two near lip; aperture white, brown tinge.

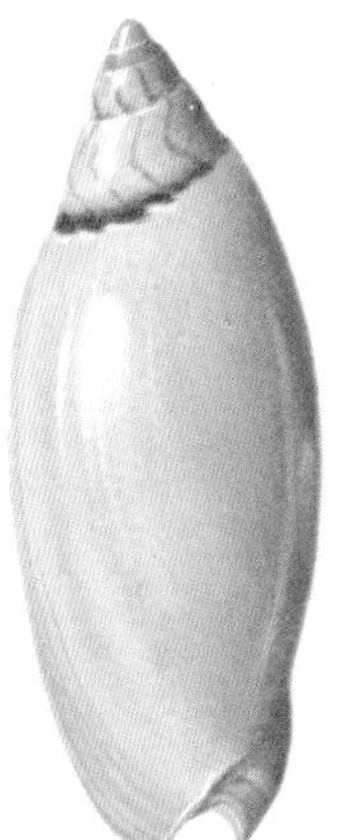

Amoria grayi

Amoria maculata

Amoria praetexta

Amoria damonii

Scaphella junonia

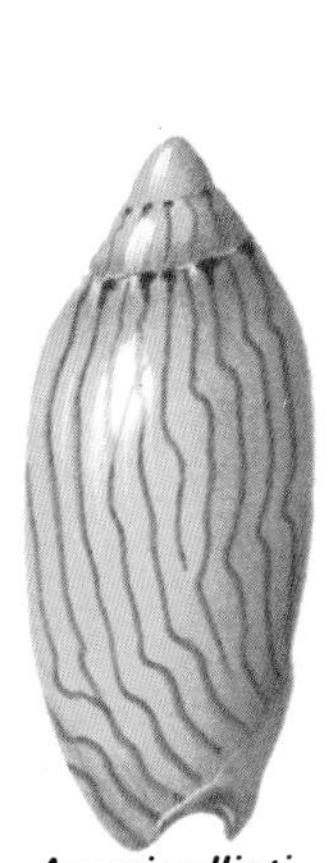

Amoria ellioti

Amoria benthalis

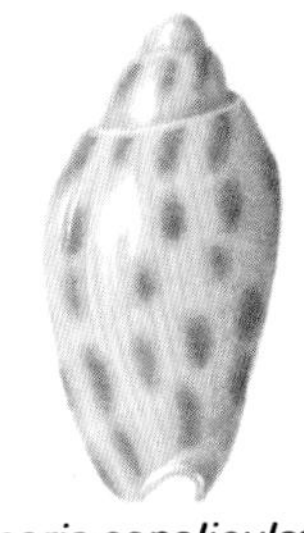

Amoria canaliculata

Amoria zebra Leach 1814 north-east Australia, 55mm. Ovate; short spire. Blunt, rounded protoconch of two-and-a-half whorls, three subsequent whorls. Indented suture; body whorl inflated at shoulder. Thickened lip; columella with four plaits, extended, curved at base; anterior canal; narrow, deep, oblique siphonal canal; fasciole with low ridge. Protoconch red; cream-yellow; wavy, axial, chocolate or orange-brown lines; aperture white; interior tinged pale brown.

A. undulata Lamarck 1804 south-east and south Australia, Tasmania, 90mm. Variable. Typically ovate with short, slightly concave spire. Small, slightly pointed protoconch of four whorls; three subsequent whorls. Smooth suture; round shoulder; a little concave between suture and shoulder. Thickened, crudely bevelled lip; straight columella, four strong plaits, sometimes small plaits between; anterior canal a shallow notch; deep, narrow siphonal canal; fasciole with low ridge. Yellow-cream to white; sparse, brown blotches in four broken bands; axial, wavy, dark brown lines; inside lip white; interior, columella yellow-pink.

A. molleri Iredale 1936 north and south-west Australia, 100mm. Elongate; moderate, slightly concave spire. Sharply pointed protoconch of four whorls; three subsequent whorls. Smooth; faint growth striae; long, narrow aperture; bevelled inside lip with sharp, white ridge, becoming obsolete at ends; four, strong columellar plaits, sometimes weak plaits between, callous hump behind last plait; deep, narrow siphonal notch. Shiny pink-brown; paler below suture and at lip; inside lip pink; interior pink-brown; columella pink; fasciole weak, white.

Cymbiolista hunteri Iredale 1931 central east Australia, 175mm. Light; low spire. Protuberant, conical protoconch of three whorls; three-and-a-half subsequent whorls with angular shoulders carrying short, sharp spines, twelve on body whorl. Slightly concave between shoulder and suture; a little inflated below shoulder; long, narrow aperture. Simple lip; four strong columellar plaits; deep, narrow siphonal notch. Commonly pale flesh; brown lines, especially across shoulder; three or four spiral bands of blue-grey marks generally edged on left with brown. Deep water form is peach-coloured, marks on bands orange-brown. Interior brown, darker farther from lip; columella pink; fasciole strong, white.

Neptuneopsis gilchristi Sowerby 1898 South Africa, 200mm. Very high spire. Protoconch of two whorls; the first larger, conical and pointed off-centre; second narrower, shorter; six subsequent whorls. Inflated; constricted suture; fine, close, spiral striae. Wide, semicircular aperture; thickened, bevelled, slightly recurved lip; smooth, extended columella forming open siphonal canal; narrow parietal area with evenly curved edge, callous. Pink-white; thin, brown periostracum (illustrated); protoconch, aperture, parietal area pink-white.

Teramachia tibiaeformis Kuroda 1931 south Japan, 80mm. Light, elongate; very high spire. Small protoconch of two whorls; ten subsequent whorls closely axially ribbed except anteriorly, ribs becoming obsolete on penultimate whorl. Constricted at suture; narrow aperture. Flaring, semicircular lip; bevelled, recurved edge. Slightly concave, smooth, extended columella. Grey-brown; darker brown band below suture on later whorls; aperture pink-grey, lighter at lip.

Ampulla priamus Gmelin 1791 Portugal, Spanish Atlantic coast, 80mm. Light, globose; moderate spire. Bluntly rounded protoconch of two-and-a-half whorls; three subsequent whorls. Inflated with indented suture; large

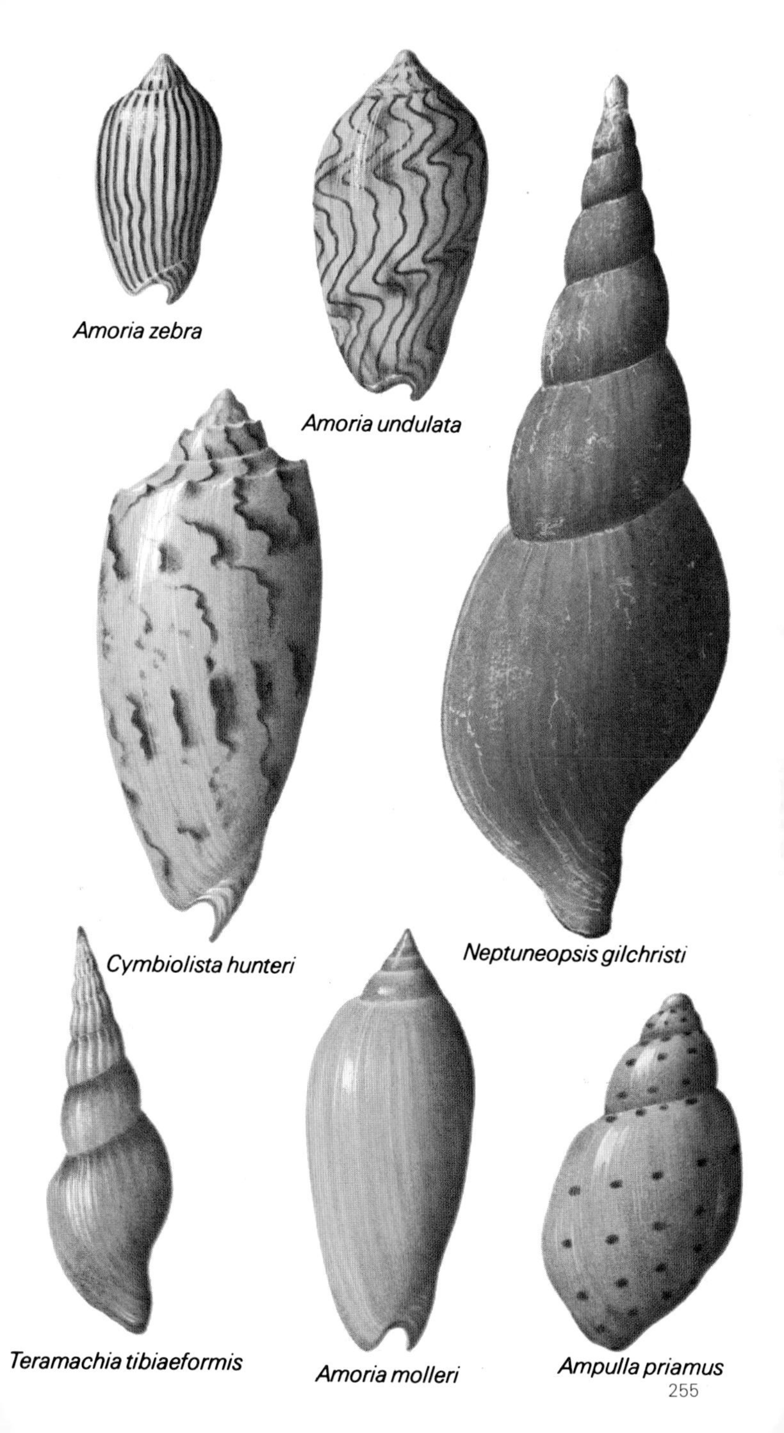

Amoria zebra

Amoria undulata

Cymbiolista hunteri

Neptuneopsis gilchristi

Teramachia tibiaeformis

Amoria molleri

Ampulla priamus

aperture; flared, unevenly curved, simple lip; concave, smooth columella; wide, lightly callous parietal area; indistinct fasciole; wide, barely indented siphonal notch. Pink-brown; seven spiral rows of red-brown spots; inside lip edge pale, interior medium brown; columella pink; parietal area pale yellow-pink with red-brown spots.

Family: Marginellidae

The marginellas are a large family found worldwide in warm and tropical seas. There is one large member of the family, *Afrivoluta pringlei*, about 120mm; two or three are about 70mm; two or three are between 40 and 50mm; a number between 20 and 30mm; and a great many smaller.

Afrivoluta pringlei Tomlin 1947 east South Africa in deep water, 120mm. Light, rather delicate, with short, blunt spire and large, rounded protoconch. Body whorl is long and bi-angulate with fine growth lines. Coarse callus, very heavy on parietal area, covers most of spire, especially sutural area and much of anterior end of body whorl. Narrow aperture; thickened lip with edge recurved, straight along middle part; columella has four, very strong plaits anteriorly, one nearest anterior end smaller than other three. Light tan; darker brown periostracum; thin glaze overall.

Marginella glabella L. 1758 north-west Africa, 45mm. Solid with short, straight-sided spire and small, rounded protoconch. Body whorl a little inflated and roundly, though noticeably, shouldered. Moderately wide aperture; heavily thickened lip is recurved and faintly dentate; columella has four strong plaits. Cream with pale brown-pink overlay, stronger in two spiral bands; base colour shows through as small dots; irregular blotches of brown-pink below suture; whole covered with thin glaze; inside lip and columella white; interior pale brown-pink.

M. desjardini Marche-Marchaud 1957 Ivory Coast, west Africa, 50mm. Solid, elongate, with short, slightly convex spire. Small, rounded protoconch and rounded shoulders sloping to end of siphonal canal. Lip much thickened, recurved and finely dentate; columella has four plaits. Pale apricot; three darker bands; small white spots rather sparsely scattered over surface; short, white streaks radiating from suture; aperture white; interior with pink tinge.

M. pseudofaba Sowerby 1846 Senegal, west Africa, 30mm. Solid with moderately high spire. Shoulders of each whorl are sharply angled, with axially pinched knobs, about fourteen on body whorl. Knobbed shoulders of earlier whorls show above suture; some of knobs extend almost to siphonal canal but most cover only about one-third of body whorl. Narrow aperture; lip is thickened, recurved and dentate within; columella has four strong plaits; shallow anal canal; siphonal canal a little extended and recurved, sharply at lip. Ivory white; green-grey clouding, generally on two spiral bands; profusely spotted with black in spiral rows, about ten on body whorl; spots also generally aligned axially; inside lip, interior and plaits white; end of siphonal canal grey.

Persicula cingulata Dillwyn 1817 west Africa, 20mm. Flat spire, smooth, with body whorl a little inflated posteriorly. Narrow aperture; thickened lip extends slightly beyond apex. Convex columella has about seven small plaits, largest anteriorly and smallest posteriorly. Small, callous knob on parietal area opposite posterior end of lip. White or pink-cream; about twelve spiral red lines, two or three of which bifurcate away from lip; aperture white.

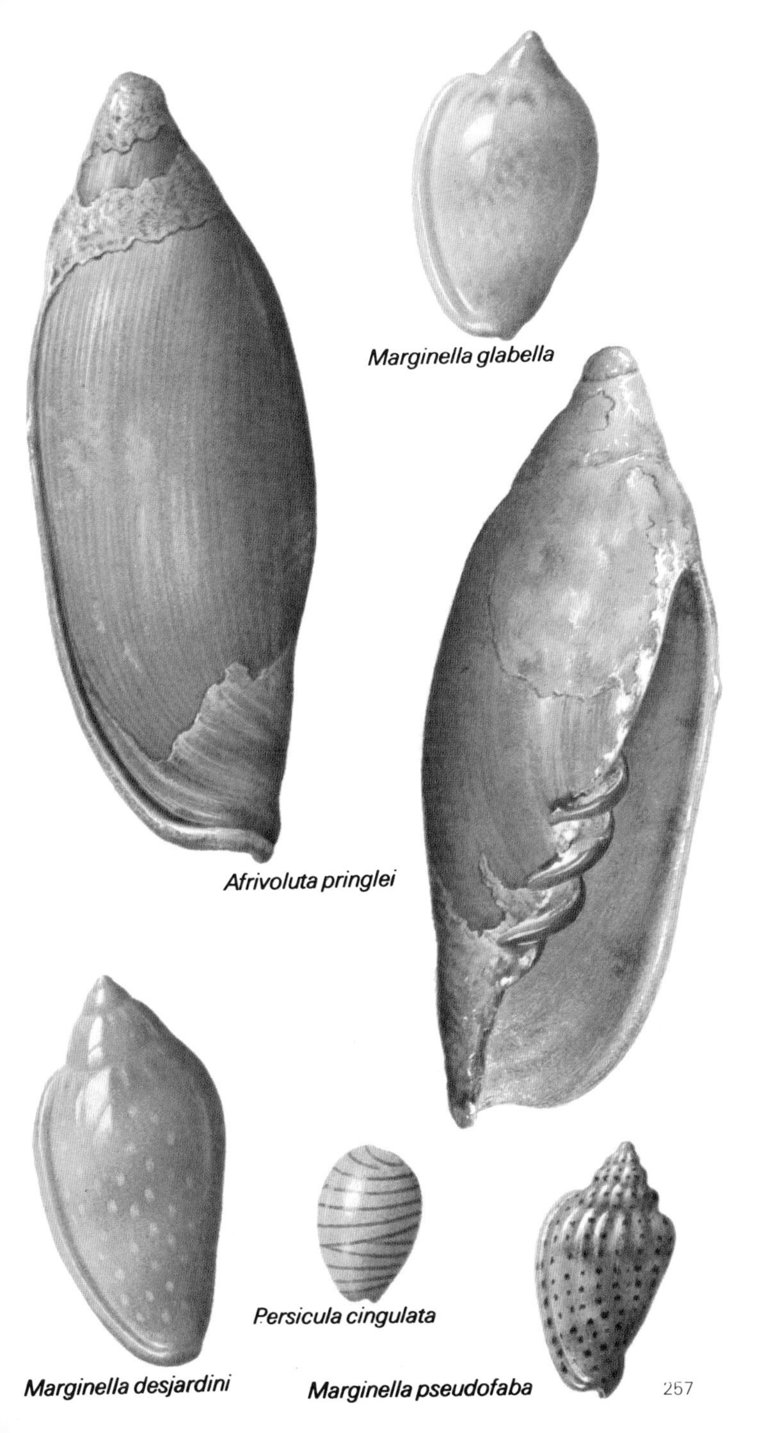
Marginella glabella
Afrivoluta pringlei
Persicula cingulata
Marginella desjardini
Marginella pseudofaba

Marginella rosea Lamarck 1822 east South Africa, 50mm. Smooth with moderate spire and about two-and-a-half shouldered whorls. Small rounded protoconch and thickened, recurved lip. Columella has four plaits. White; pale pink, generally reticulate pattern; row of darker pink-brown blotches between shoulder and suture; paler one at anterior of body whorl; lip white, dark pink-brown spots on outer edge; interior and columella white.

M. mosaica Sowerby 1846 east South Africa, 30mm. Similar in shape to above but slightly lower spire, more thickened lip, slightly extended and angled at shoulder. Creamy white flecked with pale grey-brown; spiral rows – about twelve on body whorl – of dark grey-brown spots and short dashes; aperture white.

M. piperata Hinds 1844 east South Africa, 25mm. Similar in shape to *M. rosea,* though smaller. White; slight pale brown clouding; profusely speckled with tiny axial dashes of red-brown; longer dashes of red-brown between shoulder and suture; outer edge of lip with dark brown, smudged spots; aperture white. There are a number of named forms with different colours and patterns.

M. ventricosa G. Fisher 1807 Indonesia and Malaya, 25mm. Smooth, shiny, with low spire and moderately wide aperture. Thickened, recurved lip runs round siphonal canal to join fasciole, ending posteriorly in callous area on spire. Columella has five plaits. Whole exterior is highly glazed. Creamy grey; outer edge of lip darker with light brown-grey rim; inside lip and plaits white. A well-known synonym is *M. quinqueplicata* Lamarck.

M. adansoni Kiener 1834 north-west Africa, 30mm. Rather slender; shoulders with axially pinched knobs, about sixteen on body whorl. Narrow aperture; thickened, recurved lip with angular shoulder, thirteen teeth within. Columella with four plaits, recurved at anterior end. Tan; irregular darker brown marks on shoulder and more or less conspicuous wavy axial brown lines; lip white with dark brown spots; interior and columellar plaits white.

Persicula persicula L. 1758 west Africa, 20mm. Slightly sunken spire and a little inflated at shoulder of body whorl. Narrow aperture; thickened, recurved lip, extends posteriorly beyond apex. Convex columella with about nine plaits, strong anteriorly becoming smaller until barely perceptible towards posterior. Large, blunt, callous knob at top of parietal wall. Pale cream; profuse pink-tan spots, especially crowded in three spiral bands; spire pink or red, creamy glaze; aperture and parietal knob white.

P. cornea Lamarck 1822 north-west Africa, 30mm. Graceful, slender, with very low spire covered with callus. Narrow aperture; lip thickened and slightly concave centrally, extending posteriorly slightly above apex. Columella with about ten small plaits, becoming smaller posteriorly. Cream; three slightly darker spiral bands with a pink tinge; aperture white; bands showing faintly through inside of lip.

P. elegans Gmelin 1791 Malay Peninsula, 25mm. Shaped like *M. ventricosa* but chubbier and having six strong plaits. Pale grey; many darker grey spiral bands, uneven in width and distance apart, all broken by many fine axial lines of base colour giving uneven, reticulated colour effect; lip from callous area where it joins spire, round end of siphonal notch and on to lower end of parietal area, orange-brown; interior and plaits white.

Marginella mosaica

Marginella piperata

Marginella rosea

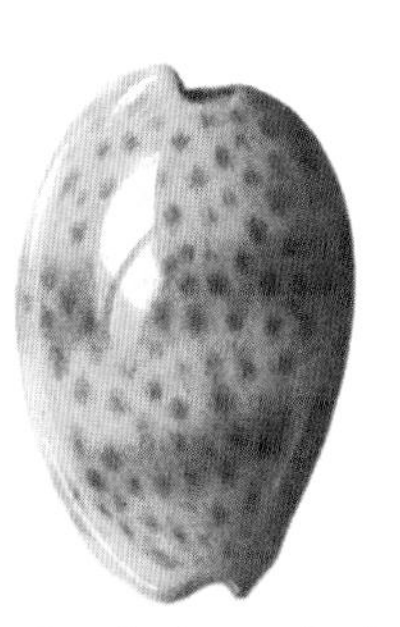

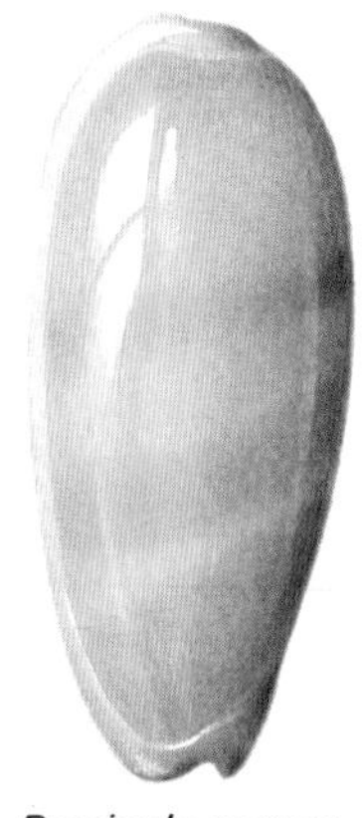

Persicula persicula

Persicula cornea

Marginella ventricosa

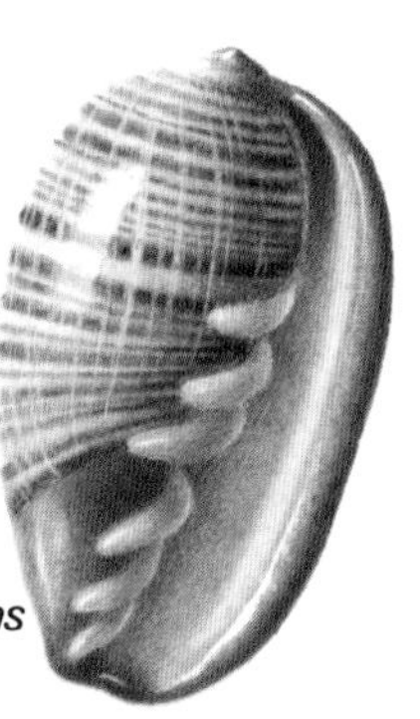

Persicula elegans

Marginella adansoni

Marginella bullata Born 1778 Brazil, 70 mm. Sunken, glazed spire; apex just proud of surrounding area; smooth bar growth lines under glaze. Aperture widening anteriorly; thickened, recurved lip, slightly concave centrally, obsoletely dentate within; four columellar plaits; strong, glazed fasciole. Pale apricot; faint, darker, spiral bands; grains trapped under glaze give spotted appearance in places; outer edge of lip darker; inner edge, plaits and fasciole shiny white; interior white with pale apricot tinge.

M. ornata Redfield 1870 South Africa, 25 mm. Chubby; moderate spire; sloping shoulder. Wide aperture; thickened, recurved lip is slightly angular at shoulder; four columellar plaits. Rose-brown; broad, central, pale band, narrow one either side, one on shoulder; central area may be stippled; lip white, dark rose-brown dots and dashes; interior mauve; columella, siphonal notch white.

M. marginelloides Reeve Philippines, 12 mm. Angular shoulders. Strong, axial ribs on all whorls. Finely spirally lirate. Narrow aperture; thickened lip with strong, internal, finely dentate ridge; columella and parietal area crossed by small ridges. Off-white to pale grey; dark brown blotch posteriorly within aperture, showing through on the shoulder on the dorsal side, especially between the ribs nearer, though not on, the lip.

M. philippinarum Redfern 1848 Philippines, 15 mm. Slender, small, solid with short spire. Smooth and shiny. Narrow aperture, wider anteriorly; thickened, turned-in, bevelled lip; almost straight columella with four plaits. Slightly concave lip makes shell look a little bent. Pale tan or red-tan; three, broken, spiral bands of darker brown with grey flecks; suture white; lip cream externally, white on edge and within; interior red-tan; columella and plaits white.

M. avena Kiener 1834 Caribbean, 12 mm. Similar shape to above. Cream; three indistinct pale tan bands.

M. cleryi Petit de la Saussaye 1836 north-west Africa, 20 mm. Fusiform; high spire. Slightly inflated; roundly shouldered. Narrow aperture; thickened, recurved, dentate lip; four columellar plaits. Green-cream; anterior end of body whorl with grey tinge; two, grey bands on body whorl, one just showing on spire whorls; axial, slightly wavy, sometimes bifurcating, dark chocolate lines at suture; lip, lower columella, plaits, interior white.

M. margarita Kiener 1834 Mozambique, 5 mm. Smooth, very shiny with low spire. Thickened lip, slighly concave centrally, some eighteen, small teeth on inner edge; four columellar teeth, middle two larger. Beautiful, shiny, very pale rose-cream; lip, aperture white.

M. apicina Menke 1828 Caribbean, Florida, 10 mm. Similar shape to above but less slender. White, cream to pink or banded pink; apex tan.

Prunum labiata Kiener 1841 Gulf of Mexico, 40 mm. Smooth, shiny; short spire; round shoulders. Thickened, dentate lip joins spire in thick, callous area, extending over spire almost to protoconch; four columellar plaits. Pale cream-pink; three, indistinct, darker bands; upper side of lip and callus near spire with strong yellow tinge; inside lip, interior, plaits white.

Persicula lilacina Sowerby 1846 Brazil, 20–25 mm. Sunken spire; slightly inflated body whorl; narrow aperture. Very thickened lip running to sunken apex; four columellar plaits; callous parietal area. Pale rose-grey; three broad darker bands; upper edge of lip orange; inner edge of lip, interior, columella and parietal area lilac; deep interior white.

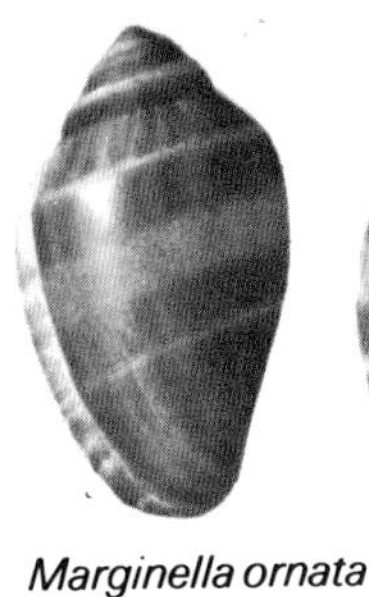

Marginella ornata

Marginella marginelloides

Marginella avena

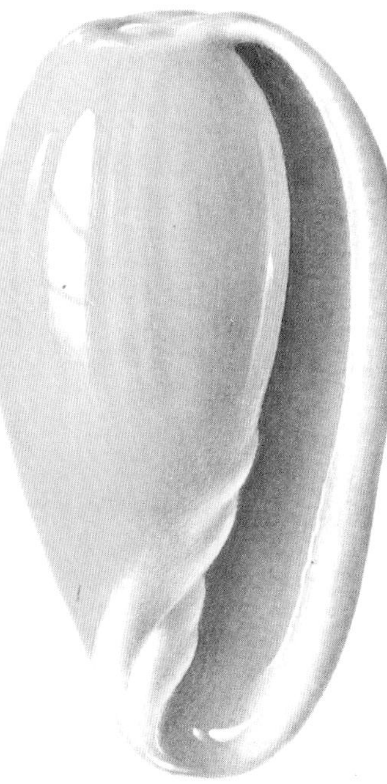

Marginella philippinarum

Marginella apicina

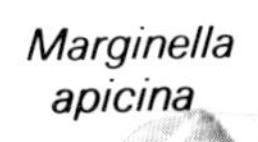

Marginella cleryi

Marginella margarita

Marginella bullata

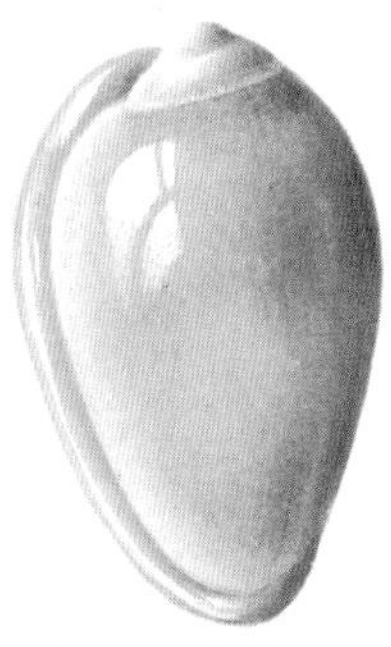

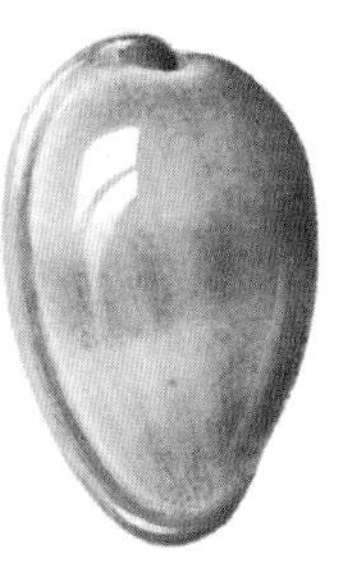
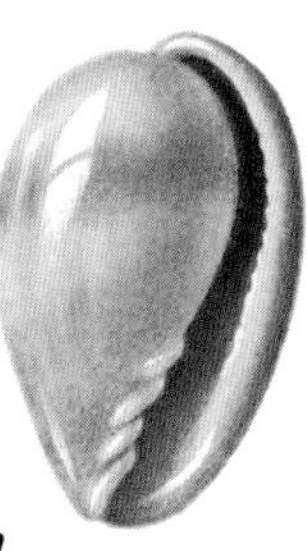

Persicula lilacina

Prunum labiata

Superfamily: Conacea

Family: Cancellariidae

The nutmeg shells live in sand in tropical areas, mostly in west America.

Trigonostoma scalata Sowerby 1832 Indo-west Pacific, 30mm. Shallow, wide channel from angular shoulder to slightly constricted suture. Weak, spiral cords; strong, axial ribs cross shoulder. Wide aperture; simple lip, tooth posteriorly, lirate within; three strong, one weak columellar plaits; lirate, narrow parietal shield. Brown-cream; sometimes brown banding; lip, columella white; interior brown.

T. scalariformis Lamarck 1822 Indo-west Pacific, 25mm. Similar to above but fewer, stronger axial ribs, nine on body whorl forming crenulations at shoulder. Weak, spiral striae cross ribs. Lip lirate within; three columellar plaits, shield almost closing umbilicus. Brown, grey or cream; white between shoulder and suture; white spiral band on body; narrow aperture, columella white; interior as exterior.

T. crenifera Sowerby 1832 Japan, 30mm. Similar to above but about fourteen ribs on body whorl. Three columellar plaits; obsolete tooth posterior of aperture; narrow, open umbilicus is partly covered by narrow columellar shield. Ivory white; brown clouding; central, spiral, pale band.

T. breve Sowerby 1832 tropical west America, 20mm. Flat or slightly concave between sharply angled shoulder and suture. Axial ribs form coronations from shoulder to suture; three, spiral ridges cross ribs forming nodules; fine striae between ridges; obsolete, axial riblets give cancellate surface. Dentate lip; three columellar folds; large umbilicus. Grey-white; brown flecks.

Cancellaria cassidiformis Sowerby 1832 tropical west America, 40 mm. Globose; short spire; sharply angled shoulders. Oblique, axial ribs end in short spines. Deeply indented suture. Faint, spiral ridges cross coarse growth lines. Thickened lip, lirate within; three, white columellar plaits; heavily glazed parietal area. Off-white to tan; spiral white band forward on body whorl; may be spiral, red lines; lip, parietal glaze yellow-brown; interior white.

C. spengleriana Deshayes 1830 west Pacific, 55mm. High spire; shouldered. Spiral ridges; axial ribs; short, blunt spines. Lip strongly lirate within; two or three columellar folds, extended fasciole; no umbilicus. Flesh-tan; pink-brown flecks on shoulders; aperture cream-white.

C. reticulata L. 1767 south-east USA, 35mm. Globose; inflated; sharp, narrow shoulders. Reticulated by axial and spiral ridges. Finely dentate lip, lirate within; one very strong, one weaker plait, spiral ridges posteriorly; strong fasciole; closed umbilicus. White; sparse, axial, brown streaks; two, broken, light brown, spiral bands; interior, plaits white.

C. asperella Lamarck 1822 west Pacific, Indian Ocean, 40mm. Similar to above; less stout; narrower shoulders. Coarse, spiral and axial threads; rough growth lines. Lirate lip; three columellar plaits; callus, parietal area, columellar shield almost close umbilicus. Cream or pale yellow-brown; two or three, indistinct, pale brown, spiral bands; aperture white.

C. similis Sowerby north-west Africa, 35mm. Solid; globose; slightly constricted suture. Cancellate; thin, spiral ridges; narrow, axial ribs. Scalloped lip, lirate within; three, curved columellar plaits, largest posteriorly; narrow umbilical chink usually open; narrower columellar shield; strong fasciole. White or pale grey; two brown-grey bands; aperture white.

Trigonostoma scalata

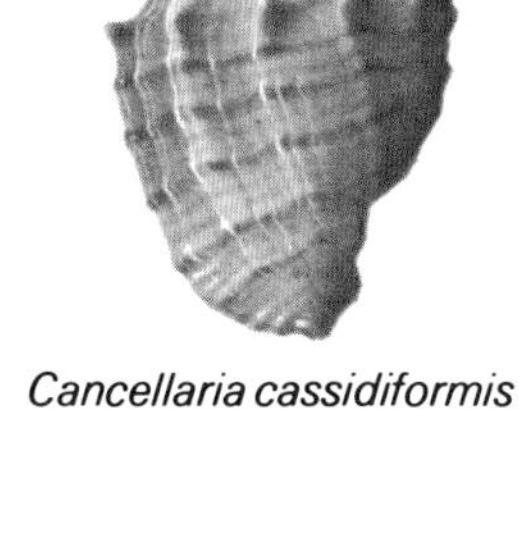
Cancellaria cassidiformis

Cancellaria reticulata

Cancellaria spengleriana

Cancellaria asperella

Cancellaria similis

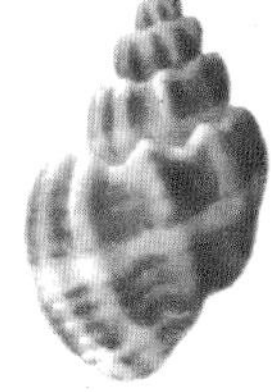
Trigonostoma scalariformis

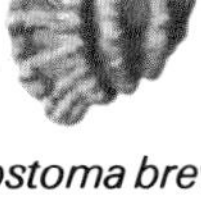

Trigonostoma breve

Trigonostoma crenifera

Family: Conidae

Genus: *Conus*

For many shell collectors this is the most popular genus of all. The enormous variety of colours and patterns, as well as shapes within the general conical base shape, have a special fascination. The cones are carnivorous and most live in tropical waters. Only a few live in subtropical waters and they are small and less colourful. They live mainly in the intertidal area, between the reef and shore, in crevices in rock and coral, and in the sandy areas around the reef. They have a soft 'skin' covering the periostracum, which can be removed by soaking in a common household bleach for a few hours, and some have a small operculum. The sting with which they kill their prey can prove fatal to human beings in a few cases. *Conus geographus* is most notorious in this respect, but the *marmoreus* and *textile* groups are also dangerous. There are some 1,500 species of named cones, divided into some thirty subgenera.

Subgenus: *Conus*

Conus marmoreus L. 1758 Indo-Pacific, 100mm. Flat, noduled spire. Black or dark chocolate with white patches overall.

Closely related are: *C. pseudomarmoreus* Cross 1875 which lacks the nodules on the spire and is endemic to New Caledonia; *C. bandanus* Hwass 1792 which has two indefinite bands where there are fewer white markings, giving the effect of black bands, and is found in the Maldive Islands, East Indies, Philippines and Melanesia to Hawaii; *C. nicobaricus* Hwass 1792 on which the banding is more marked than on *C. bandanus*, from East Indies and Philippines; *C. nocturnus* Solander 1786 which has broader and more conspicuous black bands and is found in the Philippines; and *C. vidua* Reeve 1843, the white markings forming a band at the shoulder and on the body whorl, also endemic to the Philippines. There are a few other, less common members of the group with the black or dark brown base colour and white markings, all from the same area of east Indian and west Pacific Oceans.

Subgenus: *Rhombus*

C. imperialis L. 1788 Indo-Pacific, 100mm. It has an almost flat but noduled spire. White with two brown bands; body whorl is encircled by black and brown dots and dashes.

The subgenus also includes: *C. zonatus* Hwass 1782 from the Maldives; *C. fuscatus* Born 1778 from Mauritius, which is a form of *C. imperialis*, as is *C. viridulus* Lamarck 1810 from east Africa.

Subgenus: *Lithoconus*

C. leopardus Röding 1798 almost throughout the Indo-Pacific, 220mm. One of the largest and heaviest of the cones, it has a flat spire though more raised than its near relative *C. litteratus*, and rounded shoulders, the edges of the shoulders of the earlier whorls showing as ridges on the spire. White encircled with close-set markings of shades of blue and purple forming lines, some of dots, some of short axial lines, some of joined double dots, all slightly smudged. The tip is white, the surface rather dull, and the large, old, heavy specimens are often pitted and scarred.

C. litteratus L. 1758 Indo-Pacific, 120mm. It has a flat spire with a rounded shoulder and is slightly waisted. White with more or less conspicuous bands of yellow, and overall, close-set bands of rectangular black or dark chocolate spots, larger and sometimes running together axially towards the shoulder.

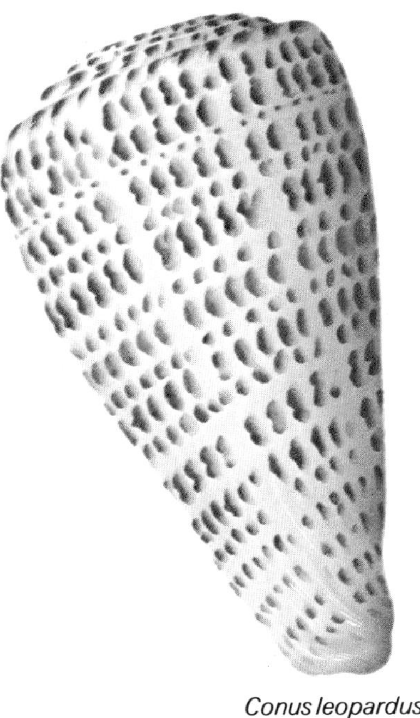

Conus leopardus

Conus marmoreus

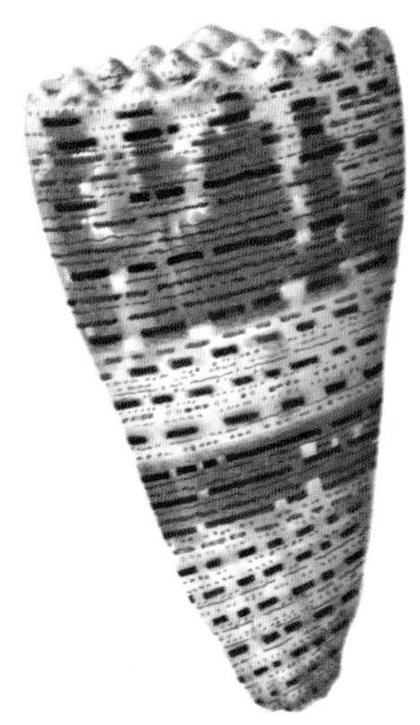

Conus imperialis

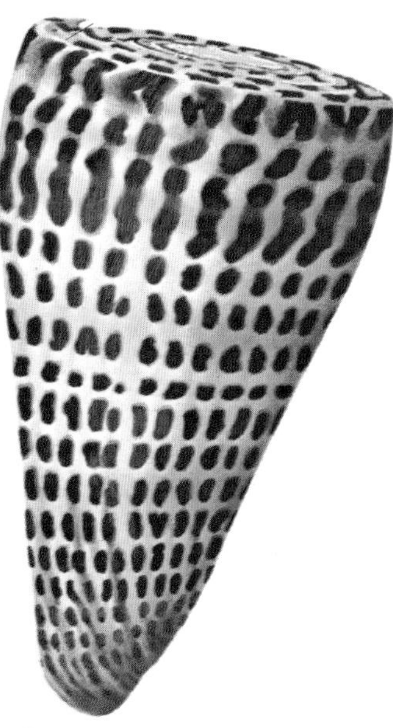

Conus litteratus

C. eburneus **Hwass 1792** throughout Indo-Pacific, 70mm. Varies a good deal over the large area. It has rounded shoulders and a flattened top with a small, pointed spire. There are two striae below the suture which show on the top when not worn away. White, sometimes with faint yellow bands, and squarish black or dark brown dots encircling the body whorl and spire. Specimens without the dark markings have been found. A small form of *C. eburneus*, *C. crassus* Sowerby 1857, is endemic to Fiji; and *C. polyglotta* A. Adams 1874 with heavier and larger black markings is found in the Philippines.

C. tessulatus **Born 1778** Indo-Pacific, 70mm. Variable in shape and colour, sometimes slightly waisted. It has a rather flat top and a short, very pointed spire with concave sides. The white background is covered with bands of squares and oblongs of red or orange, much more profuse than in *C. eburneus*; tip is often tinged with violet.

C. caracteristicus **G. Fischer 1807** Indian Ocean and Philippines, 50 mm. A chunky shell. White with two bands of wavy, brown-red lines, one on the upper part of the body whorl and one near the base; similar markings show on the apex and at the base.

There are two west African members of the subgenus: *C. pulcher* Lightfoot 1786 which can grow bigger than any other cone and is cream with brown and red markings, in a pattern similar to others of the subgenus; and *C. papilionaceus* Hwass 1792, which is smaller but somewhat similar in colour and pattern.

Subgenus: *Virroconus*

C. ebraeus **L. 1758** Indo-Pacific, 50mm. A small, chunky, rather striking shell with a moderately elevated and coronated spire. Its white background is covered with three broad bands of black chevron-shaped markings, and black markings at the top and on the spire.

C. chaldaeus **Röding 1798** range similar to that of *C. ebraeus*, though it is less common, and not always found in the same places, 25mm. It is smaller and more angular than *C. ebraeus* and has thick, black, wavy lines forming two broad bands. The lines often join up forming a 'Y' or inverted 'Y' or 'W' marking.

C. coronatus **Gmelin 1791** Indo-Pacific, 35mm. A variable species with a number of subspecies and very closely related species. It varies from slim with a well-elevated spire, to round and chunky. The colour is also very variable, blue, grey, green, beige, with dark brown-red or olive dots and mottling, with coronations on the shoulder and spire. Allied species are: *C. abbreviatus* Reeve 1843 from Hawaii; *C. miliaris* Hwass 1792, Indo-Pacific; and *C. taeniatus* Hwass 1792, Red Sea.

C. musicus **Hwass 1792** Indo-Pacific, 20mm. It has a rather flat spire. White, with a pale blue-grey band on the middle of the body whorl and at the base, with dark blue at the very tip. It is encircled with lines of dark brown dots and dashes, two of which border the central band; and larger dark spots on the shoulder which show on the spire.

C. lividus **Hwass 1792** East Indies and Pacific, 50mm. Yellow-green with a white band and white coronated spire and purple tip.

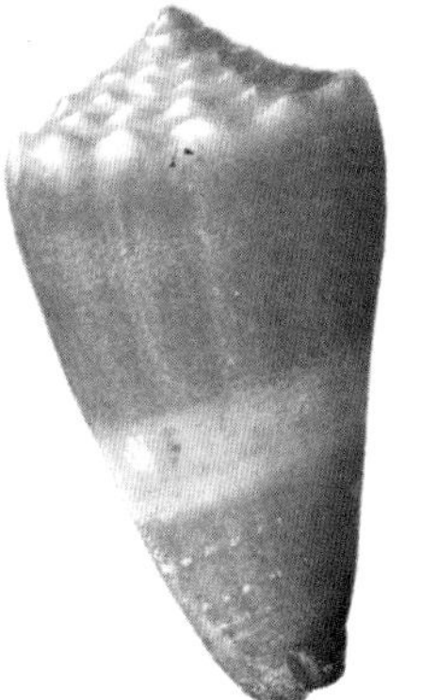
Conus lividus

Conus ebraeus

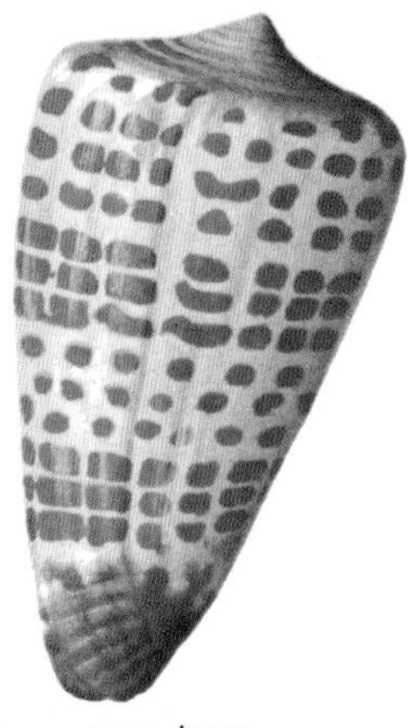
Conus tessulatus

Conus caracteristicus

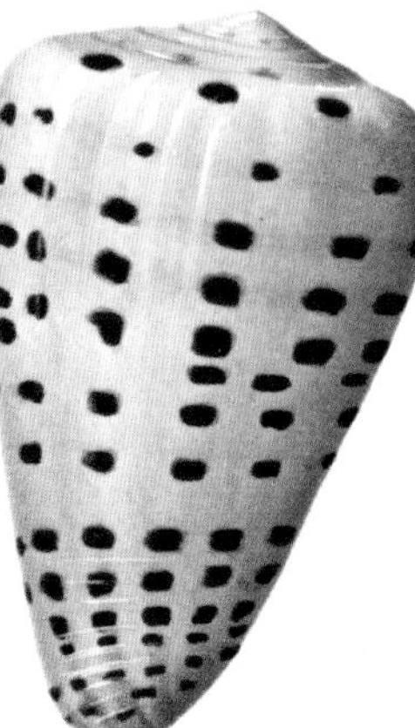
Conus eburneus

Conus coronatus

Conus coronatus

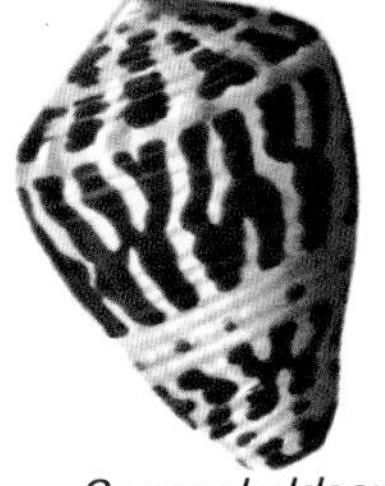
Conus chaldaeus

Conus musicus

C. piperatus **Dillwyn 1817** Indian Ocean, 30mm. Light blue-grey with indistinct, pale brown banding and a violet tip; dark brown spots between some of the coronations on the shoulder and spire.

C. distans **Hwass 1792** Indo-Pacific, 100mm. A large shell for the subgenus, it is slightly waisted. Light brown with a paler band at the waist and a darker tip, and shows rather conspicuous growth lines. Spire coronated and lighter.

Other members of the subgenus closely allied include: *C. mus* Hwass 1810 from the West Indies; *C. scitulus* Reeve 1849 from South Africa; *C. sponsalis* Hwass 1792, Pacific and Philippines; *C. nux* Broderip 1833, east Pacific; and *C. ceylanensis* Hwass 1792, Indo-Pacific.

Subgenus: *Puncticulis*

C. arenatus **Hwass 1792** Indo-Pacific, 75mm. A solid, heavy shell, slightly inflated at the shoulders, with a twist at the base of the columella forming a plait or fold. The spire is slightly elevated and coronated. Its background colour of white is speckled with small brown dots which tend to form two or three dark bands.

C. pulicarius **Hwass 1792** Pacific and East Indies, 65mm. In shape similar to *C. arenatus*. White with faint yellow blotches and marked with square-shaped, black or dark chocolate spots.

C. stercusmuscarum **L. 1758** Indo-Pacific, 50mm. A more slender member of the subgenus than the preceding two, with no coronations or very slight coronations on the shoulders. Its white background is clouded with grey and it is profusely spotted with small black and some brown dots, so close in places as to form black patches.

C. zeylanicus **Gmelin 1791** east Africa and Mauritius, 50mm. Shape somewhat similar to *C. arenatus* but with rounded shoulders and without coronations, and with a small, pointed spire. Its pinky background colour is covered with red-brown dots, darker and smaller blotches tending to form two bands. It has small, white areas mostly between the bands and on the top of the shell.

Subgenus: *Stephanoconus*

C. regius **Gmelin 1791** Florida to the West Indies, 50mm. A common member of this subgenus, it is usually a chestnut brown with blue-white markings overall (not illustrated), but has other colour variations in white and pale brown, and sometimes granulated (as illustrated). A variety *C. citrinus* Gmelin 1791 is yellow-brown only.

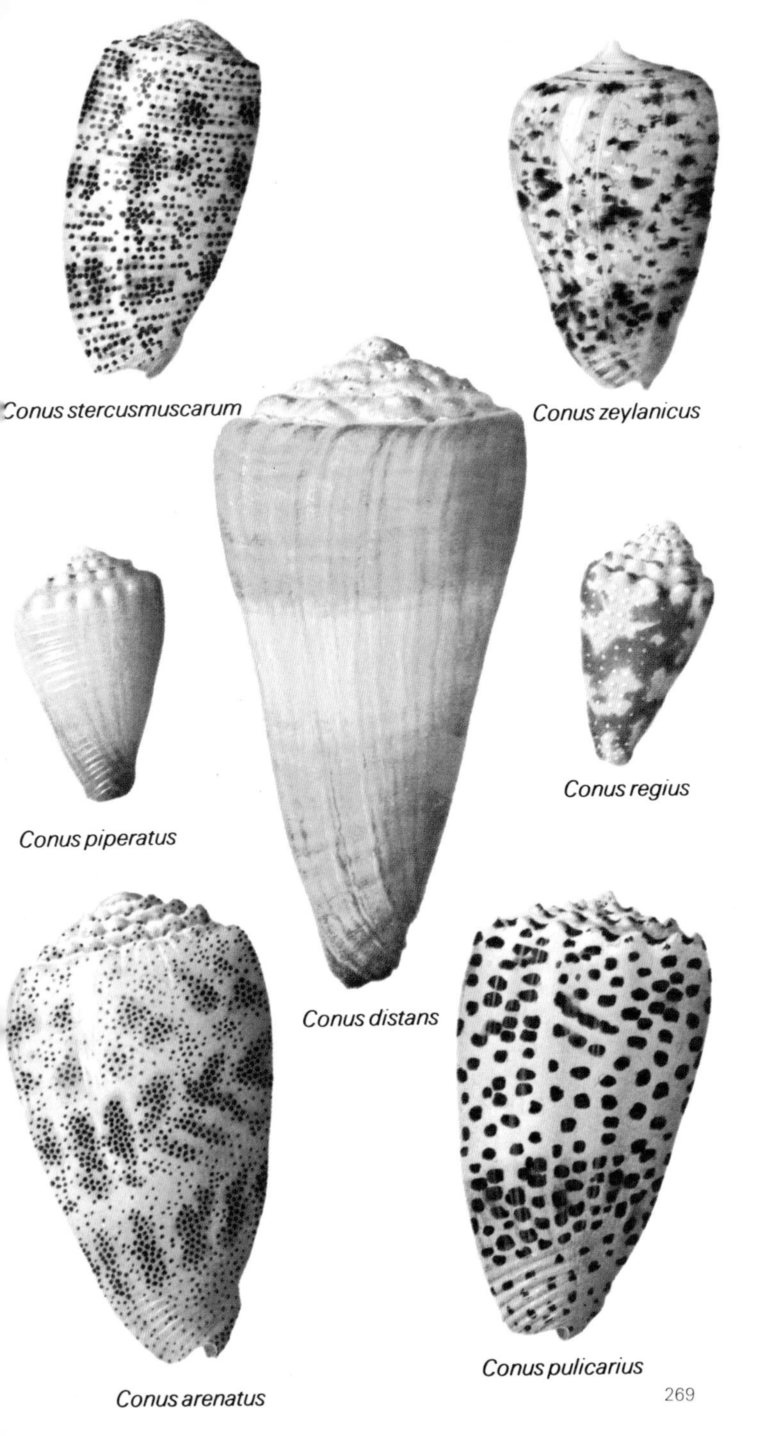
Conus stercusmuscarum
Conus zeylanicus
Conus regius
Conus piperatus
Conus distans
Conus pulicarius
Conus arenatus

C. brunneus **Wood 1828** Pacific coast of Central America and Galapagos Islands, 50mm. A beautiful, rich, mahogany brown with sparse, white dots and splashes, mostly in an indiscrete band on the body whorl. Strong crenulations on the shoulder and spire, which are also white.

C. princeps **L. 1758** Pacific coast of Mexico and Panama, 60mm. Pinky red with dark brown, axial streaks which cross the shoulder between the slight coronations. A variety, *C. p. lineolatus* Valenciennes 1832, has its dark markings as fine lines; and *C. apogrammatus* Dall 1910 has no markings.

C. varius **L. 1758** Indo-Pacific, 40mm. This shell is covered with small granulations giving it a very rough feel and appearance. Its crenulated spire, rising straight from the shoulder, forms slightly less than a right angle at the apex. Its white background colour has dark brown blotches and sparse dots. A variety from Queensland, Australia, *hwassi* Weinkauff 1874, is illustrated.

Also in the subgenus are: *C. klemae* Cotton 1953, Western Australia; *C. aurantius* Hwass 1792, a rare shell from the East Indies and Philippines area; and *C. gladiator* Broderip 1833, from the Gulf of California and Ecuador.

Subgenus: *Chelyconus*

C. purpurascens **Sowerby 1833** Pacific coast of Mexico and Panama, 50mm. A very variable shell but generally heavy and rather squat. Basically a blue-black colour, but with white-blue cloudy markings and some brown markings.

C. achatinus **Gmelin 1791** Indian and east Pacific Oceans, 55mm. Variable throughout its range but invariably handsome. It has convex sides, is lightly striated, has rounded shoulders and a moderately high spire. The background colour of pale blue has dark brown clouding and the striations are a darker brown, generally as dots across the blue and unbroken lines across the brown.

C. catus **Hwass 1792** Red Sea across the Indian and Pacific Oceans to Hawaii, 40mm. A common shell and naturally very variable, it is squat with cloudy brown or olive-yellow markings on a pale or white background. Two specimens are illustrated, one markedly granulated which is from the Tuamotu Archipelago in the south Pacific, and the other almost smooth from east Africa.

Other members of the subgenus include: *C. ranunculus* Hwass 1792 from the West Indies, *C. fulmen* Reeve 1843 from Japan, *C. monachus* L. 1758 from the Pacific, *C. nigropunctatus* Sowerby 1857 also from the Pacific—all these have some resemblance to *C. achatinus*; and *C. orion* Broderip 1833 and *C. vittatus* Hwass 1792, both from the west Central American coast and both with a central band of white with dark markings on a variable background of white, orange and lilac to brown, with dark markings on the spire.

Subgenus: *Leptoconus*

C. sculletti **Marsh 1962**, 40mm. This recently discovered species from deep water off Cape Moreton, Queensland, though rare, is included both as an example of the fact that new species continue to be discovered, and also for its unusual and attractive shape. It is narrow for its length with sharp shoulders, and is slightly waisted near the tip. It has a creamy white background encircled with tan bands, dots and flame-like markings.

Conus brunneus

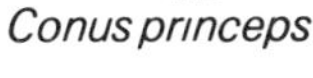

Conus princeps

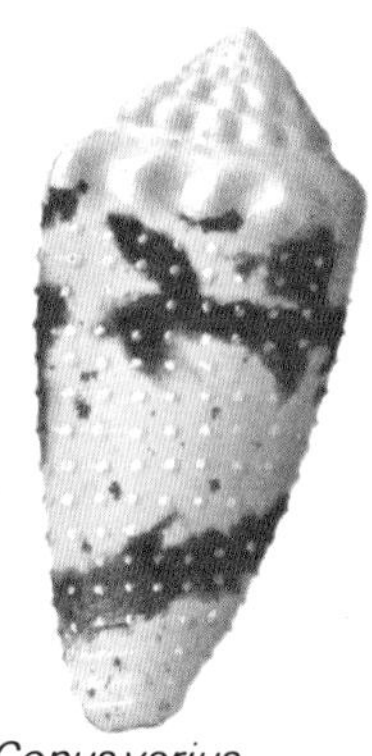

Conus varius form *hwassi*

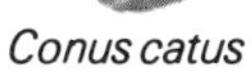

Conus catus

Conus catus

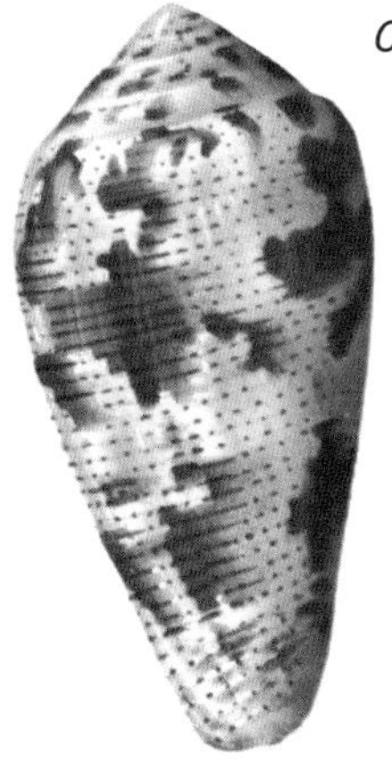

Conus achatinus

Conus sculletti

Conus purpurascens

C. thalassiarchus Sowerby 1834 Philippines and north Borneo, 90mm. Like most of the subgenus, a very handsome shell. It has a rather flat top and a small spire. The creamy white background is closely patterned with wavy, axial lines, usually light but sometimes dark brown, and generally with cream bands at the shoulder and waist, with a blue-black tip and dark markings on the top.

C. ammiralis L. 1758 east Indian and west Pacific Oceans, 60mm. To me, the most beautiful of the cones, though somewhat variable. It is straight-sided with a high, concave spire. The body whorl is a rich red of varying depths of colour with three bands of pale yellow, one at the shoulder, one in the middle, and one at the tip; the whole is covered with pale white tent marks, varying from about 5mm across down to minute; the shoulder is white with red-brown markings and the top has areas of pale red-brown crossed with dark, axial lines.

C. generalis L. 1767 Indo-Pacific, 90mm. A common species and very variable. Its rather flat top carries a sharp, concave spire. It is darker or lighter brown, with white markings, mostly forming more or less a band at the waist and also at the shoulder and tip; sometimes with dark brown, axial lines; top is white with dark brown streaks. Closely allied are *C. spirogloxus* Deshayes 1863 and *C. maldivus* Hwass 1792 from the west Indian Ocean, both usually showing more white markings, the latter with a much flatter spire.

C. regularis Sowerby 1833 Pacific Central America, 50mm. A common member of the subgenus. Its blue-white background is covered with purple-grey markings reminiscent of Arabic script, and it is encircled with rows of small red dots.

C. sozoni Bartsch 1939 Florida, 75mm. A high, pointed spire, straight sides and a very narrow tip give this handsome shell an unusual appearance. The lip is markedly convex. Its white background is encircled with bands of pale yellow-brown and many rows of dots of darker brown which form flame-like markings on the spire.

C. recurvus Broderip 1833 Pacific Central America in deep water, 60mm. Shaped like *C. sozoni* but with a flatter top, it has a white background colour with large, dark brown, flame-like markings.

C. floridanus Gabb 1868 Florida, 40mm. Similar in shape to the two preceding species though smaller. It is white with yellow to pale brown cloudy markings, and an indefinite white band with a row of dots through the centre.

C. jaspideus Gmelin 1791 south-east coast of America and West Indies, 25mm. High spire. Generally like *C. sozoni* in shape but with strong spiral grooves. Cream with irregular dark or light brown markings. Illustrated is a small form of *C. jaspideus, pigmaeus* Reeve 1844, about 12mm, from the same area.

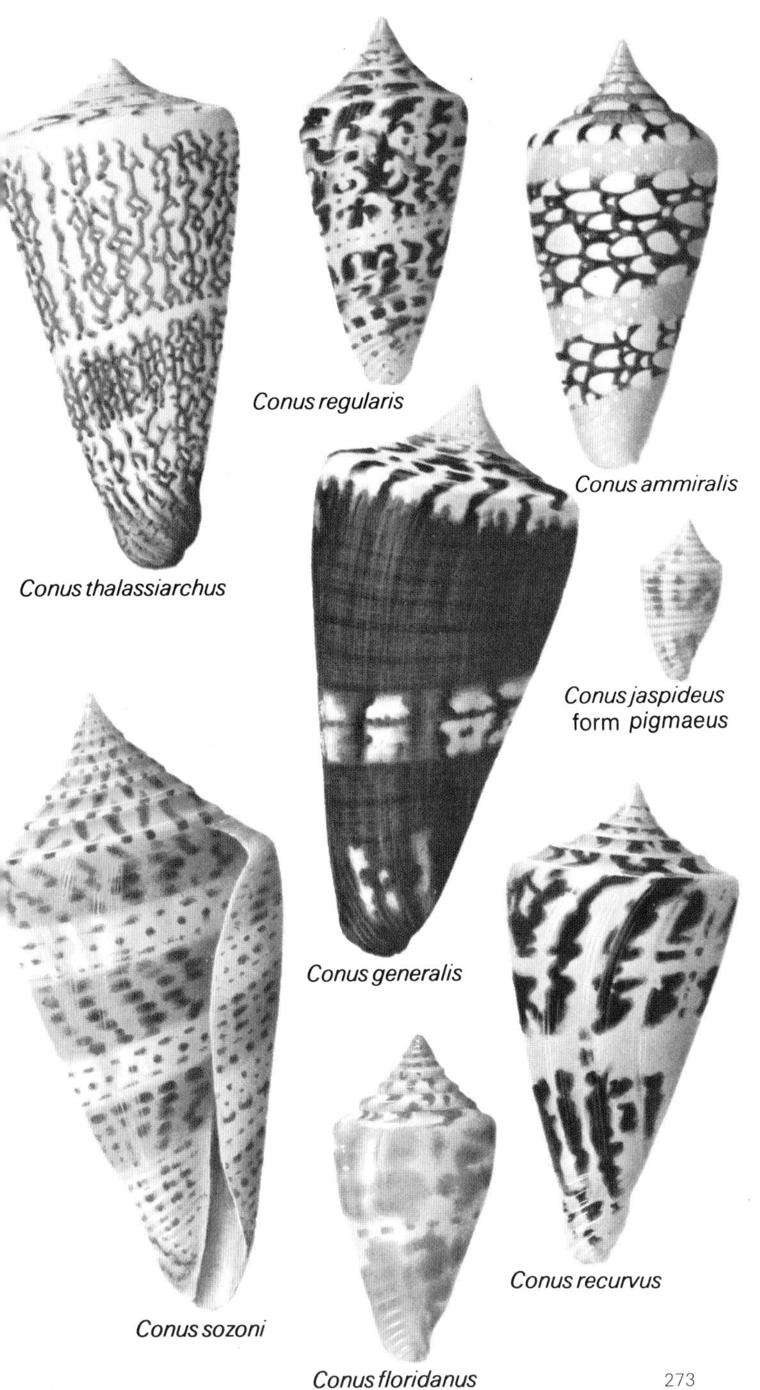

Conus thalassiarchus

Conus regularis

Conus ammiralis

Conus jaspideus form pigmaeus

Conus generalis

Conus sozoni

Conus recurvus

Conus floridanus

C. floridensis **Sowerby 1870** Florida, 40mm. This is a colour form of *C. floridanus* (see page 272) with spiral rows of dots and dashes, usually with much darker clouding than the specimen illustrated. The background colour is also usually darker.

C. spurius **Gmelin 1791** Florida and the Caribbean, 70mm. Has a rather flat top with a sharp spire. Its white background is encircled with rows of red dashes which sometimes run together vertically and horizontally. In *C.s. atlanticus* Clench 1942, the dashes do not run together, and sometimes some of the spots are blue-purple.

Other members of the subgenus include: *C. amadis* Gmelin 1791, Indian Ocean; *C. clarus* Smith 1881, Western Australia; *C. monile* Hwass 1792, Indian Ocean; *C. acuminatus* Hwass 1792, east Africa; *C. nobilis* L. 1758, Philippines; *C. virgatus* Reeve 1849, Central America; and *C. clerii* Reeve 1844, Brazil.

Subgenus: *Dauciconus*

C. augur **Solander 1786** east Africa, 70mm. Solid, heavy, rather unlike any other of this subgenus. Flat spire. Oatmeal colour, profusely banded with rows of very small dark purple-brown spots and two bands of irregular blotches of the same colour.

C. daucus **Hwass 1792** Caribbean, 50mm. Normally a red-yellow overall, but in its variety *C. luteus* Krebs 1864 it is much more yellow and shows a faint pale band. The specimen illustrated is a rare red form.

C. planorbis **Born 1780** Pacific Ocean, 75mm. Like others of the subgenus the shell varies considerably in colour pattern and shape. It varies from a plain brown-yellow through bands of darker or lighter shades of this colour (as illustrated) to specimens with much more striking banding and dark markings, not unlike *C. striatellus*. The tip is purple.

C. striatellus **Link 1807** Indian Ocean, 75mm. A synonym is *C. pulchrelineatus* Hopwood 1921. The name *C. lineatus* Hwass 1792 is also frequently used, but incorrectly as the name is preoccupied. Very variable but often a handsome shell. Generally white, with brown banding and dark flame markings on the shoulder and spire.

C. circumactus **Iredale 1929** Pacific Ocean, 40mm. Variable. Light brown, the shoulder and banding being white or with a pale mauve tinge; brown flame markings on or across the back and shoulders on to the rather flat spire.

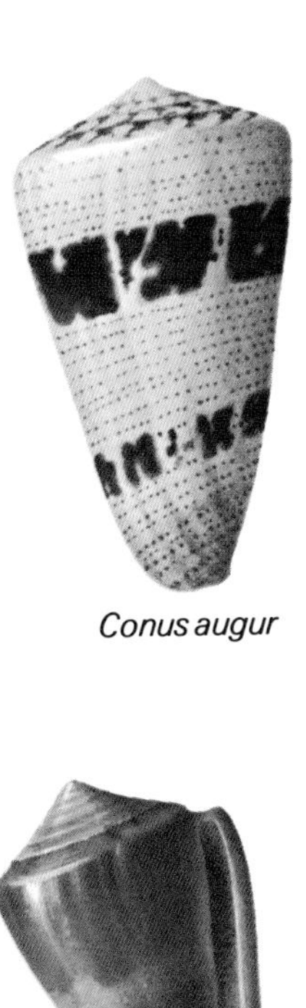

Conus augur

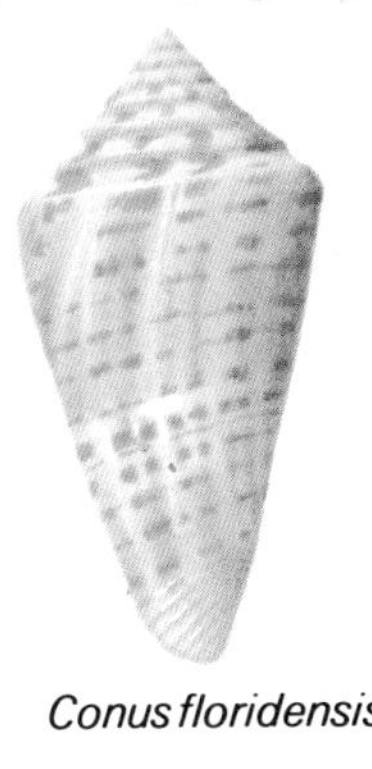

Conus floridensis

Conus planorbis

Conus daucus

Conus striatellus

Conus striatellus

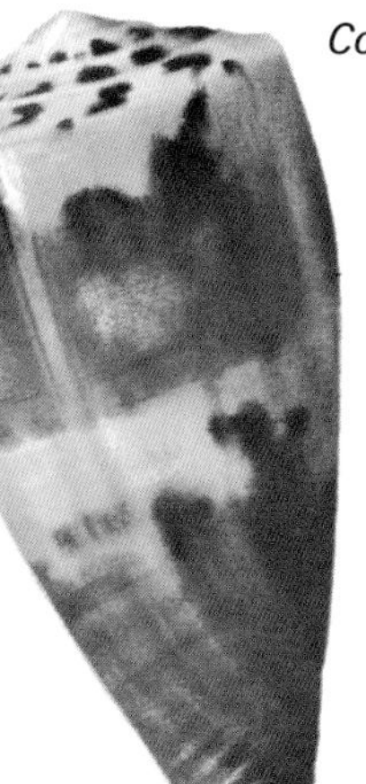

Conus circumactus

Conus spurius

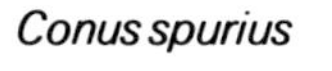

C. litoglyphus Hwass 1792 Indo-Pacific, 65mm. One of the most striking of the subgenus and less variable than most. The red-gold background has an irregular central band of pure white and another white band at the shoulder streaked with the background colour, which also flecks the spire; anteriorly the body whorl has about five rows of granules; tip is purple. The specimen illustrated is a small one.

Also in the subgenus are: *C. fulmineus* Gmelin 1791, north-east Australia; *C. furvus* Reeve 1843, Philippines; and *C. vitulinus* Hwass 1792, Indo-Pacific.

Subgenus: *Pionoconus*

Mostly narrow shells which are striate anteriorly.

C. magus L. 1758 80mm. The cone collector's nightmare. Found over wide area of Indo-Pacific centred on the Philippines, it has a very wide variety of forms in shape, colour and pattern. Some have specific rank, while others are only varieties. Among them are *C. ustulatus* Reeve 1844, *C. raphanus* Hwass 1792 and *C. circae* Sowerby 1858.

C. suturatus Reeve 1844 Pacific from East Indies to Hawaii, 30mm. Chubby, little shell with a rather flat top and sharp spire. Pale pink with three, broad, faint, yellow-brown bands and a violet tip.

Subgenus includes: *C. mercator* L. 1758, west Africa; *C. ximenes* Gray 1839 and *C. perplexus* Sowerby 1857, west Central America; *C. erythraeensis* Reeve 1843, Red Sea; *C. pertusus* Hwass 1792, Pacific; *C. consors* Sowerby 1833, Singapore; and *C. mozambicus* Hwass 1792 and *C. simplex* Sowerby 1857/8, South Africa.

Subgenus: *Phasmoconus*

A small group from central and south-west Pacific plus one from the Indian Ocean and one from South Africa. They have an elevated striate spire, a body whorl usually of one colour and are deeply grooved anteriorly.

C. carinatus Swainson 1822 Philippines, 80mm. A species within the *C. magus* complex, despite its subgeneric placing. Brown-yellow with indiscrete white band and white blotches below shoulder.

Subgenus includes: *C. radiatus* Gmelin 1791, Philippines, New Guinea; *C. infrenatus* Reeve 1848, South Africa; *C. keatii* Sowerby 1858, Red Sea, Seychelles.

Subgenus: *Rhizoconus*

Large shells with broad shoulders, sometimes sharply angled, sometimes rounded; often nearly as wide as they are long.

C. vexillum Gmelin 1791 Indian Ocean and central and south Pacific, 180mm. One of the large cones, somewhat variable in shape and markings. May have rounded or sharply angled shoulders. Basically brown with white bands on body whorl and shoulders usually broken up to form white blotches; sometimes streaked axially with black.

C. capitaneus L. 1758 Indo-Pacific, 65mm. Brown often with yellow or green tinge; white band on body whorl and shoulder; black and white chequered effect on spire spills over shoulders; lines of black spots encircle body whorl and border central white band but disappear towards purple tip.

C. rattus Hwass 1792 most of Indo-Pacific, 40mm. Variable. Shoulders either rounded or angled. Body whorl brown or green-brown with band, more or less discrete, of blue-white, horizontal flecks, blue-white shoulder band broken by blotches of body whorl colour; spire chequered with both colours.

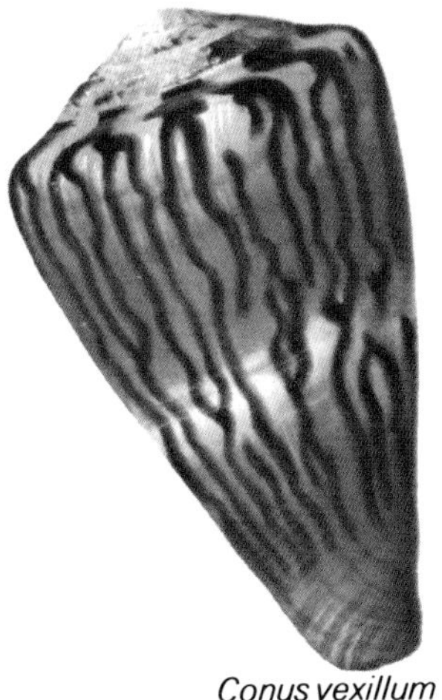

Conus vexillum

Conus rattus

Conus capitaneus

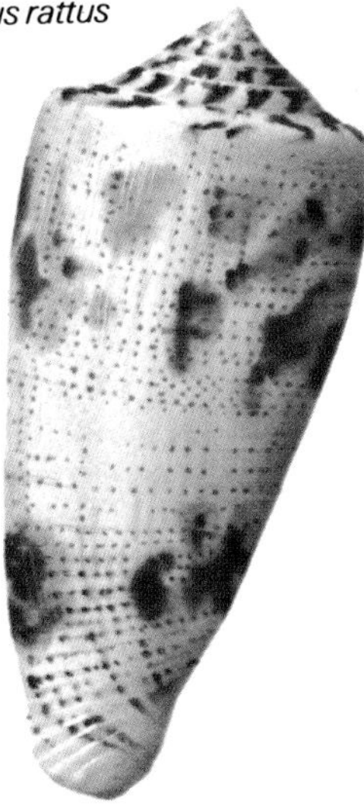

Conus magus

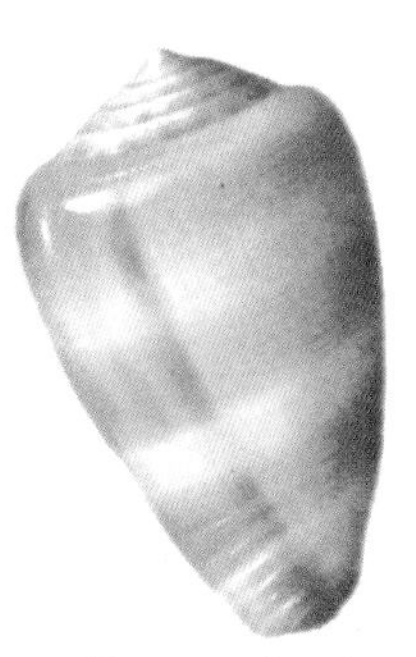

Conus suturatus

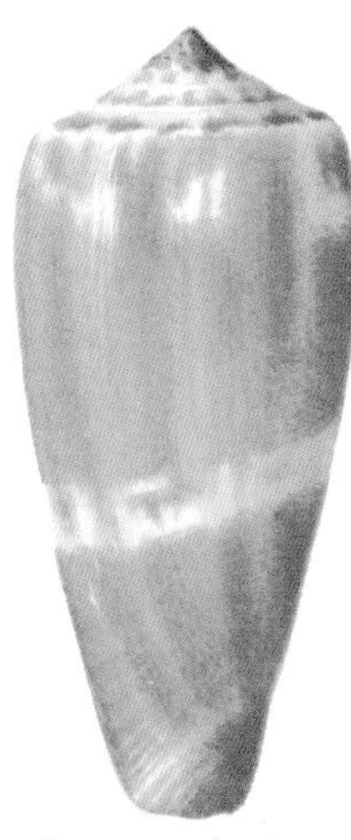

Conus carinatus

Conus litoglyphus

C. miles **L. 1758** Indo-Pacific, 90mm. Much less variable than others of the subgenus. White background colour is covered with fine, axial, wavy, light brown lines running from the apex to the base. It has a fairly narrow, dark band on the body whorl within a broader, very light brown band. The anterior quarter or fifth of the body whorl is almost black, bordered by progressively lighter bands up to one-third of its length.

Others in the subgenus include: *C. mustelinus* Hwass 1792, Indo-Pacific; *C. capitanellus* Fulton 1938, Japan; *C. namocanus* Hwass 1792 and *C. laevigatus* Sowerby 1857, Indian Ocean; and *C. trigonus* Reeve 1848, north-west Australia.

Subgenus: *Virgiconus*

Rather narrow shells. Body whorl slightly striate, especially anteriorly. One colour with purple tinge at the base.

Conus virgo **L. 1758** Indo-Pacific, 150mm. A heavy shell, straight-sided, with a dull, almost matt surface. It is pale yellow with a purple tip, and is a sand dweller.

C. emaciatus **Reeve 1849** Indo-Pacific, 50mm. Has concave sides and the body whorl is finely ridged overall. It is pale yellow with a purple tip and is a reef dweller.

C. flavidus **Lamarck 1810** Indo-Pacific, 65mm. Usually a yellow-brown with a narrow white band and a pale top and shoulders. It has a purple tip and interior.

Subgenus: *Gastridium*

Thin, light shells with faint striae and a coronated spire. The lower half of the columella is concave making a very wide aperture anteriorly.

C. geographus **L. 1758** Indo-Pacific, 150mm. Its background brown colour is clouded with some white tent marks and blotches; darker brown blotches, some forming two broken bands on body whorl. It has a poisonous sting known to have been fatal to man.

C. obscurus **Sowerby 1833** Indo-Pacific, 65mm. Shaped like a miniature *C. geographus* but shoulder often has small coronations. It is red-brown with very pale purple markings, roughly forming two bands. The Hawaiian variety is *C. halitropus* Bartsch, Rehder and Greene 1953.

Subgenus: *Regiconus*

Large shells, cylindrical in shape with rounded shoulders and a turbinate spire. Their brown background colour is covered with white tent marks.

C. aulicus **L. 1758** Indo-Pacific, 125mm. A very handsome shell, solid and somewhat inflated. Its dark or light brown background colour is covered with small and/or large white tent marks.

Others in the subgenus include: *C. auratus* Hwass 1792 from Polynesia; and *C. aureus* Hwass 1792 from the east Pacific and Indian Oceans.

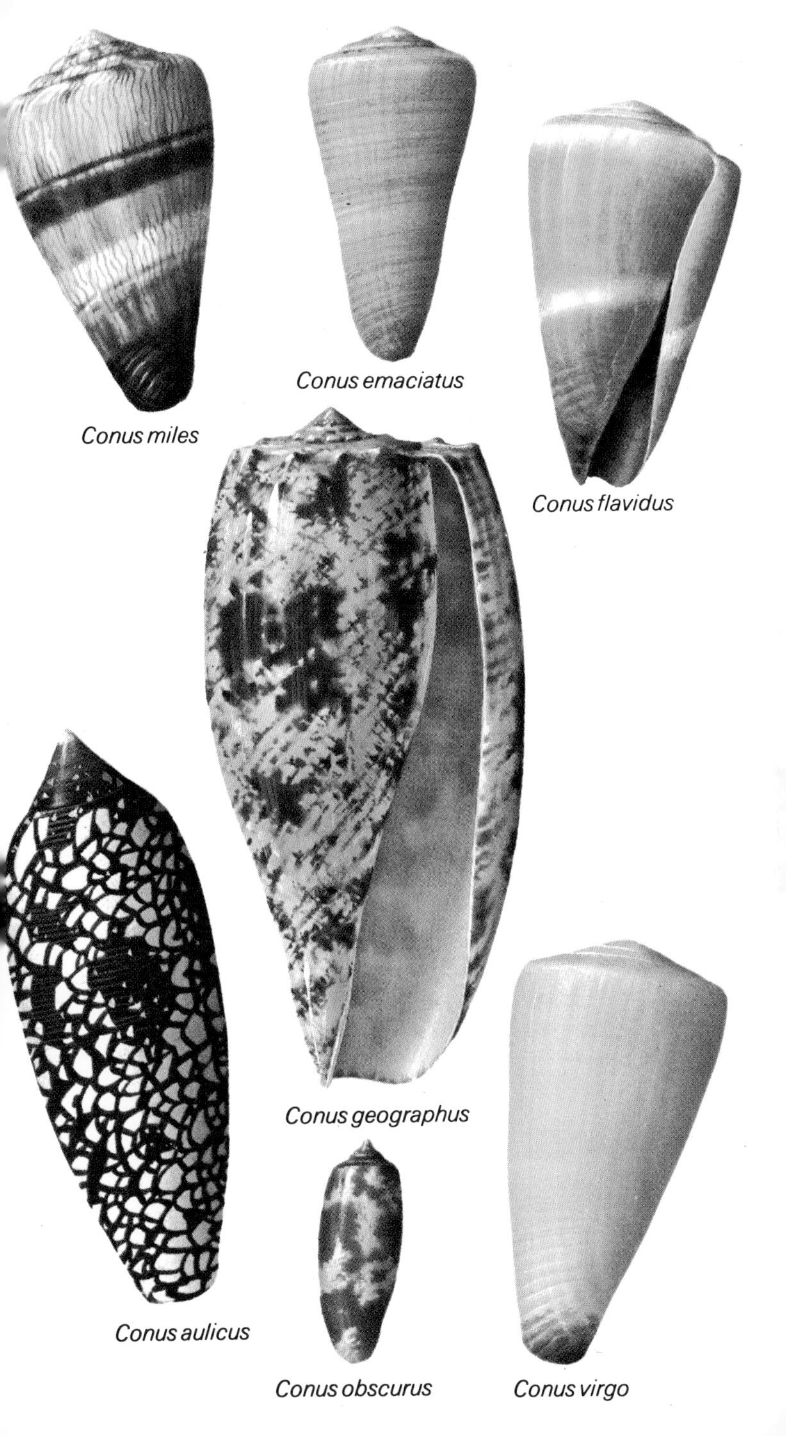
Conus miles
Conus emaciatus
Conus flavidus
Conus geographus
Conus aulicus
Conus obscurus
Conus virgo

Subgenus: *Darioconus*

The sides of the spire are more or less concave and the apex has a very small, sharp point. Markings are similar to those of the *Regiconus*.

C. omaria **Hwass 1792** Pacific Ocean and the East Indies, 80mm. Variable within the overall subgenus description both in shape of spire, colour, size and number of tent marks.

C. episcopus **Hwass 1792** is the long and narrow form of *C. omaria*, 90mm. Another synonym of *C. omaria* for those with a high spire and red-tinged tent marks is *C. magnificus* Reeve 1843.

C. pennaceus **Born 1758** Indo-Pacific, from the Red Sea to Hawaii, 50 mm. Variable both in shape, colour and tent marks. The shell illustrated is a much lighter brown than the average.

Others in the subgenus include: *C. stellatus* Kiener 1849; *C. elisae* Kiener 1849; and *C. praelatus* Born 1792 from east Africa.

Subgenus: *Cylinder*

Cylindrical with a pointed, sharp spire. Background colour golden yellow to red-brown, with dark wavy axial lines and white tent marks overall.

C. textile **L. 1758** Indo-Pacific, 80mm. Tends to be rather inflated below the shoulder and tapers to a narrow base. The two illustrations show the wide variation in colour and size of tent marks. The latter leave indiscriminate patches of the background colour and axial lines. Closely related are: *C. cholmondelyi* Melville 1800, east Africa; *C. archiepiscopus* Hwass 1792, east Australia westward across the Indian Ocean to east Africa; *C. verriculum* Reeve 1843, Indian Ocean; *C. scriptus* Sowerby 1858, east Africa; and *C. natalis* Sowerby South Africa.

C. victoriae **Reeve 1843** Western Australia to East Indies, 40mm. A light, small shell with very fine, white tent marks leaving two bands of the background red-brown colour, and axial lines. Closely related are: *C. abbas* Hwass 1792, Indian Ocean; *C. complanatus* Sowerby 1866, Western Australia; *C. panniculus* Lamarck 1810, East Indies; and *C. dalli* Stearns 1873, west Central America. They are generally smaller than the preceding members, though otherwise similar.

C. tigrinus **Sowerby 1858** Australasia and Melanesia, 50mm. A heavy shiny shell with small tent marks and no axial lines. Other members of the subgenus closely related to *C. tigrinus* are *C. pyramidalis* Lamarck 1810, east Africa; and *C. legatus* Lamarck 1810 from the Admiralty Islands. Both are solid and have a high gloss.

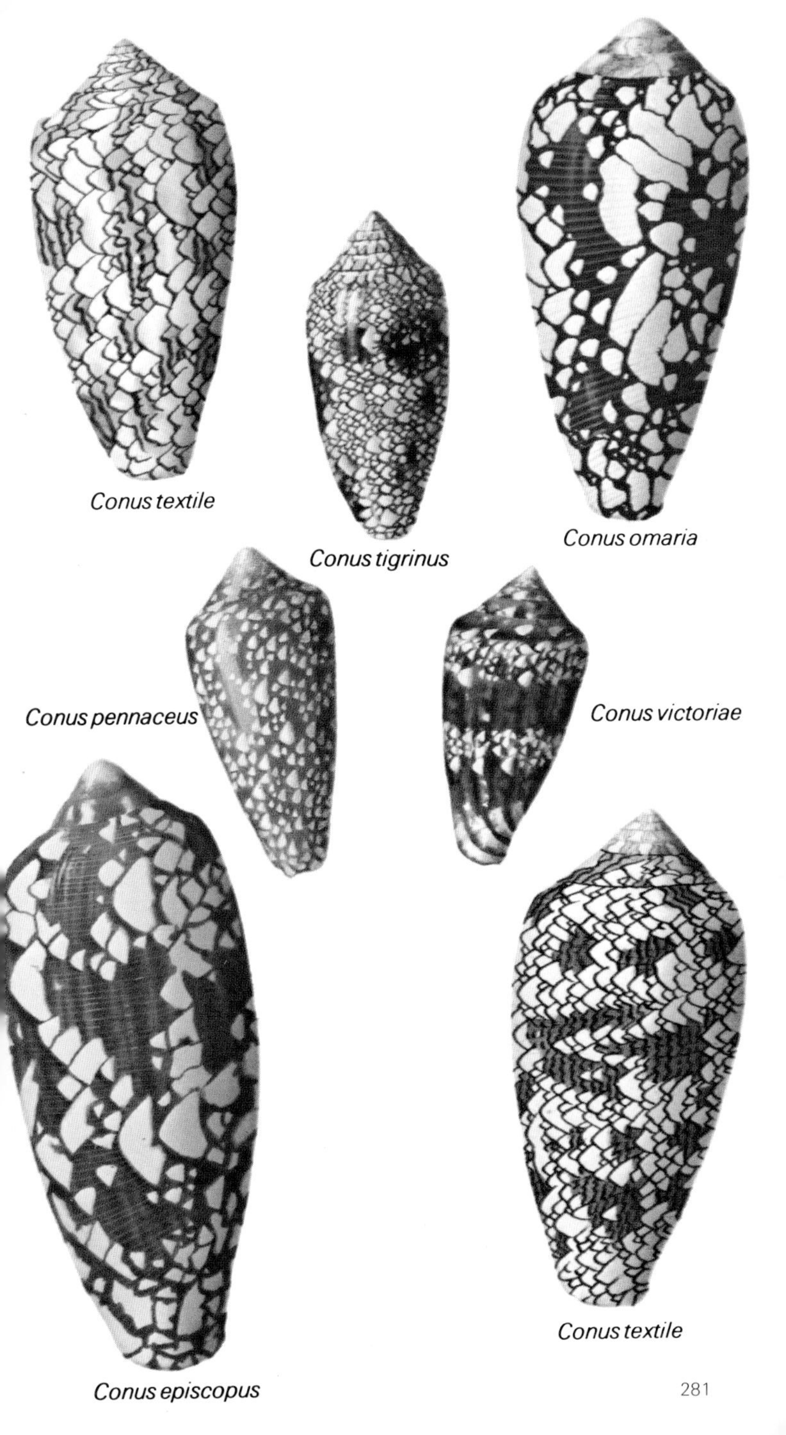

Conus textile

Conus tigrinus

Conus omaria

Conus pennaceus

Conus victoriae

Conus episcopus

Conus textile

C. retifer Menke 1829 Japan, the north Pacific to east Africa, 45mm. Solid and rather pyriform. The background tan colour has heavy, dark, axial lines and rather sparse, small tent marks, and an indiscrete darker band.

Also in the subgenus is *C. lucidus* Wood 1828 from the west Central American coast.

Subgenus: *Cleobula*

Heavy pyriform shells, of one colour or with lines of dots or dashes. Growth marks tend to be conspicuous on the body whorl.

C. suratensis Hwass 1792 Philippines, 85mm. Flesh-coloured encircled with many lines of fine, dark, purple dots and dashes.

C. genuanus L. 1758 west Africa, 70mm. Less wide at the shoulder in relation to its length than most of the subgenus. It is a most handsome shell with a clouded pink-mauve background colour and two light brown bands. The encircling lines are made up alternately of small dots and dashes, and larger dashes, dots and blotches of black with white spaces between. Below the lower brown band the lines are only of small dots and dashes.

C. glaucus L. 1758 Philippines and New Guinea, 45mm. A small member of the subgenus. Pale grey with many encircling lines of dark brown dashes.

C. figulinus L. 1758 Indo-Pacific, 80mm. Typically shaped. Brown; darker on the spire; many encircling, unbroken, darker brown lines.

C. betulinus L. 1758 Red Sea, Indian and Pacific Oceans, 150mm. The largest of the subgenus. Its brown-yellow background colour is overlaid with lines of sparse purple-brown dots.

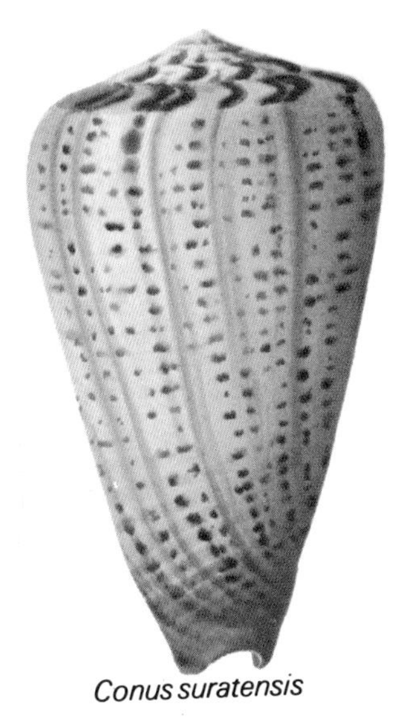
Conus suratensis

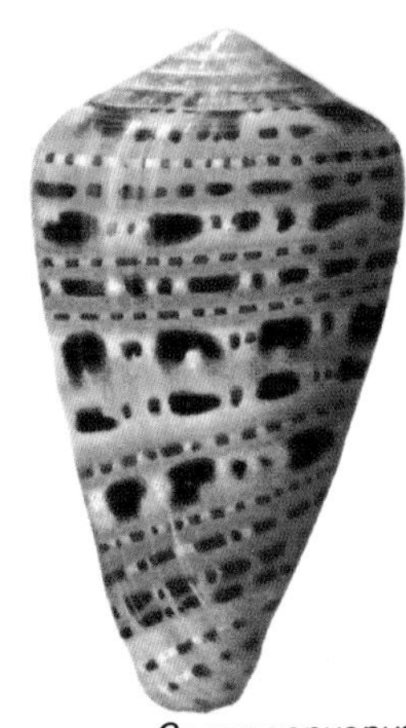
Conus genuanus

Conus glaucus

Conus retifer

Conus figulinus

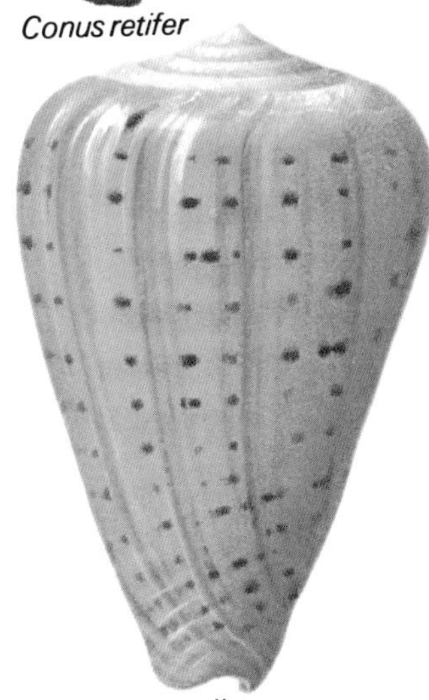
Conus betulinus

C. quercinus **Solander 1786** Indo-Pacific, 75mm. A heavy, solid and common shell. It is bright pale yellow to light tan, the colour often varying on the same shell both axially and horizontally.

Also in the subgenus are: *C. patricius* Hinds 1843 and *C. fergusoni* Sowerby 1873 from west Central America; and *C. loroisii* Kiener 1847 from East Indies and Sri Lanka.

Subgenus: *Strioconus*

Members of this group have channelled whorls on the spire and a tapered shoulder, striated body whorls and are colourful and brightly patterned.

C. striatus **L. 1758** Indo-Pacific, 120mm. Common and variable within the subgeneric description. Pink-white with blotches of purple-grey or brown.

Others in the subgenus are: *C. gubernator* Hwass 1792 from east Africa; *C. terminus* Lamarck 1810 from the Red Sea; *C. floccatus* Sowerby 1839 from central Pacific; and *C. epistomium* Reeve 1844 from Mauritius.

Subgenus: *Textilia*

Swollen, smooth shells; striae on the rather flat tops. Mostly uncommon.

Conus spectrum **L. 1758** East Indies, east, north and west Australia, 65 mm. Synonym *C. pica* A. Adams and Reeve 1848. White with brown or purple-brown markings covering the shell. These are sometimes as blotches, as in the illustration of the small form *pica,* and sometimes also with axial, wavy lines.

Others in the subgenus are: *C. adamsonii* Broderip 1836 – perhaps better known as *C. rhododendron* – long one of the cones most coveted by the collector, from the Phoenix Islands; *C. bullatus* L. 1758 from mid-Pacific to the Indian Ocean; *C. nimbosus* Hwass 1792 from the north Indian Ocean; *C. conspersus* Reeve 1844 from the Philippines and Australia; and *C. peronianus* Iredale 1931 from New South Wales.

Subgenus: *Floraconus*

Short stocky shells from south-east, south and south-west Australia and South Africa.

Conus anemone **Lamarck 1810** Victoria, south and Western Australia, 50mm. Often known in its banded form as *C. novaehollandiae* Adams 1854 from Western Australia. Variable in shape and colour pattern, but basically with dark brown and pale blue, much intermingled.

Conus rosaceus **Dillwyn 1817** south-east Africa, 40mm. A beautiful red colour form of *C. tinianus* Hwass 1792 with a band of white, and darker red blotches.

Others in the subgenus are: *C. singletoni* Cotton 1945 and *C. segravei* Gatliff 1891 from Victoria to Western Australia; *C. aplustre* Reeve 1843, *C. wallangra* Garrard 1961 and *C. papilliferus* Sowerby 1834 from New South Wales; *C. compressus* Sowerby 1866 from south Australia; *C. cyanostoma* A. Adams 1954 and *C. coxeni* Brazier 1875 from Queensland; and from South Africa *C. tinianus* Hwass 1792 – of which *C. rosaceus, C. inflatus* Sowerby 1833, *C. aurora* Lamarck 1810 and *C. caffer* Krauss 1848 are all colour forms.

Subgenus: *Hermes*

Narrow, often spindle-shaped shells, with short spires and lirate body whorls – the lirae being finely granular.

Conus nussatella **L. 1758** Indo-Pacific, 70mm. Heavy for its size; short but acuminate spire. White-brown blotches and many rows of fine, red-brown

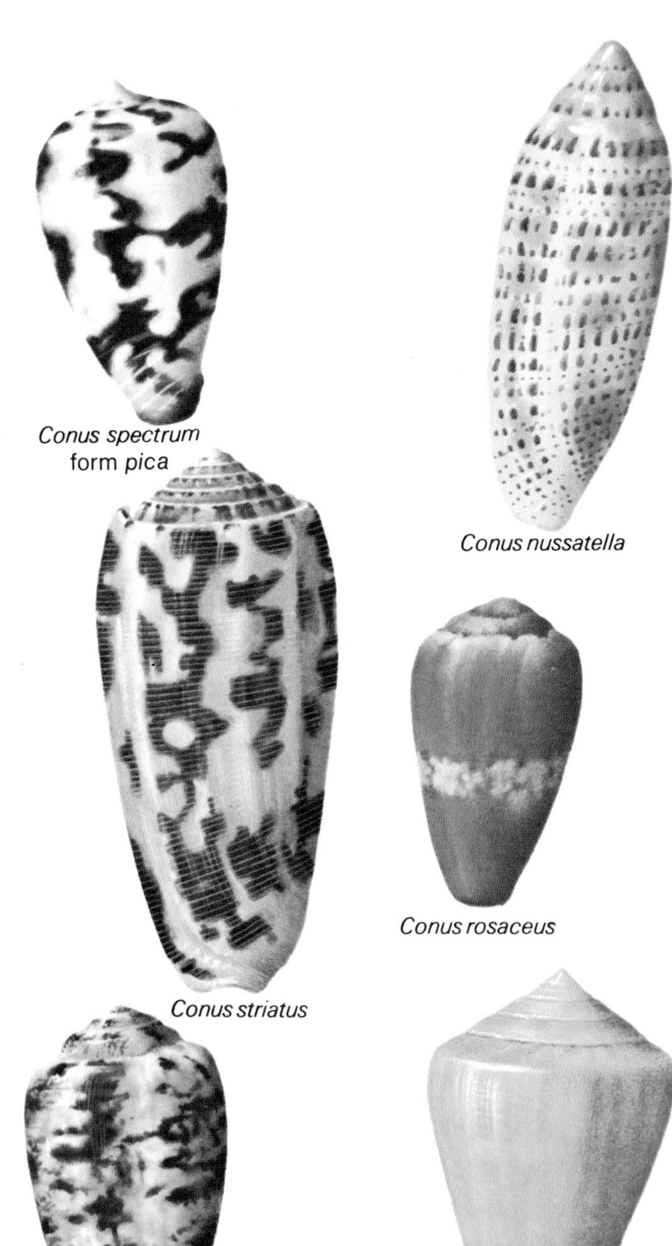

Conus spectrum form *pica*

Conus nussatella

Conus rosaceus

Conus striatus

Conus anemone

Conus quercinus

dots and vertical dashes arranged in vertical and horizontal rows.

C. terebra **Born 1780** Indian, west and central Pacific Oceans, 100mm. Synonym *C. clavus* L. 1758. A large, heavy shell tapering to a narrow base and with coarse, horizontal ridges. White with two purple or pale yellow bands, a purple tip and purple lines below the sutures.

Others in the subgenus include: *C. tendineus* Hwass 1792, South Africa and its islands; *C. lautus* Reeve 1844, South Africa; *C. luteus* Sowerby 1833, north half of Australia; *C. auricomus* Hwass 1792 from the Indian Ocean – the only member of the subgenus to carry tent marks; *C. granulatus* L. 1758 from the Caribbean; *C. circumcisus* Born 1778 from the East Indies; and *C. aurisiacus* L. 1758, a most handsome shell also from the East Indies.

Subgenus: *Tuliparia*

Somewhat similar to the subgenus *Gastridium*. The shell is thin, swollen and smooth with a faintly coronated spire. The posterior half of the columella is concave, giving a wide lower half to the aperture.

Conus tulipa **L. 1758** central Pacific to Australia, 60mm. It is sky blue with dark brown clouding, and encircling lines of tiny brown dots.

The only other member of the subgenus is *C. borbonicus* H. Adams 1808 from Mauritius to French Polynesia.

Subgenus: *Lautoconus*

Conus ventricosus **Gmelin 1791** Mediterranean, east Atlantic and the Red Sea, 40mm. A very variable shell with many synonyms, but the only cone found in the Mediterranean Sea. Chunky, usually rather drab with brown markings over a pale background colour.

Also in the subgenus is *C. californicus* Reeve 1844, off the Californian coast. Both these species occur in relatively cold waters for members of *Conus*.

Subgenus: *Leporiconus*

Somewhat similar to *Hermes*, but more pyriform. The body whorl bears coarse, granular lirae.

Conus mitratus **Hwass 1792** Indian Ocean to Queensland, 30mm. Spindle-shaped and aptly named as it looks superficially like a mitre shell. It has a very high spire, the whole being yellow-brown covered with three bands of smudged, dark brown, square blotches, which show on the whorls of the spire.

C. glans **Hwass 1792** Indonesia and west Pacific, 50mm. Rather variable but generally stubby and solid. Purple tip and streaks over a white background colour and sometimes with cream-brown banding.

Others in the subgenus are: *C. tenuistriatus* Sowerby 1858 from East Indies; *C. coccineus* Gmelin 1791, a beautiful red-orange shell from south-west Pacific; *C. cylindraceus* Broderip and Sowerby 1830, Indo-Pacific, but rare; and *C. scrabriusculus* Dillwyn 1817, south-west Pacific.

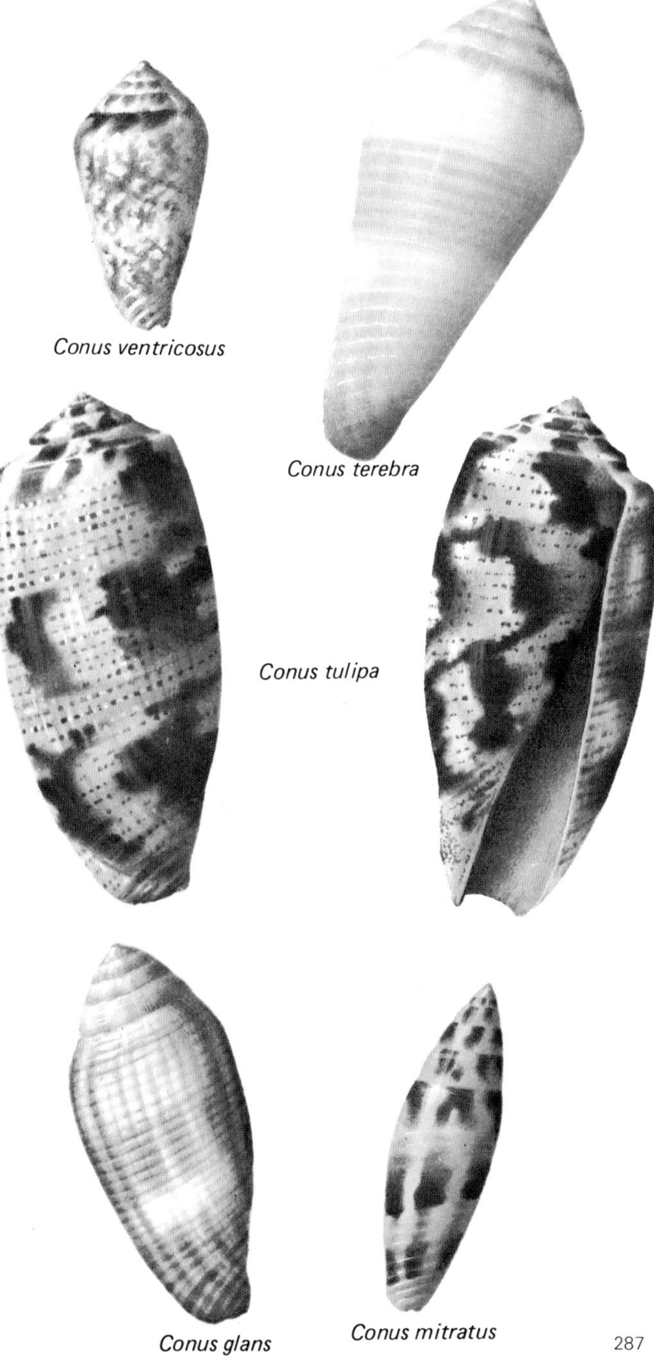
Conus ventricosus

Conus terebra

Conus tulipa

Conus glans

Conus mitratus

Subgenus: *Endemeconus*

Acute spire with a raised ridge on the sharply angled shoulder which forms a spiral ridge on the spire. Tapers to a narrow base.

Conus ione **Fulton 1938** Japan, 65mm. White encircled with rows of orange-brown dots and two bands of orange-brown blotches. The dots become lines across the shoulder and on the spire.

Conus sieboldi **Reeve 1848** Japan, 80mm. Larger than *C. ione* and much longer in relation to its shoulder width. It has the characteristic spire of the subgenus and is pure white with a few brown blotches, mainly forming a central band on the body whorls and a more indefinite band below the shoulder.

Others in the subgenus are: *C. howelli* Iredale 1929 from New South Wales; and *C. villepinii* Fischer and Bernardi 1857 from the Gulf of Mexico.

Subgenus: *Conasprella*

A high pointed spire and a sharply angled shoulder, tapering to a narrow base. The body whorl is encircled with grooves and fine axial striations.

Conus cancellatus **Hwass 1792** Japan, 45mm. Slightly swollen below the shoulder, the body whorl being white. From the shoulder to the preceding whorl on the spire it is concave, so forming a spiral groove up to the apex. This groove has brown blotches.

C. acutangulus **Lamarck 1810** Philippines to Hawaii, 25mm. A solid, little shell, sharply pointed at both ends. The spire is about one-third the length of the shell. White flecked with brown, forming spiral broken bands.

C. sowerbii **Reeve 1849** Philippines, 30mm. Spire almost as long as the body whorl. Brown encircled with rows of white dots and dashes.

Also in the subgenus are: *C. austini* Rehder and Abbott 1951 from east Central America; *C. kieneri* Reeve 1849 from Japan; and *C. verrucosus* Hwass 1792 from the Caribbean.

Subgenus: *Asprella*

Sharp spire; edges of the whorls ridged; sharply angled shoulders. Body whorl tapering to a point and encircled with grooves.

Conus orbignyi **Audouin 1831** Japan, 70mm. The shoulder angle carries small tubercles which on the earlier whorls show on the spire. It is deeply grooved and is a medium brown with some dark brown banding.

Others in the subgenus are: *C. sulatus* Hwass 1792 from south-east Asia; *C. australis* Holten 1820 from North China Sea; *C. laterculatus* Sowerby 1870 from Australia and East Indies.

Subgenus: *Dyraspis*

Contains one species.

Conus dorreensis **Peron 1807** Western Australia, 25mm. Better known by its synonym *C. pontificalis* Lamarck 1810. It is unlike any other cone, being short, stubby and heavy for its size. The shoulder and spire are heavily coronated. The tip and coronations are a dull white; the remainder green.

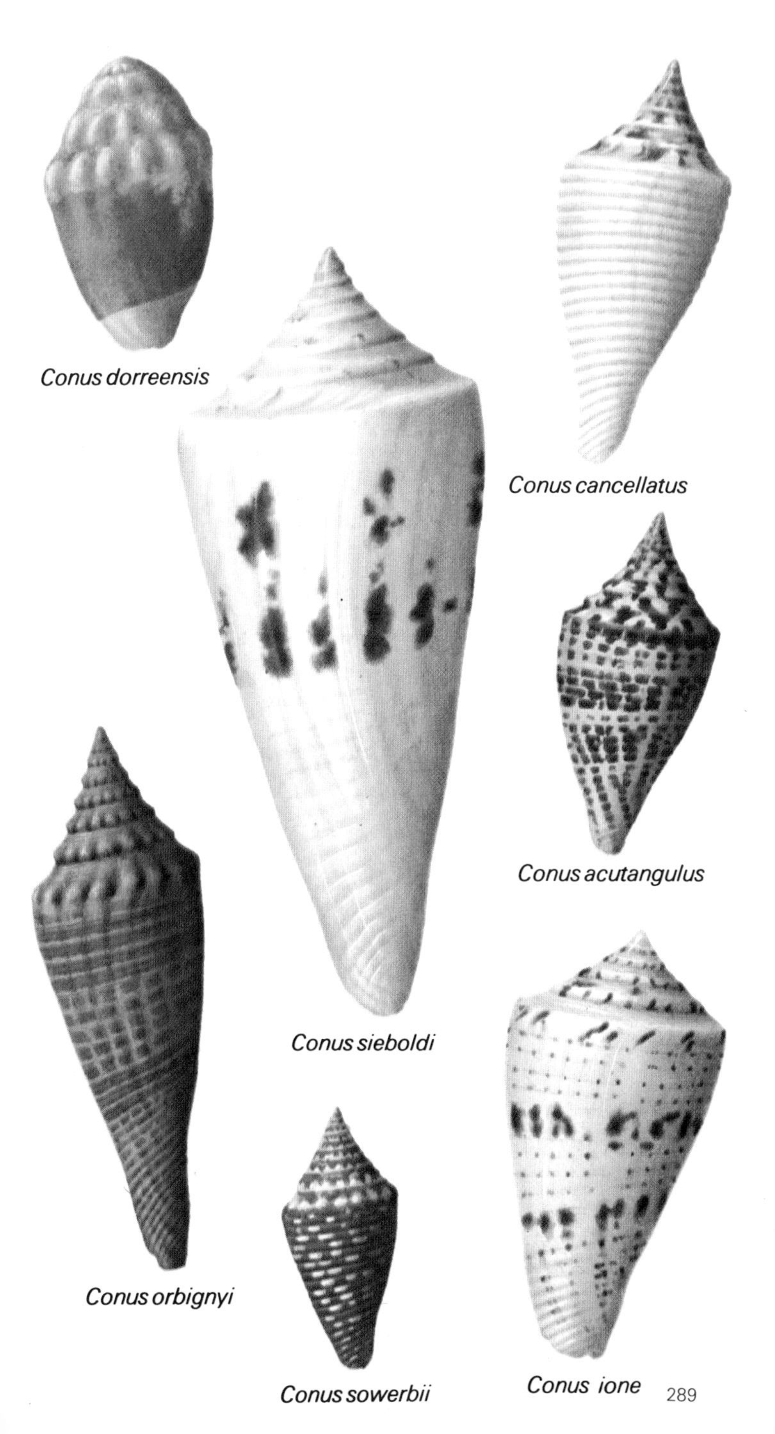
Conus dorreensis
Conus cancellatus
Conus sieboldi
Conus acutangulus
Conus orbignyi
Conus sowerbii
Conus ione

Family: Terebridae

The auger shells, of which there are some 150 species, are all long and narrow with a high, pointed spire and many whorls. They have rather small apertures with a simple lip and usually a plait on the columella. The sculpture is variable from smooth to ribbed, curved or noduled. They have thin horny opercula. The animals live in sand and are carnivorous, inhabiting tropical and semi-tropical seas. They often move just below the surface of the sand, leaving well-defined trails, and a handful of sand scooped up at the end of such a trail will usually reward the collector. They have no periostracum, and moving through sand they are clean and shiny when taken alive.

Terebra crenulata L. 1758 Indo-Pacific, 120mm. Solid. Shoulders with small nodules, about fifteen on body whorl, and a slight constriction below the nodules which becomes obsolete on the later whorls; early whorls with axial plicae giving way to fine growth lines anteriorly. Almost smooth columella; short but strong fasciole. Uneven oatmeal colour; nodules white; three or four spiral rows of red-brown dots on body whorl, two showing on other whorls; small fine streaks of the same colour between the nodules.

T. subulata L. 1767 Indo-Pacific, 150mm. Slender with some twenty-five whorls; two rows of small nodules divided by a fine groove immediately below the suture on early whorls; these become obsolete on later whorls. Otherwise smooth and shiny apart from fine growth lines. Twisted columella; small fasciole. Cream with three spiral rows of squarish, dark brown blotches on the body whorl, two of which show on the others.

T. areolata Link 1807 Indo-Pacific, 120mm. Less slender than the preceding species. About twenty whorls, earlier axially plicate, later with faint growth lines. A slight constriction below the suture forms a second 'shoulder' on each whorl and divides the whorls into approximately one-third above and two-thirds below the constriction. Aperture less 'square' than that of *T. subulata*; small but strong fasciole. Cream with four spiral rows of squarish, medium brown blotches on body whorl, three showing on other whorls; the squares in the anterior row are about four times the size of the others.

T. dimidiata L. 1758 Indo-Pacific, 120mm. Fairly solid and smooth and shiny. About twenty whorls, early whorls axially plicate, later ones with fine growth striae. A groove below the suture as in *T. areolata.* Rather elongate aperture and lip a little flared anteriorly. Columella with a weak plait and rather straight; strong fasciole. Orange-red with wavy white streaks, often bifurcating posteriorly.

T. guttata Röding 1798 Indo-Pacific, 140mm. Solid with about twenty-one whorls. A ridge and a fine groove bordering it below the suture in the early whorls giving way to blunt, very low knobs on later whorls, and a similar row on anterior end of the body whorl. Rectangular aperture; weak fasciole. Cream brown; early ridge and later knobs white.

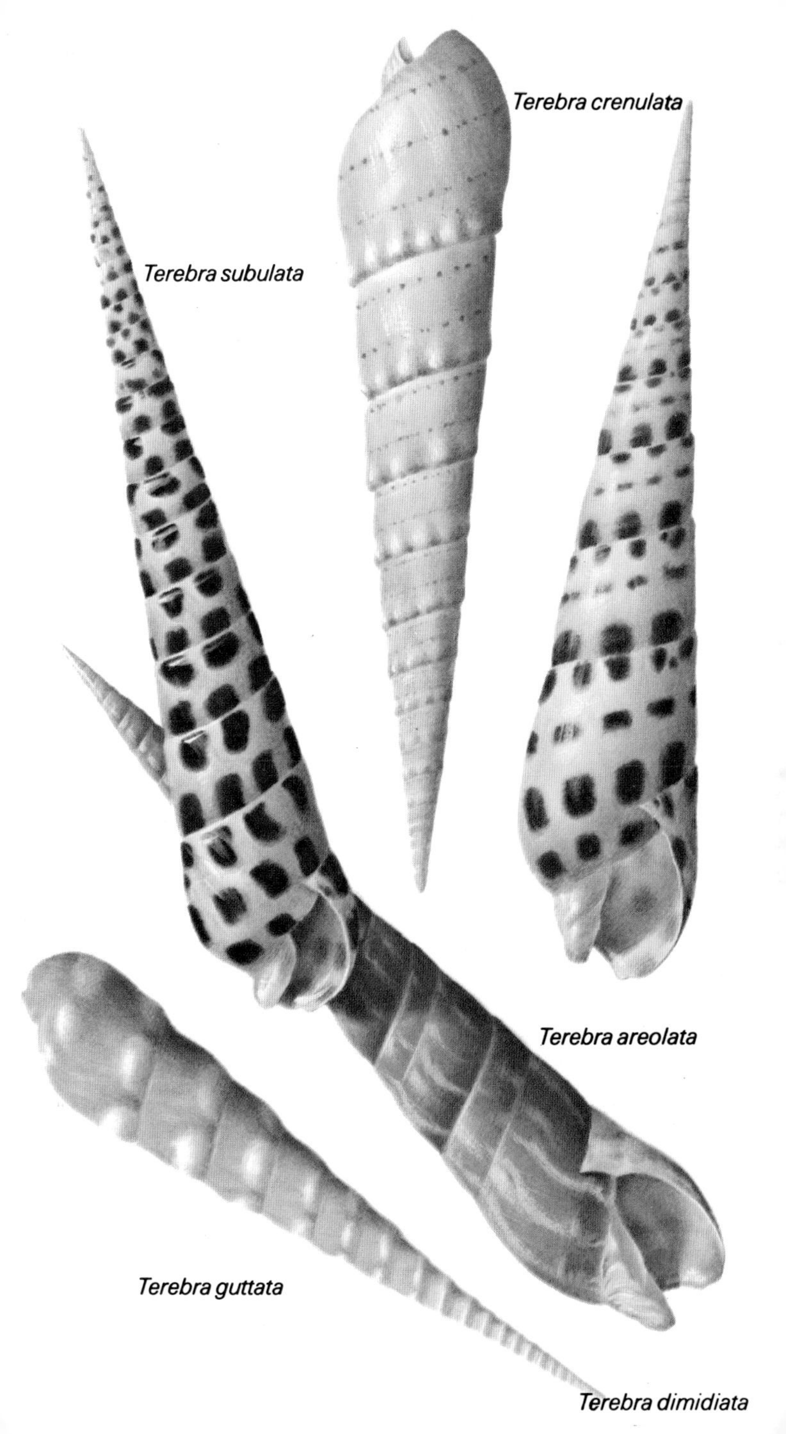
Terebra crenulata
Terebra subulata
Terebra areolata
Terebra guttata
Terebra dimidiata

Terebra commaculata Gmelin 1791 north and east Indian Ocean and west Pacific, 80mm. Long and narrow with twenty-five or more whorls. Two bands of nodules below the suture are divided by a narrow groove; below these are smaller, spiral ridges – about eighteen on the penultimate whorl – and smaller, rather curved axial riblets, giving a somewhat cancellate effect. Rectangular aperture; parietal wall at almost 90° to columella. White with axial, rather rectangular, brown flame marks, about six per whorl anteriorly, reducing to three in about a dozen whorls beyond which they become indistinct.

T. variegata Gray 1834 tropical west America, 85mm. Quite a stout, strong shell with an overall malleate appearance. Below the suture a band of uneven, rather coarse nodules, below which are weak, uneven, axial riblets, with or without spiral grooves. Aperture is scimitar-shaped and columella recurved with two plaits. Blue-grey with axial, brown dashes split by a white band which may just show above the suture on early whorls; subsutural band white with square brown blotches.

T. maculata L. 1758 Indo-Pacific, 250mm. The largest of the genus, strong and heavy. About eighteen whorls, the early ones axially plicate, the later ones smooth except for growth lines. Rather wide aperture; parietal wall at about 120° to the columella, which has a weak fold; small but strong fasciole with a central groove. White with spiral bands – about five on body whorl – of pale tan, axially aligned, rectangular blotches and two spiral bands of irregular, purple-brown blotches on the posterior half of each whorl – that part which shows above the suture.

T. robusta Hinds 1844 Baja California, Panama and the Galapagos, 140mm. More or less narrow. Early whorls with nodulose, subsutural band and axial plications, all of which become obsolete on later whorls and are replaced by faint spiral striae and axial growth lines. Somewhat rectangular aperture; twisted, recurved columella. White with creamy clouding; about four spiral rows of axially aligned, dark brown rectangular blotches which may merge to form two or even one row on spire whorls.

T. strigata Sowerby 1825 Gulf of California and Galapagos, 120mm. The largest of the west American augers, it is solid and heavy. Angle at apex is wide, though not as great as in *T. maculata*. Early whorls with slightly wavy, axial plications and subsutural groove; plications disappear well before the body whorl and groove becomes obsolete. Coarse growth lines; body whorl rather long, as is the aperture; smooth columella is only at a slight angle to the parietal wall; rather coarse fasciole. Creamy white with dark chestnut axial flames.

Duplicaria duplicata L. 1758 Indo-west Pacific, 90mm. Moderately narrow with many whorls; axial plicae on early whorls become flattish ribs on later whorls which may also be slightly inflated. Sharply defined subsutural groove, giving a broad subsutural band. Wide aperture with a deep narrow notch; slightly recurved columella; fasciole with a strong ridge. Variable in colour; may be as illustrated – blue-grey, rusty brown clouding and faint, short, axial streaks of darker grey – or unicoloured, dark brown, orange-pink, cream or white, or varieties of these with other markings, but usually with a rather opaque shine.

Terebra commaculata

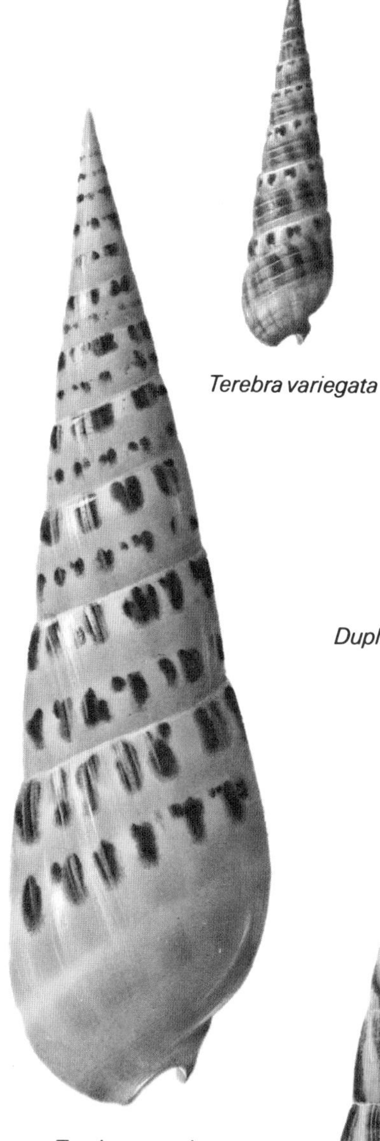

Terebra variegata

Terebra maculata

Duplicaria duplicata

Terebra robusta

Terebra strigata

Terebra monilis Quoy and Gaimard 1832 west Pacific, 50mm. Subsutural row of nodules bounded by sharply incised line; rest of surface smooth but uneven; rectangular aperture. Pale orange tan; nodules white. Note: W. O. Cernohorsky has pointed out that *monilis* is preoccupied and should be replaced.

T. nebulosa Sowerby 1825 Indo-Pacific, 75mm. Subsutural groove; slightly wavy, tight-set, axial grooves; much smaller, spiral grooves; rectangular aperture. White; irregular, squarish, orange-red blotches; band of same colour at anterior end of body whorl; aperture same.

T. ornata Gray 1834 Gulf of California and Galapagos, 80mm. Groove below suture; early whorls nodulose above suture; columella twisted with two plicae. Brown-cream; spiral rows of squarish, dark brown spots on body whorl, two or sometimes three showing on earlier whorls; top row between groove and suture.

T. felina Dillwyn 1817 Indo-Pacific, 90mm. Groove below suture becoming obsolete on last whorls, as does fine, axial ribbing. Cream or white; row of well-spaced, small, brown spots above suture; row of small spots anteriorly on body whorl.

T. succincta Gmelin 1791 Pacific and Indonesia, 30mm. Groove below suture, nodulose between; axially ribbed; fasciole with low ridge. Dark brown; shiny nodules a shade lighter.

T. pertusa Born 1780 Indo-Pacific, 75mm. Nodulose band below suture; early whorls axially ribbed; on last whorls ribs become obsolete; spiral, incised lines in interstices on early whorls giving cancellate structure with obsolete ribs on last whorls. Pale yellow to orange; sutural band white; short purple streaks between many of nodules; light band anteriorly on body whorl.

T. anilis Röding 1798 Philippines to Samoa, 75mm. Row of strong, oblique nodules; second row of small nodules below suture; weak, axial ribbing is cut by spiral grooves, about seven on penultimate whorl. Tan; rows of nodules lighter.

Duplicaria bernardi Deshayes 1857 east Australia, 40mm. Deeply indented suture and subsutural groove; axial ribs overall. Red-brown; blue-white band on centre of body whorl shows just above sutures on earlier whorls; ribs on sutural band, especially near suture, blue-white.

Hastula lanceata L. 1767 Indo-Pacific, 60mm. Slender, graceful; axial ribs on early whorls become obsolete anteriorly. Narrow aperture; columella concave posteriorly, recurved anteriorly; weak fasciole. Shiny white; thin, wavy, axial, red-brown lines, broken on body whorl by white band.

H. diversa E. A. Smith 1901 Pacific, 30mm. Axial ribs. Purple-brown to orange with white band below suture on which there are dark brown spots, about eight on body whorl; narrow, white band at posterior end of body whorl.

Impages hectica L. 1758 Indo-Pacific, 80mm. Smooth; callous near suture; smooth columella; fasciole with central groove. White or cream; slightly broken purple band below suture; columella brown.

Impages confusa Smith north-west Pacific, 20mm. Growth lines and weak ribs below suture. Combination of purple and waxy white, from nearly all purple to nearly all white; when purple, a small white area with tiny purple spots below suture.

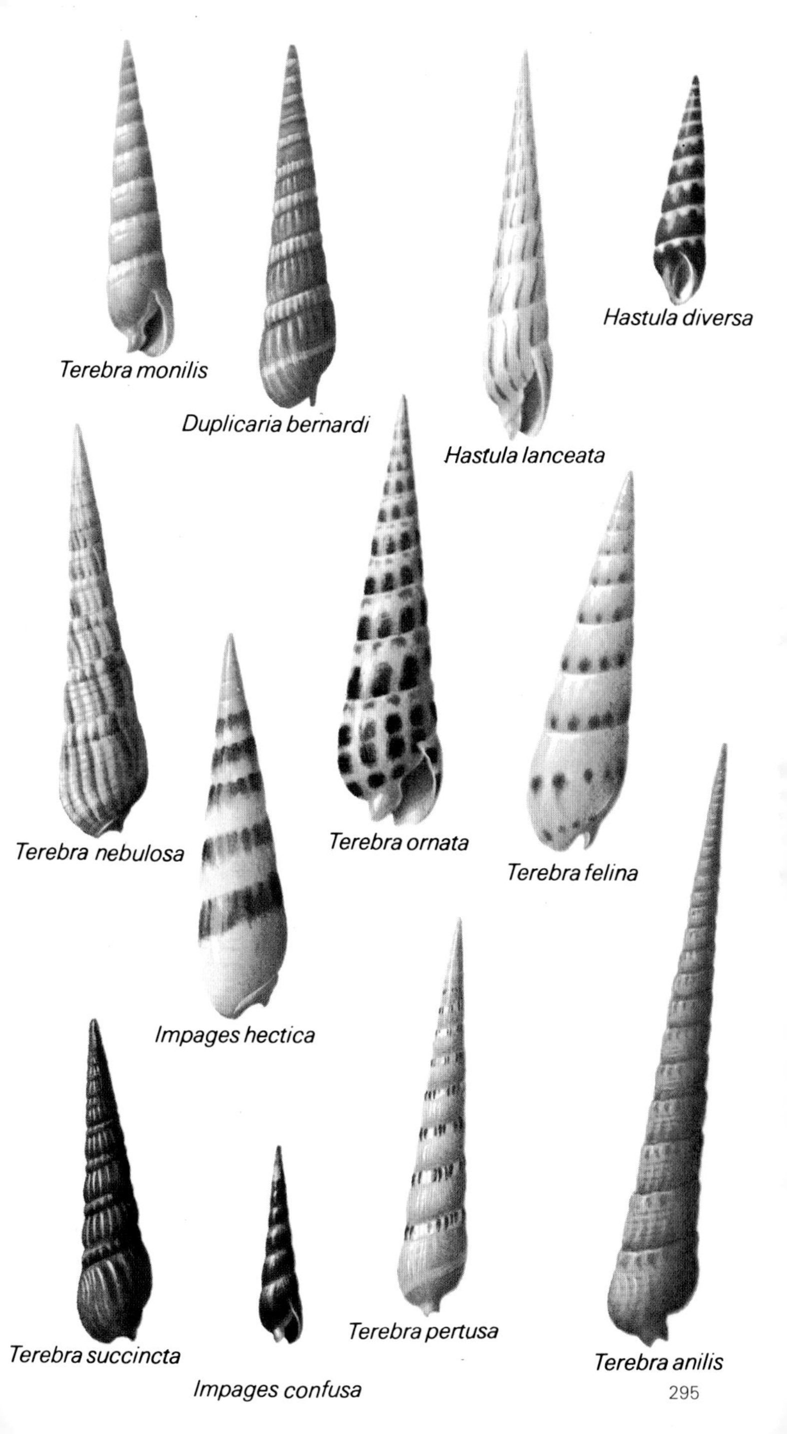
Terebra monilis
Duplicaria bernardi
Hastula lanceata
Hastula diversa
Terebra nebulosa
Terebra ornata
Terebra felina
Impages hectica
Terebra succincta
Impages confusa
Terebra pertusa
Terebra anilis

Family: Turridae

The turrids form one of the biggest families of molluscs; some 1200 species are known. Found in all seas in deep and shallow water. Many are very small. Generally have a turret-shaped spire, extended siphonal canal, and a notch, slit or sinus in the posterior half of the lip. Like the Conidae and Terebridae they use a dart-like tooth to harpoon and poison their prey.

Micantapex lühdorfi **Lischke 1872** Japan, 40mm. Moderate spire; short, broad siphonal canal. Shoulder has rounded nodules, row of smaller ones below suture; about three, fine, spiral ridges in concave area between rows of nodules. About eight, larger, spiral ridges from shoulder to top of siphonal canal, twelve, small ones from there to tip of canal; sinus at end of shoulder ridge; thin lip. Light brown; interior, columella white.

Turris crispa **Lamarck 1816** Madagascar to Fiji, including India, Japan, Indonesia, Philippines, north Australia, 150mm. High spire; long siphonal canal. Posterior end of aperture is marginally nearer anterior end than apex of spire. Sculptured with spiral ridges of varying height and width, that ending in sinus flat, others pointed; fine, lamellate, spiral threads in interstices. White or cream profusely covered with axial, dull chocolate dashes, tending to align themselves especially on body whorl and siphonal canal; in older specimens axial dashes tend to form wider blotches and lose the axially lined effect; aperture white. Shell is variable and gives rise to subspecies: *T.c. variegata* Kiener 1839, India; *T.c. yeddoensia* Joussaume 1883, Japan; *T.c. intricata* Powell 1964.

Lophiotoma acuta **Perry 1811** Indo-Pacific, 65mm. Synonym *tigrina* Lamarck 1822. Variable, typically with high spire and long, straight canal – relatively shorter than *T. crispa*. Spirally ridged; two, strong ridges with narrow, dividing channel on shoulder protruding strongly, and moderately large ones either side of deep suture; other ridges and threads generally small and finely beaded. Sinus at end of strong shoulder ridge. White speckled with small, dark brown dots, most prominent on main ridges but with smaller specks on lesser ridges and threads; aperture creamy white.

Thatcheria mirabilis **Angas 1877** Japan, 100mm. One of the most fascinating of all shells. About eight whorls with very angular shoulders and small carina. Wide area from shoulder to suture is slightly concave and angle at suture is close to 90°; this presutural area spirals up the shell like a ramp. Body narrows sharply to wide, open siphonal canal. Wide aperture; simple, thin lip; and slightly concave and extended columella. Smooth but with very fine, spiral threads and weak growth lines. Pinky tan to tan; interior white.

Polystira albida **Perry 1811** Gulf of Mexico and West Indies, 100mm. High spire and long siphonal canal. Strong, spiral ridges and threads; largest ridge is at shoulder, ending at short sinus. White; thin, brown periostracum.

Turricula javana **L. 1767** north Indian Ocean and South China Sea, 75 mm. High spire and moderately long. At times rather twisted siphonal canal. Shoulder carries obliquely set nodules, longer than wide. Two, small, spiral ridges below suture. Strong spiral threads from shoulder to end of siphonal canal. Wide sinus from shoulder to suture. Light tan to dark purple-brown; nodules lighter.

Drillia suturalis **Gray** south-east Asia, 35mm. High spire, short siphonal canal. Shoulders with slightly oblique nodules, longer than wide. Small ridge below suture; strong, spiral threads overall. Brown; nodules, threads white.

Micantapex lühdorfi

Turris crispa

Lophiotoma acuta

Thatcheria mirabilis

Polystira albida

Turricula javana

Drillia suturalis

Order: Cephalaspidea

Superfamily: Acteonacea

Family: Hydatinidae

Thin, fragile with sunken apex. Large, inflated body whorl. Sand dwellers.

Hydatina velum Gmelin 1790 South Africa and Mauritius, 50mm. Very fragile with flat spire. Suture a deep groove; fine growth lines; wide aperture; thin lip; smooth, slightly callous columella. Waxy white; two, very broad, spiral bands of oblique, axial, close-set, brown lines, both edged on each side with an unbroken, darker brown line; gap between the two bands is narrow.

H. albocincta Van der Hoeven 1839 Indo-west Pacific, 60mm. Very thin, fragile; concave spire. Very large aperture; lip extends beyond apex; smooth, thinly callous columella. White or very pale brown; four, broad, spiral bands of curved, axial, brown lines; bands may be edged with narrow lines of brown on white, or white on brown; aperture white; outer colour shows through.

Superfamily: Philinacea

Family: Scaphandridae

Scaphander lignarius L. 1758 Mediterranean, 70mm. Thin, moderately strong; sunken apex. Body whorl contracting towards apex and expanding anteriorly; spiral incised lines; fine growth lines. Thin lip, flared anteriorly; smooth, thinly callous columella. Pale yellow-brown; darker periostracum.

Superfamily: Bullacea

Family: Atyidae

Thin shells with deeply sunken spire. Sand dwellers.

Atys cylindricus Hinds 1779 Indo-Pacific, 30mm. Thin, fragile, sub-cylindrical with sunken spire. Lip extends beyond apex of shell. Body whorl with incised spiral lines at each end, smooth in middle; fine growth lines. Aperture narrow posteriorly, wide anteriorly; thin lip; smooth columella. White or very light brown and may have a few, axial, brown hair lines.

Family: Bullidae

Bubble shells, related to Atyidae, have a deep narrow umbilicus at the apex.

Bulla ampulla L. 1758 Indo-Pacific, 60mm. Thin, moderately solid, globose; expanded body whorl. Simple lip, extended posteriorly beyond apex, slightly constricted centrally and expanded anteriorly. Columella, reversed 'S' shape, smooth and thinly callous. Cream; blotches of dark purple-brown; overall clouding of light brown, rarely with two spiral bands; aperture, columellar callus white.

B. striata Bruguière 1792 Mediterranean, 30mm. Thin, delicate, rather narrow. Aperture narrow posteriorly, wide anteriorly. Incised spiral lines at ends of body whorl; smooth centrally bar fine growth lines. Thin lip, slightly constricted centrally, extends beyond apex. Brown-grey; darker, smudged dots and dashes; aperture and smooth, thin columellar callus white.

B. punctulata A. Adams 1850 Pacific, 30mm. Shape as *B. ampulla* but smaller, more cylindrical. Cream; clouding of brown or grey in two to four spiral bands, generally spotted with squarish chocolate dots smudged to the left and bordered to the right by white spots; aperture, columellar callus white.

Atys cylindricus
Hydatina velum
Bulla ampulla
Bulla striata
Hydatina albocincta
Scaphander lignarius
Bulla punctulata

Class: Bivalvia, Lamellibranchia or Pelecypoda

Molluscs which have two valves or shells. This and the following five pages give just a few examples of this class of molluscs to show something of the variety of shape, size and coloration within the class.

Family: Pinnidae

Large, fragile shells. The animals live in sand or muddy sand, point anchored into the substrate by a series of hairs called the byssus. Most are long in relation to width, as illustrated, but others are more rounded.

Pinna incurva Gmelin 1791 north-east Indian Ocean and South China Sea to north Australia, 300mm. Thin, fragile and translucent. Inner layer of the shell is partly nacreous.

Family: Isognomonidae

Flattened shells. Hinges have rows of pits into which the joining ligaments fit. They are found in the Indo-Pacific and Atlantic Oceans.

Isognomon isognomum L. 1758 Indo-Pacific, 120mm. Both valves very flat. Shape very variable. Blue-grey, usually heavily encrusted; interior pearly white. Strongly anchored by a byssus to mangrove roots, coral and so on.

Family: Pteriidae

The wing oysters. Flattened and fragile with hinge projected both ways. Pearly interior. They are edible and rarely produce pearls. They live in tropical to temperate regions and anchor themselves to the bottom.

Pteria levanti Dunker Indo-Pacific, 80mm. Rather inflated on the ventral side. Edge very fragile. Outline like a bird. Light brown with a fine periostracum.

Family: Arcidae

The ark shells are usually rather oblong with a long, straight hinge on which there are very many, small, subequal, fine teeth. The beaks are hooked and separated from each other. Most anchor themselves by a byssus of hairs. They live worldwide and are edible.

Trisidos tortuosa L. 1758 Indo-Pacific, 120mm. This shell is unusual and interesting as it is twisted so that the extremities are at 90° to each other. Many fine radial threads. White.

Anadara maculosa Reeve Indo-Pacific, 60mm. It is a more normal ark shell. Strongly ribbed and solid with a thick periostracum as shown in the illustration.

Family: Glycymeridae

Solid, equivalved and rather circular in outline, with a porcellaneous interior. Always have many teeth on the curved hinge, larger laterally. They have a thick periostracum and are not attached by a byssus. They are found worldwide and are generally edible.

Glycymeris glycymeris L. 1758 the Dog Cockle, British and European waters, up to 60mm. Indistinct radial sculpture. Cream with dark blotches of red-brown, often in zigzag pattern; variable.

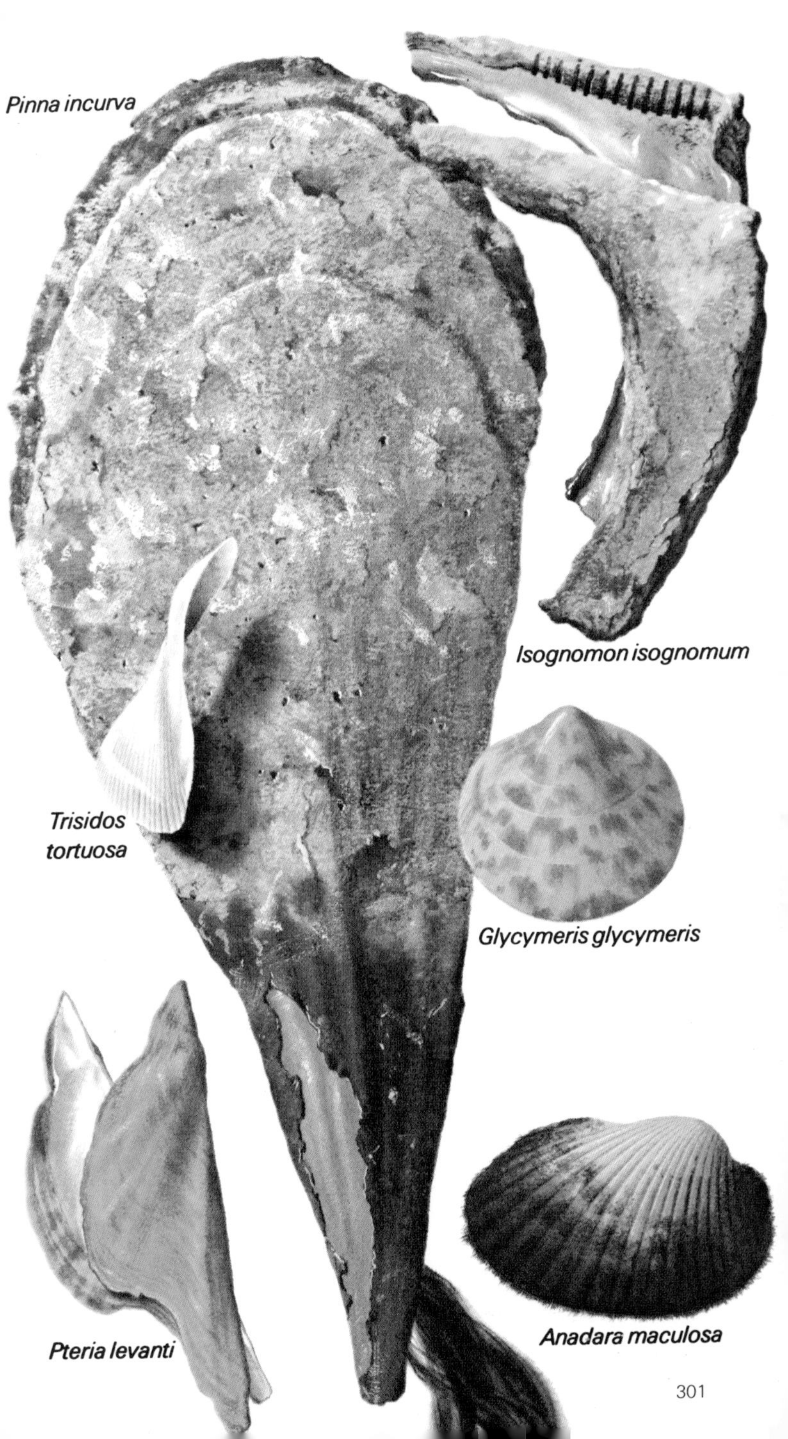
Pinna incurva
Isognomon isognomum
Trisidos tortuosa
Glycymeris glycymeris
Pteria levanti
Anadara maculosa

Family: Pectinidae

The scallop shells include many species, often very colourful. They are usually regular in outline either with the right valve convex and the left flat or slightly concave, or more or less equivalved. They are generally sculptured with radiating ribs which sometimes carry small spines. The beak generally has an 'ear' on each side, the anterior larger. They are found worldwide and some species are highly prized as food. They are one of the most popular molluscs to appear in art and used as the symbol of the Royal Dutch/Shell Group are recognised worldwide. They swim, apparently haphazardly, by a rapid opening and shutting of their valves, especially to escape their main predators, starfish.

Decatopecten striatus Schumacher Indo-Pacific, 25mm. Solid, inflated and somewhat triangular, with about five rather large ribs and many fine radial cords. Unequal ears. White usually marked with pink-brown.

Pecten tranquebaricus Gmelin 1790 South China Sea, 40mm. About twenty radial ribs; not very inflated; uneven ears. White, sometimes with red clouding, and irregular, generally horizontal, black banding.

Chlamys swifti Bernardi 1858 north Pacific, 100mm. Large and solid with five broad ribs which are nodulose where they cross heavy, concentric ridges and fine radial riblets. Somewhat inflated with very unequal ears. Attached to rocks by a byssus. Left valve usually shades of purple-red, darker at the nodules; right valve much paler or white with pink bands.

Aequipecten opercularis L. 1758 the Queen Scallop, Europe, 80mm. Very rounded with about twenty ribs. Ears almost equal. Colour variable, yellow, orange, brown, red, pink or purple, and often blotched or spotted.

Family: Limidae

The file clams are generally ovate and somewhat compressed. The anterior side is straighter and the ear smaller. The posterior side is more rounded and the valves usually gape. They are generally radially ribbed and white in colour. The Limidae swim quite rapidly by opening and closing their valves and using the long filaments which extend from the margin of the animal.

Lima sowerbyi Deshayes Indo-Pacific, 40mm. It has about twenty, radial, scaly ribs. White.

Family: Mytilidae

The mussels are generally rather elongate, oval or roundly triangular, and fairly thin-shelled. The valves are equal, inflated and have a thick periostracum. The interior is nacreous. They are found worldwide, usually attached by a byssus to rock or other material and often in thick clusters. Some bore into soft rock or coral and some into quite hard material such as limestone. Some species are eaten, especially in the Mediterranean.

Lithophaga teres Philippi Indo-Pacific, 60mm. Despite boring into rock and coral, it is very thin and fragile. Rather cylindrical with the beaks near the anterior end. It is white with a dark brown periostracum, as illustrated.

Perna viridis L. South China Sea, 50mm. A solid shell with a rich, deep green periostracum, a little lighter near the beak where the underlying white of the shell shows through.

Mytilus perna L. 1758 east coast of South Africa, 70mm. South Africa's common mussel. Beaks pointed and a little down-turned. Purple-brown; interior blue or yellow.

Decatopecten striatus

Pecten tranquebaricus

Lima sowerbyi

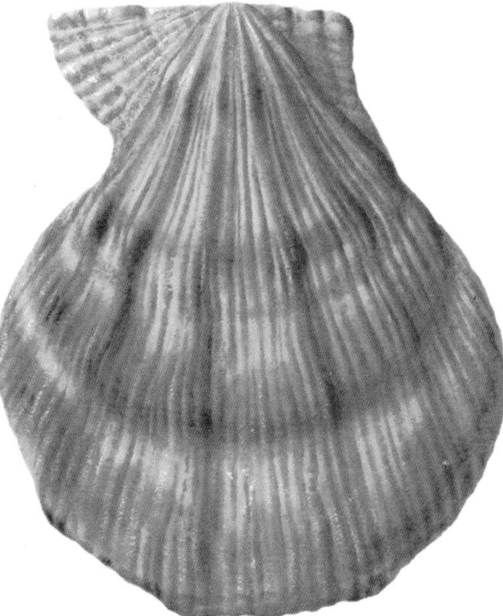
Chlamys swifti

Lithophaga teres

Mytilus perna

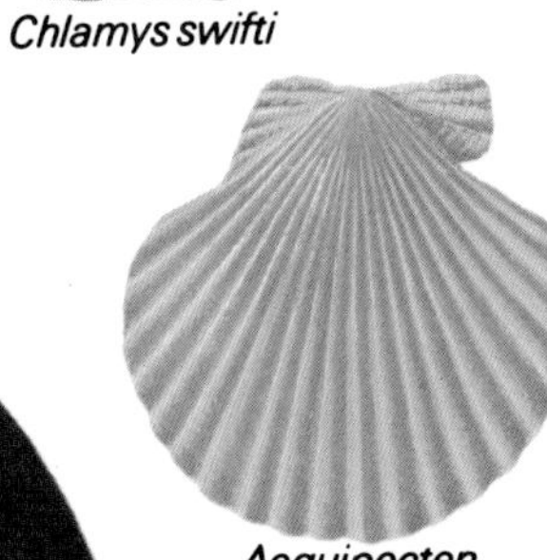
Aequipecten opercularis

Perna viridis

Family: Cardiidae

The cockles are usually ovate and nearly globular, with radial ribs sometimes carrying lamellae or short spines. The hinge has two central teeth and there are usually one or two lateral teeth on each side of the hinge. They are very active and move about with their long, strong foot. They are found worldwide. Like other burrowing bivalves they provide food for many predators.

Discors lyratum Sowerby 1841 Indo-Pacific, 50mm. Solid, swollen with strange sculpture in that it has radial riblets, stronger posteriorly, and clear, oblique ridges restricted to the anterior half. The colour is also unusual, a rich, slightly purple-red periostracum; white spots more frequent near the white beaks; interior white and yellow; pink in the umbo area.

Cardium costatum L. 1758 west Africa, 100mm. Has about ten, strong, hollow, elevated, quite sharp, radiating ribs ending abruptly at the edge of the valves. Posterior end of the shell—the corselet—is rather flattened, has smaller, very rough-topped ribs, and does not close, but 'gapes'. White or off-white; brown or orange between ribs near umbo.

Cerastoderma edule L. 1758 the European Cockle, Europe, 60mm. Equivalved with twenty-five ribs and strong growth lines. Yellow-white to orange-brown; sometimes darker rings. Eaten by man, bird and fish.

Family: Glossidae (see also page 308)

These are even more heart-shaped in anterior view than the Cardiidae, with the beaks twisting in a spiral and turning towards the anterior.

Meiocardia moltkiana Gmelin 1791 Pacific, 40mm. Delicate and handsome, with a strong ridge or carina running from the beak to the posterior edge of the shell; small, concentric, incised lines in front of the carina, smooth behind it. Cream-white.

Family: Spondylidae

The thorny oysters are perhaps the most popular bivalves among collectors. Attractive in sculpture and colour. The surface is sculptured with radiating ribs which generally bear more or less long, spatulate or pointed spines. They cement themselves to rock or coral by the right valve which is strongly convex; the left is usually flatter. Their 'ears' are small or obsolete and they have no byssus. Some 100 species.

Spondylus regius L. 1758 Japan and south-east Asia, 200mm. One of the most splendid of the thorny oysters. It is almost equivalved with seven, strong, elevated, radial ribs bearing long, erect, strong spines. Between the large ribs it has small, prickly, radial riblets. Rose-pink, spines lighter.

S. barbatus Reeve 1856 Japan and south-east Asia, 70mm. The radial ribs carry some broad, rather spatulate spines. The right valve is flat or slightly concave. The colour varies, generally purple-red stained with rose and deep red near the beaks, or pinky mauve, as illustrated.

Family: Ostreidae

The oysters, of which there are many species, mostly edible, are found worldwide. They are generally rather dull. One of the more interesting is:

Lopha cristagalli L. 1758 the Cock's-comb Oyster, Indo-Pacific, 70mm. The edge of the shell is zigzag; surface rough like a piece of coarse leather. A dull purple. It cements itself firmly to rock or coral, as illustrated.

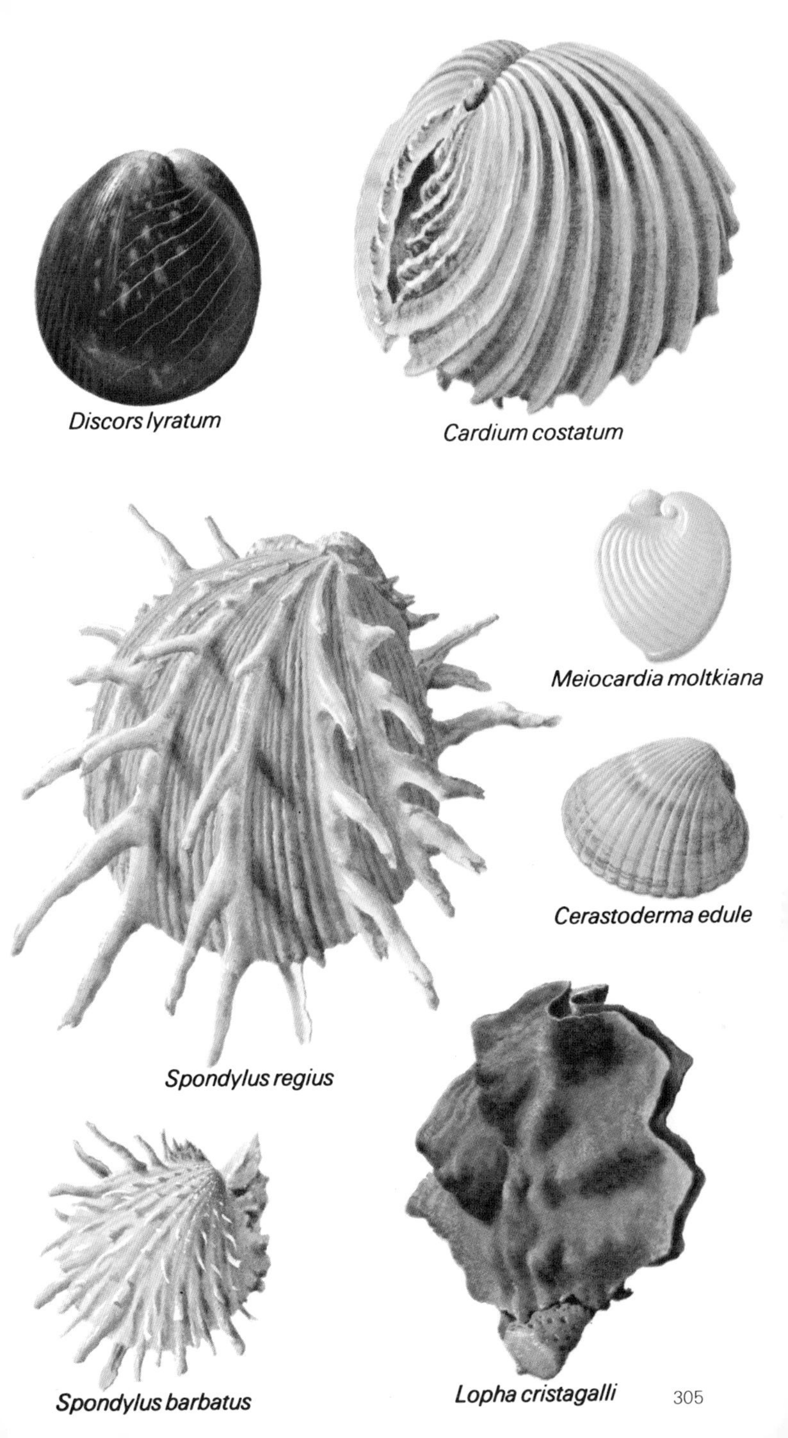

Discors lyratum

Cardium costatum

Meiocardia moltkiana

Cerastoderma edule

Spondylus regius

Spondylus barbatus

Lopha cristagalli

Family: Veneridae

The venus shells are a large family, found worldwide. They generally have a strong shell and are often colourful. They are equivalved, with the beaks more or less central and directed forwards and inwards. They live usually a little below the sediment surface. Many species are used for food.

Hysteroconcha lupanaria Lesson 1830 the Comb Venus, tropical west America, 80mm. It is one of the most spectacular of the family. A row of strong spines runs from the beak to the posterior edge of the shell on both valves. The spines increase in size away from the beak. In front of the spines are sharp, concentric ridges and the corselet is smooth. Grey, with purple spots at the base of the spines; purple radial streaks near the posterior margin.

Chione paphia L. the Key Venus, Caribbean, 40mm. It has ten, strong, coarse, heavy, concentric ridges undercut on the beak side and strongly grooved between. On the rear quarter of the shell the ridges end sharply and continue as thin flanges. Off-white or cream with red-brown zig-zag lines and dark brown escutcheon.

Paphia amabilis Philippi Indo-Pacific, 75mm. Rather elongated, concave in front of the beaks; close concentric ridges. Cream with fine, zig-zag, light brown lines and four, darker brown radial rays.

Lioconcha castrensis L. 1758 Indo-Pacific, 40mm. Solid and inflated with round outline. Smooth except for fine growth lines. Cream with some grey shading and dark brown-black, zigzag lines.

Anomalodiscus squamosus L. 1758 Indo-Pacific, 30mm. Inflated and extended posteriorly into a beak. Radially ribbed; strong concentric ridges; edges strongly crenulate. White with some yellow or yellow-brown clouding.

Family: Tridacnidae

The giant clams. The family contains six species all from the Indo-Pacific.

Tridacna gigas L. 1758 (not illustrated) is the largest living shelled mollusc. It can grow up to 1,350mm and weigh over 260kg. Often embedded in rock and coral. The brightly coloured blue, green and yellow mantles of the animals are very conspicuous.

Hippopus hippopus L. 1758, 300mm. Equivalved with strong carina from umbo to anterior edge. Rather triangular with several broad, low, strong radial ridges with close riblets between; the ribs bear a few short scales. The ribbed area in front of the carina is wide. Generally flattened but concave near the beaks. The ventral margin is strongly undulating. This genus has a very small byssal gape. Creamy white with purple-brown and white blotches.

Tridacna squamosa Lamarck 1819, 300mm. Equivalved and elongate with about five, strong, low, rounded radial ridges, each carrying curved, fluted scales growing larger further from the beak. It has a large byssal gape, typical of this genus. White, becoming lemon yellow towards the margin.

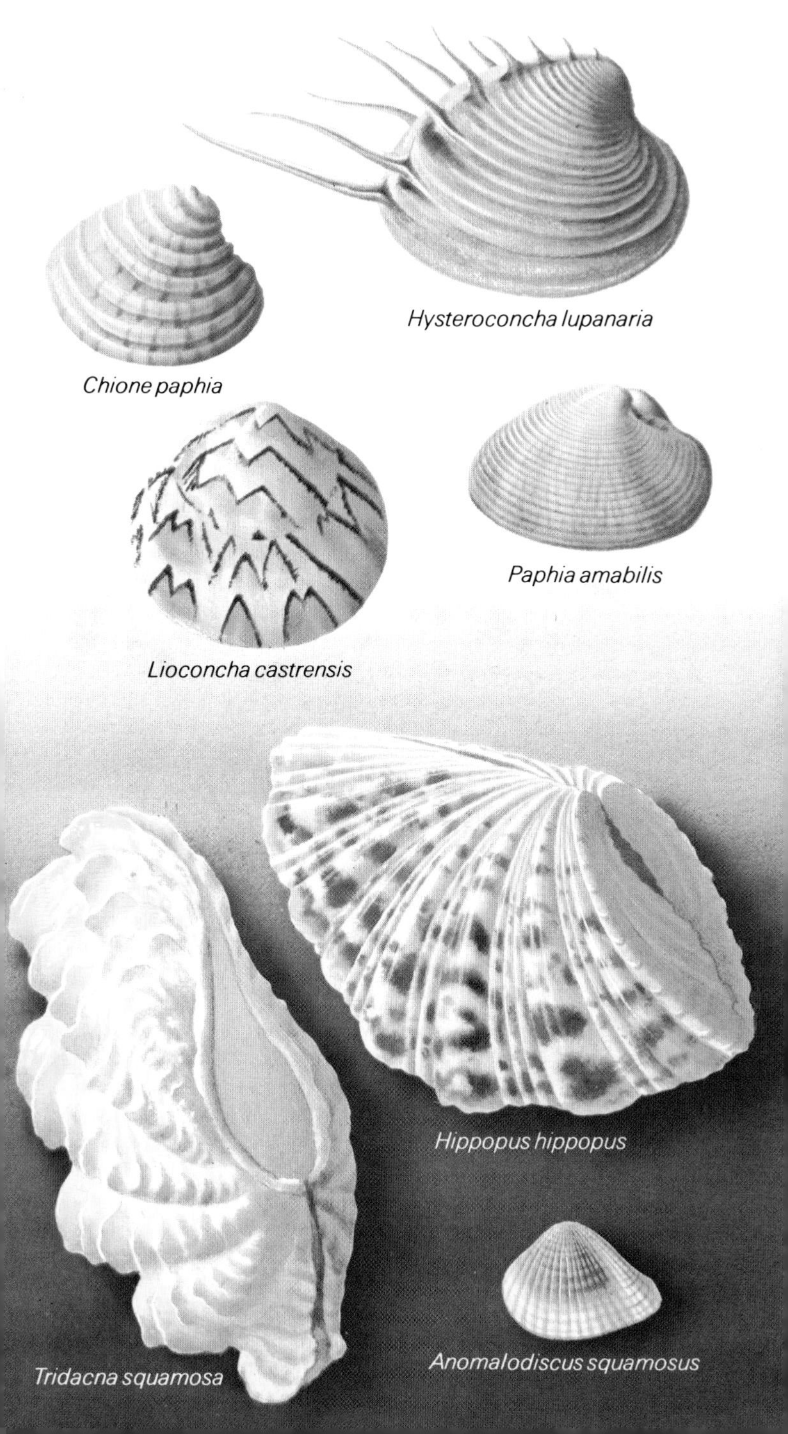
Hysteroconcha lupanaria
Chione paphia
Paphia amabilis
Lioconcha castrensis
Hippopus hippopus
Anomalodiscus squamosus
Tridacna squamosa

Family: Glossidae (see page 304)

Glossus humanus L. 1758 north-east Atlantic and Mediterranean, 100mm. The Heart Cockle, though not in fact a cockle (Cardiidae). Equivalved and solid, but not heavy; very inflated. Beaks which do not touch each other coil forward, outward and downward like a pair of horns. Off-white; thick brown or dark green periostracum.

Family: Mactridae

The otter and trough shells. Generally thin, smooth, light shells and most have a triangular-shaped pit in the hinge to which the ligament is attached. Most live burrowed well into the sand. Some are used for food, but they are not as tasty as most of the edible bivalves.

Spisula elliptica Brown 1827 north-east Atlantic, 30mm. Equivalved, oval and fairly solid. Smooth except for growth lines. Dirty white but sometimes stained with blue, green or yellow concentric rings (as illustrated). Lives in muddy sand or gravel.

Family: Lucinidae

Found worldwide, they are generally circular and often thin-shelled. White predominates as colour.

Codakia orbicularis L. 1758 the Tiger Lucina, south-east USA and the West Indies, 90mm. Almost circular. Smooth close to the beaks, otherwise with fine concentric ridges and radial riblets giving a cancellate sculpture. White; interior with a pale yellow tinge, often with a rose tinge on the hinge.

Family: Tellinidae

The tellins generally have rather flat, thin shells. The posterior end is often a little twisted to the right. They burrow quickly and deeply and are often very brightly coloured.

Tellina radiata L. 1758 south-east USA and West Indies, 100mm. Elongate, smooth and shiny with faint growth lines. Shallow groove from beak to near posterior. White or white with pink-red or yellow rays and/or concentric bands; beaks usually with red tips; interior may have a pale yellow clouding.

T. rostrata L. Malaysia, 80mm. Elongate and flattened; strongly twisted posteriorly. Pointed posterior. Concentric fine ridges becoming oblique posteriorly. Each valve has a low, spined ridge running from the beak to each extremity; below each posterior ridge is a flattened furrow. White.

Family: Donacidae

The wedge shells are generally triangular, medium to small, fairly solid shells, rather similar to the tellins. They are found worldwide.

Donax vittatus da Costa 1778 Europe from Norway to the Mediterranean, 40mm. Wedge-shaped, smooth and shiny with very fine growth lines. Somewhat inflated. Inside edge strongly but finely toothed. Brown, purple-brown or yellow, usually with white rays; interior mostly purple.

D. serra Röding 1798 South Africa, 65mm. Posterior end truncated with wavy concentric ridges. Otherwise smooth with fine growth lines and very faint radial grooves. Inside edge of shell serrated. Purple-brown, yellow-brown or white with narrow, darker, concentric bands; interior purple, white or both.

Spisula elliptica

Glossus humanus

Tellina radiata

Codakia orbicularis

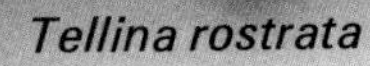

Tellina rostrata

Donax vittatus

Donax serra

Family: Solecurtidae

Oblong and rectangular, gaping at both ends. Rather similar to the tellins.

Solecurtus strigillatus L. 1758 Portugal to Senegal and Mediterranean, 150mm. Inflated, it has strong, some coarse, growth lines, and fine oblique striae. Pale tan with a pinky tinge and a few pale rays; interior white with pink tinge at the edges.

Family: Solenidae

The razor shells are long, narrow, and open at each end. They burrow a foot or more into the sand. They are found worldwide and many of the species are used for food.

Solen marginatus Pulteney 1799 north Europe to Morocco, 130mm. Straight and square-ended, gaping widely at each end. Rather delicate with growth lines becoming coarse posteriorly. Beak at anterior with furrow running just inside anterior margin. The illustration shows some of the periostracum still attached to the shell. Colour pale tan.

Family: Pholadidae

The piddocks are boring molluscs. They have thin, fragile shells, gaping at each end, with file-like ridges anteriorly enabling them to bore into rock or clay. Beaks are obscured by reflected shell margin. Inside each valve, under the beak, they have a rod or spoon-like projection, and there is a variable number of accessory plates between the valves which protect the muscles holding the two valves together – a function of the ligament in almost all other bivalves.

Cyrtopleura costata L. 1758 Angel Wings, east North America and West Indies, 200mm. The largest of the family. Elongate and rather fragile with strong radiating ribs carrying short spines, strongest anteriorly; the ribs show internally as grooves – pitted beneath the spines. The illustration shows the spoon-like projection and an accessory plate.

Family: Clavagellidae

The watering pot shells start life as small, typical-looking bivalves, but develop a shelly tube. The original valves can be seen in the illustration. The bottom of the tube is buried in the sand or mud. The base of the tube has many tiny holes, hence the name watering pot.

Penicillus penis L. 1758 Indo-Pacific, 200mm. Tube tapering from base. Smooth with indistinct growth lines. No attached foreign matter. Domed base is surrounded by a flat, thin frill of many radiating, fine tubules; dome with narrow, central slit and fine, raised grooves. Elongate shells, situated just above the fringe, not noticeably sunken. White to pale grey.

P. cumingianum Australia, 120mm. Tube somewhat uneven, not tapered, becoming slightly flared at top. Upper one-third has conspicuous growth lines and plated ruffles. Small granules of foreign matter cemented to lower two-thirds of tube; larger pieces cemented to base. Base rough, obscured by foreign matter and consists of scattered, relatively wide, elevated tubules which also form a rough fringe. Valves sunken near base of tube (see illustration) and almost equi-dimensional, wide at top. Fringes white; base of tubes orange-brown; shells white.

Country Life Guides

Previously published as **Hamlyn Guides.** All are written by acknowledged experts and are illustrated in colour throughout. Guides to the following are available.

Aquarium Fishes
Klaus Paysan
0 600 34473 8

Astronomy
David Baker
Illustrated by David A Hardy
0 600 30348 9

Birds of Britain and Europe
(New revised edition)
Bertel Bruun
Illustrated by Arthur Singer
0 600 31411 1

Countryside of Britain and Northern Europe
Editor Pat Morris
0 600 35606 X

Edible and Medicinal Plants of Britain and Northern Europe
Dr Edmund Launert
Illustrated by Roger Gorringe and Anne Davis
0 600 35281 1

Flora and Fauna of the Mediterranean Sea
Dr A.C. Campbell
Illustrated by James Nicholls and Roger Gorringe
0 600 35279 X

Freshwater Fishes of Britain and Europe
Dr P.S. Maitland
Illustrated by Keith Linsell
0 600 33986 6

Horses and Ponies of the World
Elwyn Hartley Edwards
Illustrated by David Nockels
0 600 34533 5

Minerals, Rocks and Fossils
Dr A.R. Woolley, Dr A.C. Bishop and Dr W.R. Hamilton
0 600 34398 7

Seashore and Shallow Seas of Britain and Europe
Dr A.C. Campbell
Illustrated by James Nicholls
0 600 34396 0

Shells of the World
A.P.H. Oliver
Illustrated by James Nicholls
0 600 34397 9

Trees of Britain and Europe
Dr C.J. Humphries, J.R. Press and D.A. Sutton
Illustrated by I. Garrard, T. Hayward and D. More
0 600 35278 1

Weather Forecasting
F. Wilson and S. Dunlop
0 600 39024 1

Spiders of Britain and Northern Europe
Dick Jones
0 600 35665 5

ISBN 0-600-34397-9

9 780600 343974

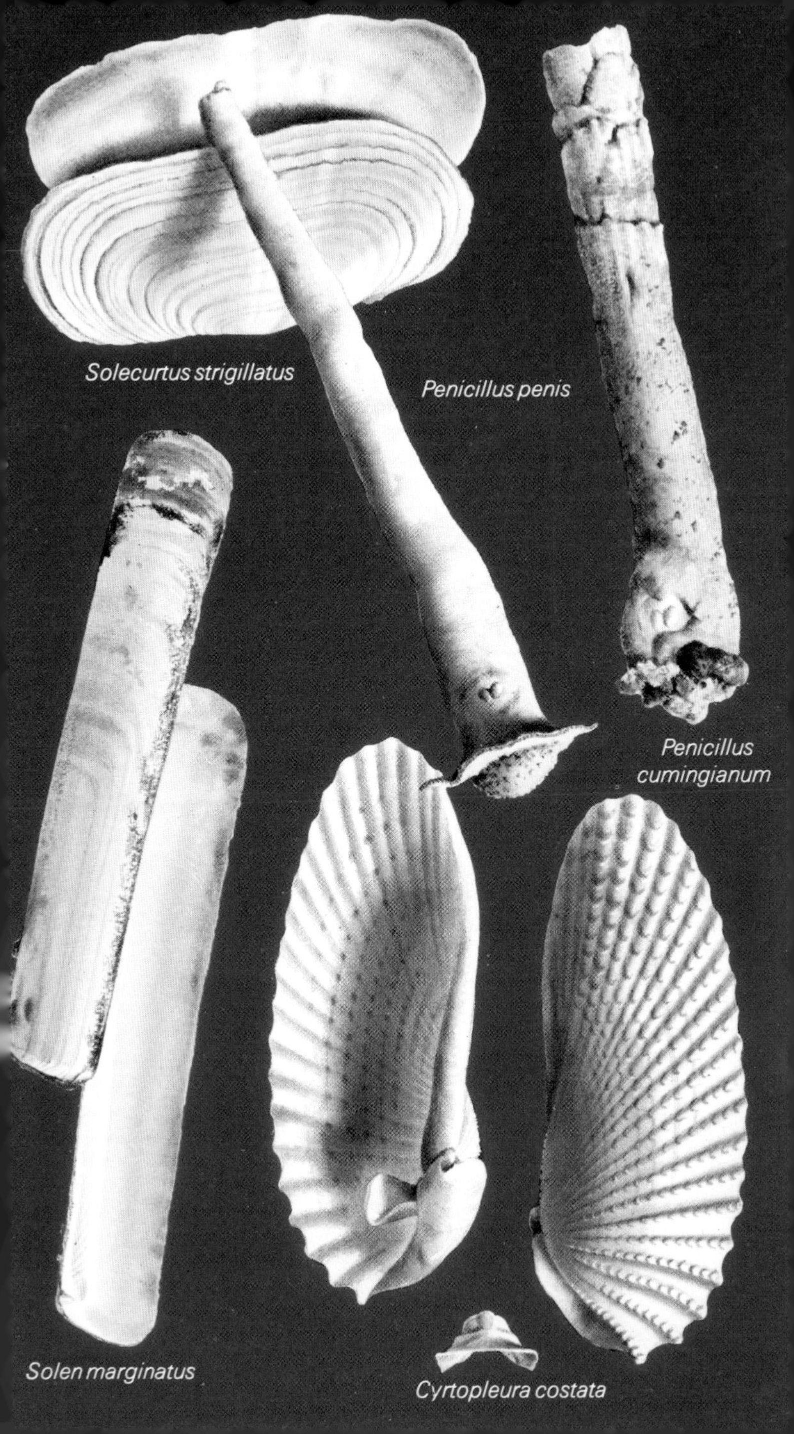

Solecurtus strigillatus

Penicillus penis

Penicillus cumingianum

Solen marginatus

Cyrtopleura costata

Class: Cephalopoda

Includes the octopus, cuttlefish, spirula, squid and nautilus. Active carnivores.

Nautilus pompilius L. 1758 the Pearly Nautilus, Indo-Pacific, 200mm. Shell coiled in flat spiral with both umbilici covered. Large aperture, nacreous interior. Smooth, apart from distinct growth striae. Area of shell facing aperture with thin black callus. Cream or white with red-brown radial bands narrowing towards umbilical area. Other species, much rarer, are very similar but have a more or less deep umbilicus. The interiors of all these shells have chambers serving as buoyancy tanks allowing the animal to change depth.

Argonauta hians Dillwyn 1817 all warm seas, 75mm. Very thin and fragile. Coiled in a flat plane and lacking umbilici. Radial ridges and angular shoulders with two rows of knobs on either side of a flat, smooth periphery. Brown; nodules darker, becoming paler towards the aperture.

The females of the *Argonautidae* all produce a somewhat similar delicate structure, which is not chambered, in which to deposit and incubate their eggs. Other species are *A. argo* L. 1758, 300mm and *A. nodosa* Lightfoot 1786, 125mm. Both white.

Spirula spirula L. 1758 all warm seas, 35mm. The only species in the superfamily Spirulacea. The shell is a free coil and has chambers like the nautilus. It is almost completely embedded within the animal, rather like a squid. The shell is very finely reticulated. White; interior nacreous.

Class: Scaphopoda

Tusk shells are hollow, tusk-shaped shells but open at both ends, often with a notch or slit at the narrow end. Some are very small and most are white. They are found in most seas and live on protozoa.

Dentalium elephantinum L. 1758 Indo-Pacific, 80mm. This species is strongly longitudinally ridged—about nine ridges—and banded with shades of green, darker at the wider end; white at the lip.

D. vernedei Sowerby 1860 Japan, 120mm. Fine, close-set, longitudinal threads. White, irregularly banded usually with light yellow or brown.

Class: Amphineura

Chitons are nearly all flattened and oval-shaped, rather like a wood louse. They have eight plates which overlap and are kept together by a surrounding girdle. They live on rocks and other hard surfaces and are vegetarians. There are about 550 species in a number of families and numerous genera. Illustrated is a typical example and also the eight plates separated. The largest, *Amicula stelleri* Middendorff 1847 west North America, grows to 250mm.

Class: Monoplacophora

These are the most primitive of the molluscs. They have no eyes or tentacles and unlike all other molluscs they have segmented bodies (see illustration). Their fragile shells look like that of a limpet, but have a series of paired muscle scars inside. The apex is close to the anterior end and it has fine, concentric, uneven ridges. They were known only as fossils until the 1950s and have only been found in very deep water. The illustration shows the inside of the shell with the animal, *Neopilina adenensis* Tebble.

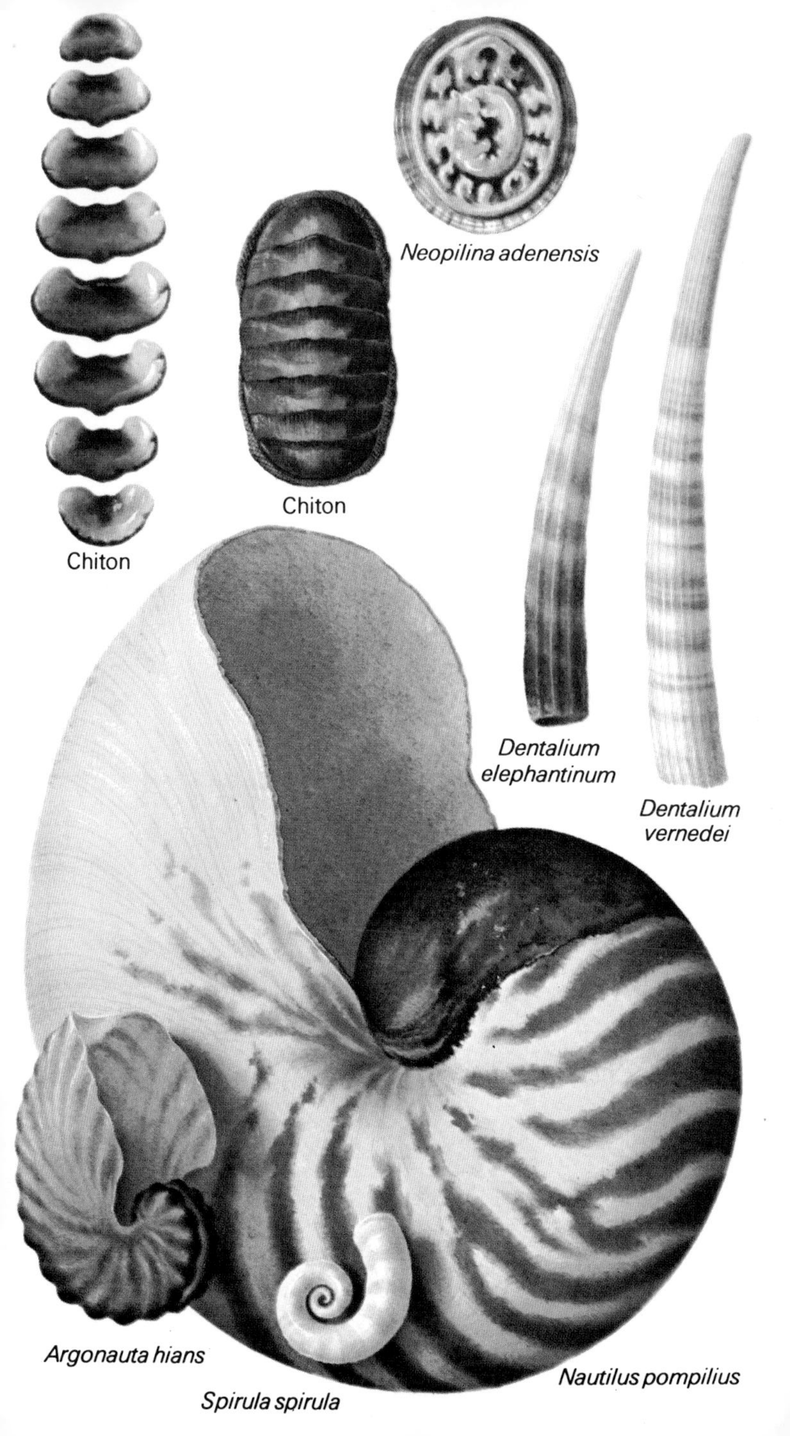
Chiton
Chiton
Neopilina adenensis
Dentalium elephantinum
Dentalium vernedei
Argonauta hians
Spirula spirula
Nautilus pompilius

Index

Page numbers in italic refer to illustrations.